CELL BIOLOGY
by the numbers

CELL BIOLOGY
by the numbers

Ron Milo
Rob Phillips

illustrated by

Nigel Orme

Garland Science
Vice President: Denise Schanck
Senior Editor: Summers Scholl
Senior Editorial Assistant: Katie Laurentiev
Production Editor: Natasha Wolfe
Cover Designer and Illustrator: Nigel Orme
Copyeditor: John Murdzek
Typesetting: Cenveo, Inc.
Proofreader: Elephant Editorial Services
Indexer: Bill Johncocks

Chapter Opening Images:
Chapter 1: Courtesy of Shutterstock.com
Chapter 2: Courtesy of Mark van Coller
Chapter 3: Courtesy of George Karbus
Chapter 4: Courtesy of Clark Little Photography
Chapter 5: Courtesy of Sean Davey
Chapter 6: Courtesy of Shutterstock.com

Cell biology painted by the numbers. The cover image shows a growing microcolony of *E. coli* where the cells are colored by age, defined as the number of divisions since the plating of a single cell. The oldest cells are red and the youngest blue. Published in: (2005) Aging and Death in *E. coli. PLOS Biol* 3(2): e58. doi:10.1371/journal.pbio.0030058. Photograph provided by Eric Stewart and Stefanie Timmermann.

First published 2016 by Garland Science

2 Park Square, Milton Park, Abingdon, Oxfordshire OX14 4RN
52 Vanderbilt Avenue, New York, NY 10017

Routledge is an imprint of the Taylor & Francis Group, an informa business

First issued in paperback 2019

ISBN 978-0-8153-4537-4 (pbk)

The Library of Congress Cataloging-in-Publication data:

Milo, Ron, author.
  Cell biology by the numbers / Ron Milo, Rob Phillips.
      p. ; cm.
    ISBN 978-0-8153-4537-4
    [DNLM: 1. Cells–ultrastructure. 2. Cellular Structures. 3. Data Interpretation, Statistical. QU 350]
    QH585.5.D38
    572—dc23
                                        2015033337

This book is dedicated to our parents,
Ada & Igal and Lee & Bob.

Table of Contents

Detailed Table of Contents

List of Estimates

Preface

> ❝ *I often say that when you can measure what you are speaking about, and express it in numbers, you know something about it; but when you cannot measure it, when you cannot express it in numbers, your knowledge is of a meagre and unsatisfactory kind; it may be the beginning of knowledge, but you have scarcely in your thoughts advanced to the state of Science, whatever the matter may be.* ❞
>
> **William Thomson (Lord Kelvin)** [Popular lectures and addresses, Vol. 1, Electrical Units of Measurement, 1883]

Though Lord Kelvin was unaware of the great strides that one can make by looking at bands on gels without any recourse to numbers, his exaggerated quantitative philosophy focuses attention on the possible benefits of biological numeracy.

One of the great traditions in biology's more quantitative partner sciences, such as chemistry and physics, is the value placed on centralized, curated quantitative data. Whether thinking about the astronomical data that describes the motions of planets or the thermal and electrical conductivities of materials, the numbers themselves are a central part of the factual and conceptual backdrop for these fields. Indeed, often the act of trying to explain why numbers have the values they do ends up being an engine of discovery.

In our view, it is a good time to make a similar effort at providing definitive statements about the values of key numbers that describe the lives of cells. One of the central missions of our book is to serve as an entry point that invites the reader to explore some of the key numbers of cell biology. We hope to attract readers of all kinds—from seasoned researchers, who simply want to find the best values for some number of interest, to beginning biology students, who want to supplement their introductory course materials. In the pages that follow, we provide a broad collection of vignettes, each of which focuses on quantities that help us think about sizes, concentrations, energies, rates, information content, and other key quantities that describe the living world.

However, there is more to our story than merely providing a compendium of important biological numbers. We have tried to find a balance between presenting the data itself and reasoning about these numbers on the basis of simple estimates that provide both surprises and sanity checks. With each vignette, we play with the interaction of two mindsets when thinking about cell biology by the numbers. First, we focus on trying to present in one place the relevant numbers for some particular biological structure or process. A second thrust is to "reason out" the numbers—to try and think about what determines their values and what the biological repercussions of those numbers might be. We are inspired by the so-called "Fermi problems" made famous as a result of the simple estimates made by Enrico Fermi on subjects ranging from the number of piano tuners in a large American city to the advantages of having double windows for thermal insulation in winter. We were interested in the extent to which it is possible to gain insights from a Fermi-inspired order-of-magnitude biology in which simple order-of-magnitude estimates serve as a sanity check on our understanding of biological phenomena.

When our hypothetical readers page to an entry of interest, be it the rate of translation or the number of genes in their favorite organism, we hope to greet them with a vignette that is at once entertaining and surprising. Rather than a dry elucidation of the numbers, as captured in our many tables, we use each vignette as a chance to tell some story that relates to the topic in question. We consider our book to be a quantitative companion to classic textbooks on molecular and cell biology and a source of enrichment for introductory and advanced courses. We thus aim to supply a quantitative component, which we consider an important complementary way of organizing and viewing biological reality. We think that knowing the measure of things is a powerful and different way to get a "feel" for the organisms and their inner life.

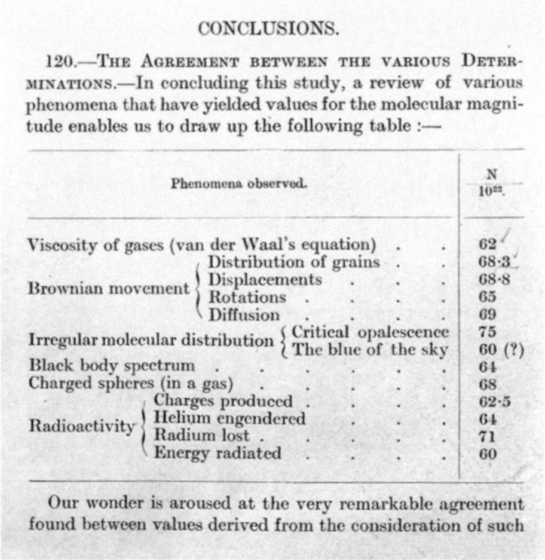
CONCLUSIONS.

120.—THE AGREEMENT BETWEEN THE VARIOUS DETERMINATIONS.—In concluding this study, a review of various phenomena that have yielded values for the molecular magnitude enables us to draw up the following table :—

Phenomena observed.		$\frac{N}{10^{22}}$.
Viscosity of gases (van der Waal's equation)		62
Brownian movement	Distribution of grains	68·3
	Displacements	68·8
	Rotations	65
	Diffusion	69
Irregular molecular distribution	Critical opalescence	75
	The blue of the sky	60 (?)
Black body spectrum		64
Charged spheres (in a gas)		68
Radioactivity	Charges produced	62·5
	Helium engendered	64
	Radium lost	71
	Energy radiated	60

Our wonder is aroused at the very remarkable agreement found between values derived from the consideration of such

Figure P-1 The many measurements of Avogadro's number. The French physicist Jean Perrin in his book *Atoms* noted the broad diversity of ways to determine "atomic dimensions" and was justly proud of the consistent picture of the world that emerged from such different approaches.

Another reason for writing this book emerged from our own research. We often want to do "quick-and-dirty" analyses to estimate time scales, rates, energy scales, or other interesting biological parameters as a sanity check to see if some observation or claim makes sense. The issue is how to make it quick. Looking for key biological numbers using the internet or flipping through textbooks is laborious at best and often futile. It is a common experience that even after hours of searching, we are left either with no result at all or a value with no reference to the experimental conditions that gave rise to that number, hence providing no sense of either the uncertainty or variability in the reported values. Our aspirations are for a biology that can boast the same kind of consistency in its data as is revealed in **Figure P-1**, which shows how in the early twentieth century a host of different methods yielded a surprisingly consistent set of values for Avogadro's number. Often in biology we are not measuring specific physical constants such as Avogadro's number, nevertheless, when measuring the same quantity under identical conditions we should find similar results. One of the points that will come up again in the first chapter is that reproducibility is required first as the basis for recognizing regularities. Then, once scientists are confident in their regularities, it becomes possible to recognize anomalies. Both regularities and anomalies provide a path to new scientific discoveries.

Our vision is that we need a sort of a "cheat sheet" for biology, just like those we got in high school for physical and chemical constants. We hope this book will serve as an extended cheat sheet or a brief version of the handbooks of the exact sciences—namely, those used prevalently in engineering, physics, and so on. Marc Kirschner, the head of the Systems Biology department at Harvard University, compared doing biology without knowing the numbers to learning history without knowing geography. Our aim is that our readers will find this book to be a useful atlas of important biological numbers with allied vignettes that put these numbers in context.

We are well aware that the particular list of topics we have chosen to consider is subjective and that others would have made different choices. We limited our vignettes to those case studies that were consistent with our mutual interests and to topics where we felt we either knew enough or could learn enough to make a first pass at characterizing the state of the art in quantifying the biological question of interest.

The organization of the various numbers in the pages that follow is based upon roughly five different physical axes rather than biological context. First, we provide a narrative introduction to both the mindset and methods that form the basis for the remainder of the book. We offer our views on why we should care about the numbers described here, how to make back-of-the-envelope estimates, and simple rules on using significant digits in writing out numbers. We then begin the "by-the-numbers" survey in earnest by examining the sizes of things in cell biology. This is followed by a number of vignettes whose aim is to tell us how many copies of the various structures of interest are found. Taking this kind of biological census is becoming increasingly important as we try to understand the biochemical linkages that make up the many pathways that have been discovered in cells. The third axis focuses on force and energy scales. The rates of processes in biology form the substance of the fourth section of the book, followed by different ways of capturing the information content of cells. As is often the case in biology, we found that our human effort at rational categorization did not fit nature's appetite for variety, and thus the last section is a biological miscellany that includes some of our favorite examples that defy inclusion under the previous headings.

Unexpectedly to us, as our project evolved, it became ever more clear that there is a hierarchy of accuracy associated with the determination of the numbers we describe. For example, our first chapter deals with sizes of components in the cell, a relatively accurate and mature outgrowth of modern structural biology with its many different microscopies. Our second chapter on the cellular census ramps up the difficulty, with many of the numbers we report coming from very recent research literature, some of which show that calibrations of different methods, such as fluorescence techniques and those based upon antibodies, are not entirely consistent. Chapter 3, which deals with energy scales of various processes within the cell, suffers from challenges as severe as ambiguities in the definition of the quantities themselves. We thought hard about how to represent in writing the uncertainties associated with values that we collected from the literature. The guidelines we follow regarding how many significant digits to use are summarized in the opening chapter. It is our hope that attention to this issue of quantitative sanitation will become the norm among students and researchers in biology.

Inspiration for the approach taken here of "playing" with the numbers has come from many sources. Some of our favorites, which we encourage our readers to check out, include: *Guesstimation* by Lawrence Weinstein and John Adam; John Harte's two books, *Consider a Spherical Cow* and *Consider a Cylindrical Cow*; Richard Burton's *Physiology by Numbers* and *Biology by Numbers*; *Why Big Fierce Animals Are Rare* by Paul Colinvaux; and Sanjoy Mahajan's fine books, *Street Fighting Mathematics* and *The Art of Insight in Science and Engineering: Mastering Complexity*. We are also big fans of the notes and homeworks from courses by Peter Goldreich, Dave Stevenson, and Stirl Phinney on "Order of Magnitude Physics." What all of these sources have in common is the pleasure and value of playing with numbers. In some ways, our vignettes are modeled after the examples given in these other books, and if we have in some measure succeeded in inspiring our readers as much as these others have inspired us, our book will be a success.

Acknowledgments

One of the great pleasures of writing a book such as ours is the many stimulating and thoughtful interactions we have had with our colleagues. The most important such interaction was our long-standing collaboration with our illustrator Nigel Orme. We hope that it will become clear to our readers that by thinking hard about how to visually represent some biological process, we can actually make discoveries about how living systems work. This is similar in spirit to an observation made by Darwin in his autobiography—namely, by formulating our thoughts about systems mathematically we can see things that are completely invisible when couched in verbal language. When we couch our thoughts in careful visual language, it also reveals new ways of understanding biology. Nigel Orme has made us clearer thinkers, better communicators, and has helped us understand more about the world we live in. We are also deeply grateful to Uri Moran, who showed outstanding devotion in searching a wide range of scientific resources and scouring the original literature to find the numbers essential for making this book scientifically meaningful. He never tired of the many iterations we made both of individual vignettes and entire book drafts, and of guiding us through his wide knowledge and peaceful manner.

A number of hearty souls actually read the entire book from cover to cover and provided insights, suggestions, criticisms, and encouragements. This group consisted of Olaf Andersen, David Bensimon, Maja Bialecka-Fornal, Ian Booth, Dennis Bray, Rachel Brockhage, Avantika Lal, Molly Phillips, Rahul Sarpeshkar, and Carolina Tropini. We are deeply grateful to all of them for their insights, which both changed and improved our book substantially and from which we have learned so much. Others who dipped widely into our manuscript, making a host of important comments, include Niv Antonovsky, Wilfredo Ayala-López, Uri Barenholz, Bill Broadhurst, Robert Endres, Arren Bar Even, Wayne Fagerberg, Shai Fuchs, Akira Funahashi, Noriko Hiroi, Ian Kay, Stefan Klumpp, Tetsuya Kobayashi, Jenny Koenig, Patricia Marsteller, Stephen Martin, Steve Quake, Doug Rees, Maya Shamir, and Zhanchun Tu. Of course, it should be clear that mistakes in both the truth or logic of what we say is our fault and not theirs.

Much of the work presented here has been featured in a number of different courses that the two of us have taught. Many students gave us motivation and feedback, and their willingness to serve as guinea pigs while we developed the ideas presented here is very much appreciated. They were members of courses focusing on the subject of Cell Biology by the Numbers given at the Weizmann Institute of Science, Caltech, and Cold Spring Harbor Laboratory in the years 2009–2015.

The two of us have been deeply lucky to have had the chance to work together over these last seven years. We are grateful to Elliot Meyerowitz, who was the matchmaker and who saw the potential for like-minded souls to have a scientific collaboration, despite their 12,000-km separation and 10-hour time difference. Jon Widom also served as a constant source of support and insight during the incipient phases of our project.

Another source of extremely important support whose impact cannot be underestimated is the financial support that it takes to carry out such an enormous undertaking as the BioNumbers website and our book that grew from it. The NIH Director's Pioneer Award on one side of the Atlantic and the European Research Council on the other gave the essential support to the labs of the authors that made this book possible. The Donna and Benjamin M. Rosen Center for Bioengineering at Caltech, under the excellent leadership of Niles Pierce and Frances Arnold, has been a source of generous and constant support, which has funded many critical parts of this project and which has made it possible for us to provide a copy of the book free of charge for those who are less fortunate than well-supported researchers. The development of the BioNumbers website itself was made possible through the vision, generosity, and expertise of Paul Jorgensen, Marc Kirschner, Daniel Kreiter, Ben Marks, Mason Miranda, Uri Simhony, Mike Springer, and Becky Ward. In addition, we are deeply grateful to the team at Garland Science who has joined us in this adventure, with special thanks to our editor Summers Scholl, who has been a steady supporter of the project, to Natasha Wolfe, who expertly served as production editor, and to John Murdzek for improving the readability and accuracy of our book through his copy editing.

Our list of specific scientific advisors is long and distinguished. These generous friends provided us insights on everything from the size of synapses to the scaling of the nucleus size to the physical limits of the senses. We are grateful to Yaarit Adamovich, Noam Agmon, Ariel Amir, Shira Amram, Tslil Ast, David Baltimore, Naama Barkai, Scott Barolo, Mark Bathe, Nathan Belliveau, Danny Ben-Zvi, Howard Berg, Michael Brenner, James Briscoe, Jan Brugues, Tabita Bucher, Richard Burton, Michael Clague, Bil Clemons, Adam Cohen, Athel Cornish-Bowden, Peter Curtin, Dan Davidi, Avigdor Eldar, Yuval Eshed, Daniel Fisher, Avi Flamholtz, Sarel Fleishman, Dan Fletcher, Michael Gliksberg, Dani Gluck, Matthew Good, Talia Harris, Yuval Hart, Elizabeth Haswell, Peter Henderson, Hermann-Georg Holzhütter, Joe Howard, KC Huang, Shalev Itzkovitz, Grant Jensen, Daniel Koster,

Eric Lander, Michael Laub, Heun Jin Lee, Avi Levy, Wendell Lim, Jennifer Lippincott-Schwartz, Jeff Marlow, Wallace Marshall, Sarah Marzen, Markus Meister, Joe Meyerowitz, Lishi Mohapatra, Phil Mongiovi, Daniel Needelman, Ronny Neumann, Tom O'Halloran, Victoria Orphan, Linnaea Ostroff, Michal Perach, Matthieu Piel, Tzachi Pilpel, Rami Pugatch, Tobias Reichenbach, Joel Rothman, Michael Roukes, James Saenz, Dave Savage, Elio Schaechter, Declan Schroeder, Oren Schuldiner, Maya Schuldiner, Eran Segal, Pierre Sens, Joshua Shaevitz, Tom Shimizu, Tomer Shlomi, Alex Sigal, Thomas Simmen, Victor Sourjik, Todd Squires, Lucia C. Strader, Lubert Stryer, Arbel Tadmor, Amos Tanay, Dan Tawfik, Julie Theriot, Elitza Tocheva, Katja Tummler, Antoine van Oijen, Alexander Varshavsky, Doug Weibel, Talia Weiss, Irving Weissman, Jeremy Young, Yossi Yovel, and Jesse Zalatan.

We would like to thank the following societies and publishers for granting permission to adapt or republish material appearing in the following figures: The American Association for the Advancement of Science for Figures 1-16, 2-7, 2-9, 2-18, 2-22, 2-24, 2-27, 3-18, 4-12, 4-37, 4-44, 5-4, and 5-10. The American Chemical Society for Figure 1-25. The American Institute of Physics for Figure 3-2. The American Physical Society for Figure 3-14. The American Society for Biochemistry and Molecular Biology for Figure 2-14. The American Society for Cell Biology for Figures 1-19 and 1-23. The American Society for Microbiology for Figures 1-14, 2-34, and 4-13. The Association for Research in Vision and Ophthalmology for Figure 2-37. The Bibliothèque de Genève for Figure 2-2. Elsevier Inc. for Figures 1-2, 1-5, 1-22, 1-26, 1-47, 2-21, 2-31, 2-35, 2-42, 4-2, 4-4, 4-14, 4-16, 4-24, 4-31, 4-34, 4-36, 4-38, 4-39, 4-46, 5-12, 6-2, and 6-9. John Wiley & Sons for Figures 1-12, 2-38, 2-43, 3-13, 4-18, 4-27, and 4-32. Macmillan Publishers Ltd. for Figures 1-6, 1-24, 1-29, 1-31, 2-13, 2-16, 2-19, 2-20, 2-29, 3-5, 3-10, 3-17, 4-30, 4-41, 5-8, 5-9, 6-3, and 6-4. The National Academy of Sciences for Figures 1-3, 1-20, 1-41, 2-6, 2-10, 2-15, 2-17, 2-36, 3-9, 3-19, and 6-6. The Royal Society of Chemistry for Figure 4-17. The Rockefeller University Press for Figure 1-46. The Society for General Microbiology for Figure 2-11. Springer Science + Business Media for Figures 1-21 and 2-12. The Trustees of the Natural History Museum, London for Figure 1-15.

In many ways that go far beyond the science, we were supported by Yoram Altman, Uri Alon, Efrat Binya Talmi, Efrat Cohen, Hernan Garcia, David Goodsell, Alon Gildoni, Alex Kizin, Jane Kondev, Ronen Levi, Yael Milo, Elad Noor, Daniel Ramot, Richard Sever, and Julie Theriot.

It is a long-held tradition that book acknowledgments end with heartfelt thanks to families. Of course, there is a reason for this and that is this: Though the lives of those who write books are so deeply enriched by the act of writing, they always know that there is something that matters much more, and that is our loved ones. Both of us value our families above all and are deeply grateful to our spouses, Hilla and Amy, our parents, Ada and Igal Milo and Lee Cudlip and Bob Phillips, and our children, Geffen, Yaara, and Rimon and Molly and Casey, for bringing magic to our lives.

The Path to Biological Numeracy

> 66 *...[I]n after years I have deeply regretted that I did not proceed far enough at least to understand something of the great leading principles of mathematics, for men thus endowed seem to have an extra sense.* 99
>
> [**Charles Darwin,** Autobiography, 1887]

WHY WE SHOULD CARE ABOUT THE NUMBERS

This introduction sets the stage for what is to unfold in upcoming chapters. If you feel the urge to find some number of interest now, you can jump to any vignette in the book and come back later to this chapter, which presents both the overall logic and the basic tools used to craft biological numeracy. Each of the $\approx 10^2$ vignettes in the book can be read as a stand-alone answer to a quantitative question on cell biology by the numbers. The formal structure for the remainder of the book is organized according to different classes of biological numbers, ranging from the sizes of things (Chapter 1) to the quantitative rules of information management in living organisms (Chapter 5) and everything in between. The goal of this section of the book is decidedly more generic, laying out the case for biological numeracy and providing general guidelines for how to arrive at these numbers using simple estimates. We also pay attention to the question of how to properly handle the associated uncertainty in both biological measurements and estimates. We build on the principles developed in the physical sciences, where estimates and uncertainties are common practice, but in our case require adaptation to the messiness of biological systems.

What is gained by adopting the perspective of biological numeracy we have called "cell biology by the numbers"? The answer to this question can be argued along several different lines. For example, one enriching approach to thinking about this question is by appealing to the many historic examples, where the quantitative dissection of a given problem is what provided the key to its ultimate solution. Examples abound, whether from the classic discoveries in genetics that culminated in Alfred Sturtevant's map of the geography of the *Drosophila* genome or Hodgkin and Huxley's discoveries of the quantitative laws that govern the dynamics of nerve impulses. More recently, the sharpness of the questions, as formulated from a quantitative perspective, has yielded insights into the limits of biological information transmission in processes ranging from bacterial chemotaxis to embryonic development and has helped establish the nature of biological proofreading that makes it possible for higher fidelity copying of the genetic material than can be expected from thermodynamics alone (some of these examples appear in a paper we wrote together[*]).

A second view of the importance of biological numeracy centers on the way in which a quantitative formulation of a given biological phenomenon allows us to build sharp and falsifiable claims about how it works. Specifically, the state of the art in biological measurements is beginning to reach the point of reproducibility, precision, and accuracy where we can imagine discrepancies between theoretical expectations and measurements that can uncover new and unexpected phenomena. Further, biological numeracy allows scientists an "extra sense," as already appreciated by Darwin himself, to decide whether a given biological claim actually makes sense. Said differently, with any science, in the early stages there is a great emphasis on elucidating the key facts of the field. For example, in astronomy, it was only in light of advanced naked-eye methods in the hands of Tycho Brahe that the orbit of Mars was sufficiently well understood to elucidate central facts, such as that Mars travels around the sun in an elliptical path with the sun at one of the foci. But with the maturity of such facts comes a new theoretical imperative—namely, to explain those facts on the basis of some underlying theoretical framework. For example, in the case of the observed elliptical orbits of planets, it was an amazing insight to understand how this and other features of planetary orbits were the natural consequence of the inverse-square law of gravitation. We believe that biology has reached the point where there has been a sufficient accumulation of solid quantitative facts that this subject, too, can try to find overarching principles expressed mathematically that serve as theory to explain those facts and to reveal irregularities when they occur. In the chapters that follow, we provide a compendium of such biological facts, often presented with an emphasis that might help as a call to arms for new kinds of theoretical analysis.

Another way to think about this quest for biological numeracy is to imagine some alien form coming to Earth and wishing to learn more about

---

[*] Phillips R & Milo R (2009) A feeling for the numbers in biology. *Proc Natl Acad Sci USA* 106:21465–21471.

what our society and daily lives look like. For example, if we could give the friendly alien a single publication, what such publication might prove most useful? Though different readers may come up with different ideas of their own, our favorite suggestion would be the report of the Bureau of Statistics, which details everything from income to age at marriage to level of education to the distributions of people in cities and in the country. The United Nations posts such statistics on their website: https://unstats. un.org/unsd/default.htm.

Hans Rosling has become an internet sensation as a result of the clever and interesting ways that he has found not only to organize data, but also to milk it for unexpected meaning. Our goal is to provide a kind of report of the bureau of statistics for the cell and to attempt to find the hidden and unexpected meaning in the economy and geography of the cell.

As an example of the kind of surprising insights that might emerge from this exercise, we ask our readers to join us in considering mRNA, the "blueprint" for the real workhorses of the cell, the proteins. Quickly, ask yourself: Which is larger, the blueprint or the thing being blueprinted? Our intuition often thinks of the blueprint for a giant skyscraper, and it is immediately obvious that the blueprint is but a tiny and flattened caricature of the building it "codes for." But what of our mRNA molecule and the protein it codes for? What is your instinct about the relative size of these two molecules? As we will show in the vignette entitled "Which is bigger, mRNA or the protein it codes for?" (pg. 43), most people's intuition is way off, with the mRNA molecule actually being substantially larger than the protein it codes for. This conclusion has ramifications, for example for whether it is easier to transport the blueprint or the machine it codes for.

Finally, we are also hopeful for a day when there is an increasing reliance in biology on numerical anomalies as an engine of discovery. As the measurements that characterize a field become more mature and reproducible using distinct methodologies, it becomes possible to reliably ask the question of when a particular result is anomalous. Until the work of David Keeling in the 1950s, no one could even agree on what the level of CO_2 in the atmosphere was, let alone figure out if it was changing. Once Keeling was able to show the rhythmic variations in CO_2 over the course of a year, then questions about small overall changes in the atmospheric CO_2 concentration over time could be addressed. Perhaps more compellingly, Newton was repeatedly confounded by the 20% discrepancy between his calculated value for the speed of sound and the results from measurements. It was only many years later that workers such as Laplace realized that a treatment of the problem as an adiabatic versus isothermal process could explain that discrepancy. The recent explosion of newly discovered extrasolar planets is yet another example where small numerical anomalies are received with such confidence that they can be used as a tool of discovery. In our view, there is no reason at all to believe that similar insights don't await those studying living matter once our measurements have been codified to the point that we know what is irregular when we see it. In a situation where there are factors of 100 distinguishing different answers to the same question, such as how many proteins are in an *E. coli* cell, there is little chance to discern even regularities, let alone having confidence that anomalies are indeed anomalous. Often, the great "effects" in science are named such because they were signaled as anomalous. For example, the change in wavelength of an oncoming ambulance siren is known as the famed Doppler effect. Biochemistry has effects of its own, such as the Bohr effect, which is the shift in binding curves for oxygen to hemoglobin as a function of the pH. We suspect that there are many such effects awaiting discovery in biology as a result of reproducibly quantifying the properties of cells and then paying close attention as to what those numbers can tell us.

THE BIONUMBERS RESOURCE

As a reminder of how hard certain biological numbers are to come by, we recommend the following quick exercise for the reader. Pick a topic of particular interest from molecular or cell biology and then seek out the corresponding numbers through an internet search or by browsing your favorite textbooks. For example, how many ribosomes are there in a human cell? Or, what is the binding affinity of a celebrated transcription factor to DNA? Or, how many copies are there per cell of any famous receptor, such as those of chemotaxis in bacteria or of growth hormones in mammalian cells? Our experience is that such searches are at best time-consuming, and often they are inconclusive or even futile. As an antidote to this problem, essentially all of the numbers presented in this book can be found from a single source—namely, the BioNumbers website (http://bionumbers.hms.harvard.edu/). The idea of this internet resource is to serve as an easy jumping-off point for accessing the vast biological literature in which quantitative data is archived. In particular, the data to be found in the BioNumbers database have been subjected to manual curation, have full references to the primary literature from which the data are derived, and provide a brief description of the method used to obtain the data in question.

As signposts for the reader, each and every time that we quote some number, it will be tied to a reference for a corresponding BioNumbers Identification (BNID). Just as our biological readers may be familiar with the PMID, which is a unique identifier assigned to published articles from the biological and medical literature, the BNID serves as a unique identifier of different quantitative biological data. For example, BNID 103023 points us to one of several determinations of the number of mRNA per yeast cell. The reader will find that both our vignettes and the data tables are filled with BNIDs, and by pasting this number into the BioNumbers website (or just Googling "BNID 103023"), the details associated with that particular quantity can be uncovered.

HOW TO MAKE BACK-OF-THE-ENVELOPE CALCULATIONS

The numbers to be found in the BioNumbers compendium and in the vignettes throughout this book can be thought of as more than simply data. They can serve as anchor points to deduce other quantities of interest and can usually be themselves scrutinized by putting them to a sanity test based on other numbers the reader may know and bring together by "pure thought." We highly recommend the alert reader to try and do such cross tests and inferences. This is our trail-tested route to powerful numeracy. For example, in Chapter 4 we present the maximal rates of chromosome replication. But we might make an elementary estimate of this rate by using our knowledge of the genome length for a bacterium and the length of the cell cycle. Often such estimates will be crude (say, to within a factor of two), but they will be good enough to tell us the relevant order of magnitude as a sanity check for measured values.

There are many instances in which we may wish to make a first-cut estimate of some quantity of interest. In the middle of a lecture you might not have access to a database of numerical values, and even if you do, this skill of performing estimates and inferring the bounds from above and below as a way to determine unknown quantities is a powerful tool that can illuminate the significance of measured values.

One handy tool is how to go from upper and lower bound guesses to a concrete estimate. Let's say we want to guess at some quantity. Our first step is to find a lower bound. If we can say that the quantity we are after is bigger than a lower bound x_L and smaller than an upper bound x_U, then a simple estimate for our quantity of interest is to take what is known as the geometric mean—namely,

$$x_{\text{estimate}} = \sqrt{x_L x_U} .$$

(0.1)

Though this may seem very abstract, in most cases we can ask ourselves a series of questions that allow us to guess reasonable upper and lower bounds to within a factor of 10. For example, if we wish to estimate the length of an airplane wing on a jumbo jet, we can begin with "Is it bigger than 1 m?". Yes. "Is it bigger than 5 m?" Yes. "Is it bigger than 10 m?" I think so, but I'm not sure. So, we take 5 m as our lower bound. Now the other end, "Is it smaller than 50 m?" Yes. "Is it smaller than 25 m?" I think so, but I'm not sure. So, we take 50 m as our upper bound. Using 5 m and 50 m as our lower and upper bounds, respectively, we then estimate the wing size as $\sqrt{5\text{m} \times 50\text{m}} \approx 15$ m, the approximate square root of 250 m². If we had been a bit more bold, we could have used 10 m as our lower bound, with the result that our estimate for the length of the wing would be ≈22 m. In both cases we are accurate to within a factor of two compared with the actual value—that is, well within the target range of values we expect from "order-of-magnitude biology."

Let's try a harder problem, which will challenge the intuition of anyone we know. What would you estimate is the number of atoms in your body? 10^{10} is probably too low—in fact, that sounds more like the number of people on Earth. 10^{20}? Maybe, but that vaguely reminds us of the exponent in Avogadro's number. 10^{80} sounds way too high, because such exponents are reserved for the number of atoms in the universe. 10^{40}? Maybe. So, $\sqrt{10^{20} \times 10^{40}} \sim 10^{30}$. A more solid calculation is given later in the book using the Avogadro constant (can you see how to do it?), but it suffices to say that we are within about two orders of magnitude of the correct order of magnitude, and this is based strictly on educated guessing. We may object to pulling 10^{20} and 10^{40} out of thin air, but this is exactly the kind of case where we have extremely little intuition and thus have nothing to start with aside from vague impression. But we can still construct bounds by eliminating estimates that are too small and too large as we did above, and somewhat surprisingly, with the aid of the geometric mean, that takes us close to the truth. We probably have to try this scheme out several times to check if the advertised effectiveness actually works. The geometric mean amounts really to taking the normal arithmetic mean in log space (that is, on the exponents of 10). Had we chosen to take the normal mean on the values we guessed, our estimate would have been completely dominated by the upper bound we chose, which often leads to extreme overestimation.

One question worth asking is: How do we know whether our estimates are actually "right"? Indeed, often those who aren't used to making estimates fear of getting the "wrong" answer. In his excellent book, *Street Fighting Mathematics*, Sanjoy Mahajan makes the argument that an emphasis on this kind of "rigor" can lead, in fact, to mathematical "rigor mortis." The strategy we recommend is to think of estimates as successive approximations, with each iteration incorporating more knowledge to refine what the estimate actually says. There is no harm in making a first try and getting a "wrong" answer. Indeed, part of the reason such estimates are worthwhile is that they begin to coach our intuition so that we can *a priori* have a sense of whether a given magnitude makes sense or not without even resorting to a formal calculation.

ORDER-OF-MAGNITUDE BIOLOGY TOOLKIT

As noted above, one of the most elusive, but important, skills is to be able to quickly and efficiently estimate the orders of magnitude associated with some quantity of interest. Earlier, we provided some of the conceptual rules that fuel such estimates. Here, we complement those conceptual rules with various helpful numerical rules that can be used to quickly find our way to an approximate but satisfactory assessment of some biological process of interest. We do not expect you to remember them all on first pass, but give them a quick look, and maybe a few of them will stick in the back of your mind when you need them.

Arithmetic sleights of hand

- $2^{10} \approx 1000$
- $2^{20} = 4^{10} \approx 10^6$
- $e^7 \approx 10^3$
- $10^{0.1} \approx 1.3$
- $\sqrt{2} \approx 1.4$
- $\sqrt{0.5} \approx 0.7$
- $\ln(10) \approx 2.3$
- $\ln(2) \approx 0.7$
- $\log_{10}(2) \approx 0.3$
- $\log_{10}(3) \approx 0.5$
- $\log_2(10) \approx 3$

Big numbers at your disposal

- Seconds in a year $\approx \pi \times 10^7$ (the approximate value of pi, a nice coincidence and an easy way to remember this value)
- Seconds in a day $\approx 10^5$

- Hours in a year $\approx 10^4$
- Avogadro's constant $\approx 6 \times 10^{23}$
- Cells in the human body $\approx 4 \times 10^{13}$

Rules of thumb

Just as there are certain arithmetical rules that help us quickly get to our order-of-magnitude estimates, there are also physical rules of thumb that can similarly extend our powers of estimation. We give here some of our favorites and you are most welcome to add your own at the bottom of our list and also send them to us. Several of these estimates are represented pictorially as well. Note that here and throughout the book we try to follow the correct notation where "approximately" is indicated by the symbol $\approx$, and loosely means accurate to within a factor of two or so. The symbol $\sim$ means "order of magnitude," so only to within a factor of 10 (or in a different context it means "proportional"). We usually write approximately because we know the property value indeed roughly but to better than a factor of 10, so $\approx$ is the correct notation and not $\sim$. In the cases where we only know the order of magnitude, we will write the value only as 10^x without extraneous significant digits.

- 1 dalton (Da) = 1 g/mol $\approx 1.6 \times 10^{-24}$ g (as derived in **Estimate 0-1**).
- 1 nM is about 1 molecule per bacterial volume, as derived in **Estimate 0-2**, 10^1–10^2 per yeast cell, and 10^3–10^4 molecules per characteristic mammalian (HeLa) cell volume. For 1 µM, multiply by a thousand; for 1 mM, multiply by a million.
- 1 M is about one per 1 nm^3.
- There are 2–4 million proteins per 1 $µm^3$ of cell volume.

Converting between daltons and grams

H atom

$m_H = 1$ Da

$$1 \text{ g of hydrogen} = N_A \times m_H \implies m_H = \frac{1 \text{ g}}{6 \times 10^{23}} \approx 1.6 \times 10^{-24} \text{g}$$

Avogadro's number hydrogen atom mass

Estimate 0-1

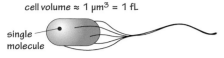

Nanomolar in *E. coli* units

cell volume ≈ 1 μm³ = 1 fL

single
molecule

$$c = \frac{1 \text{ molecule}}{1 \text{ fL}} \times \frac{1 \text{ mole}}{6 \times 10^{23} \text{ molecules}} \times \frac{1 \text{ fL}}{10^{-15} \text{ L}} = \frac{1}{6} \times 10^{-8} \text{ M} \approx 1.6 \text{ nM}$$

⟹ rule of thumb: 1 molecule per bacterial volume ≈ 1 nM

Estimate 0-2

- Concentration of 1 ppm (part per million) of the cell proteome is ≈ 5 nM.
- 1 mg of DNA fragments 1 kb long is ≈ 1 pmol or ≈ 10^{12} molecules.
- Under standard conditions, particles at a concentration of 1 M are ≈ 1 nm apart.
- Mass of typical amino acid ≈ 100 Da.
- Protein mass [Da] ≈ 100 × Number of amino acids.
- Density of air ≈ 1 kg/m³.
- Water density ≈ 55 M ≈ × 1000 that of air ≈ 1000 kg/m³.
- 50 mM osmolites ≈ 1 atm osmotic pressure (as shown in **Estimate 0-3**).
- Water molecule volume ≈ 0.03 nm³, (≈ 0.3 nm)³.
- A base pair has a volume of ≈ 1 nm³.
- A base pair has a mass of ≈ 600 Da.
- Lipid molecules have a mass of ≈ 500–1000 Da.
- 1 $k_B T$ ≈ 2.5 kJ/mol ≈ 0.6 kcal/mol ≈ 25 meV ≈ 4 pN nm ≈ 4×10^{-21} J.
- ≈ 6 kJ/mol sustains one order of magnitude concentration difference [= $k_B T \ln(10)$ ≈ 1.4 kcal/mol].

Relating solute concentration to osmotic pressure

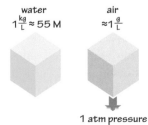

water
$1 \frac{kg}{L} \approx 55$ M

air
$\approx 1 \frac{g}{L}$

50 mM solute more than surrounding ≈
1/1000 of water conc. ≈ air conc. ≈
1 atm osmotic pressure

1 atm pressure

air density ≈ 1/1000 of water

Estimate 0-3

- Movement across the membrane is associated with 10–20 kJ/mol per one net charge due to membrane potential.
- ATP hydrolysis under physiological conditions releases $20\ k_B T \approx 50$ kJ/mol ≈ 12 kcal/mol $\approx 10^{-19}$ J.
- One liter of oxygen releases ≈ 20 kJ during respiration.
- A small metabolite diffuses 1 nm in ~ 1 ns.
- 1 $OD_{600} \approx 0.5$ g cell dry weight per liter.
- There are $\approx 10^{10}$ carbon atoms in a 1 μm^3 cell volume.

RIGOROUS RULES FOR SLOPPY CALCULATIONS

One of the most important questions that all readers should ask themselves is: Are any of the numbers in this book actually "right"? What does it even mean to assign numbers to quantities such as sizes, concentrations, and rates that are so intrinsically diverse? Cellular processes show immense variability, depending upon both the type of cell in question and the conditions to which it has been subjected. One of the insights of recent years that has been confirmed again and again is that even within a clonal population of cells, there is wide cell-to-cell variability. Hence, both the diversity and intrinsic variability mean that the task of ascribing particular numbers to biological properties and processes is fraught with the danger of misinterpretation. One way to deal with this challenge is by presenting a range of values rather than "the value." Equally important, a detailed discussion of the environmental conditions under which the cells grew and when and how the measurement was taken and analyzed is in order. Unfortunately, this makes the discussion very cumbersome and is often resolved in textbooks and journals by avoiding concrete values altogether. We choose in this book to give concrete values that clearly do not give the "full" picture. We expect, and caution the reader to do the same, to think of them only as rough estimates and as an entry point to the literature. Whenever a reader needs to rely on a number for research rather than merely obtain a general impression, he or she will need to turn to the original sources. For most values given in this book, finding a different source that reports a value that is a factor of two higher or lower is the rule rather than the exception. We find that knowing the "order of magnitude" can be very useful, and we give examples in the text. Yet, awareness of the inherent variability is critical so as not to get a wrong impression or perform inferences that are not merited by the current level of data. Variety (and by extension, variability) is said to be the spice of life—it is definitely evident at the level of the cell and should always be kept in the back of our minds when discussing values of biological properties.

How many digits should we include when reporting the measured value of biological entities such as those discussed throughout this book? Though this question might sound trivial, in fact there are many subtle issues we had to grapple with that can affect the reader's capability to use these numbers in a judicious fashion. To give a concrete example, say you measured the number of mitochondria in three cells and found 20, 26, and 34. The average is 26.666..., so how should you best report this result? More specifically, how many significant digits should you include to characterize these disparate numbers? Your spreadsheet software will probably entice you to write something like 26.667. Should it be trusted?

Before we dig deeper, we propose a useful conservative rule of thumb. If you forget everything we write below, try to remember this: It is usually a reasonable choice when reporting numbers in biology to use two significant digits. This will often report all valuable information without the artifact of too many digits giving a false sense of accuracy. If you write more than three, we hope some inner voice will tell you to think about what it means or just press the backspace key.

We now dive deeper. Significant digits are all digits that are not zero, plus zeros that are to the right of the first nonzero digit. For example, the number 0.00502 has three significant digits. Significant digits should supply information on the precision of a reported value. The last significant digit—that is, the rightmost one—is the digit that we might be wrong about, but it is still the best guess we have for the accurate value. To find what should be considered significant digits, we will use a rule based on the precision (repeatability) of the estimate. The precision of a value is the repeatability of the measurement, given by the standard deviation, or in the case of an average, by the standard error. If the above sentence confuses you, be assured that you are in good company. Keep on reading and make a mental note to visit Wikipedia at your leisure for these confusing terms, as we do ourselves repeatedly.

Going back to the example above of counting mitochondria, a calculator will yield a standard deviation of 4.0552.... The rule we follow is to report the uncertainty with one significant digit. Thus 4.0552 is rounded to 4, and we report our estimate of the average simply as 26, or more rigorously as 26 ± 4. The last significant digit reported in the average (in this case, 6) is at the same decimal position as the first significant digit of the standard error (in this case, 4). We note that a leading 1 in some conventions does not count as a significant digit (for example, writing 123 with one significant digit will be 120) and that in some cases it is useful to report the uncertainty with two digits rather than just one, but that should not bother us further at this point. But be sure to stay away from using three or more digits in the uncertainty range. Anyone further interested can read a whole report (http://tinyurl.com/nwte4l5) on the subject.

Unfortunately, for many measured values relating to biology, the imprecision is not reported. Precision refers to how much variation you have in your measurements, whereas accuracy refers to how different it is from the real value. A systematic error will cause an inaccuracy but not an imprecision. Precision you can know from your measurements, but for knowing accuracy you have to rely on some other method. You might want to add the distinction between accuracy and precision to your Wikipedia reading list, but bear with us for now. Not only is there no report of the imprecision (error) in most biological studies, but the value is often written with many digits—well beyond what should be expected to be significant given the biological repeatability of the experimental setting. For example, the average for the volume of a HeLa cell may be reported as 2854.3 μm^3. We find, however, that reporting a volume in this way is actually misleading, even if this is what the spreadsheet told the researcher. To our way of thinking, attributing such a high level of precision gives the reader a misrepresentation of what the measurement achieved or what value to carry in mind as a rule of thumb.

Because the uncertainty in such measurements is often not reported, we resort to general rules of thumb, as shown in **Estimate 0-4**. Based on reading many studies, we expect many biological quantities to be known with only twofold accuracy, in very good cases maybe to 10%, and in quite variable cases to within 5- or 10-fold accuracy. Low accuracy is usually not because of the tools of measurement, which have very good precision, but because systematic differences, say, due to growth conditions being different, can lead to low accuracy with respect to any application where the value can be used. In this book, we choose to make the effort to report values with a number of digits that implicitly conveys the uncertainty. The rules of thumb we follow are summarized in Estimate 0-4 as a work flow to infer how many significant digits should be used in reporting a number based on knowing the uncertainty or by estimating the level of uncertainty. For example, say we expect the reported HeLa cell average volume to have 10% inaccuracy (pretty good accuracy for biological data)—that is, about 300 μm^3. As discussed above, we report the uncertainty using one significant digit—that is, all the other digits are rounded to zero. We can now infer that the volume should be written as 2900 μm^3 (two significant digits). If we thought the value has a twofold uncertainty—that is, about 3000 μm^3, we would report the average as 3000 μm^3 (one significant digit).

How to report a value with the appropriate number of significant digits

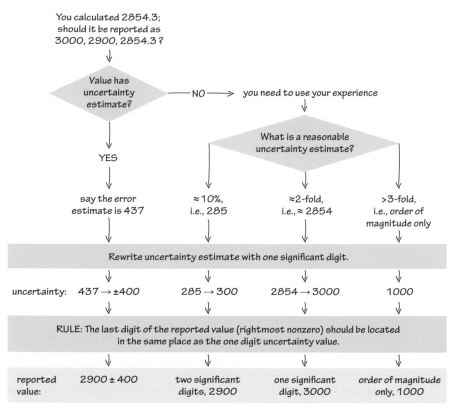

We prefer the convention where a leading 1 does not count as a significant digit—i.e., 1234 with one significant digit is 1200; for order of magnitude round in log space—i.e., 3000 → 1000; 4000 → 10000

Estimate 0-4

Finally, if we think there are very large imprecisions, say, to a factor of five or 10, we will resort to reporting only the order of magnitude (that is, 1000 µm³), or better still, we will write it in a way that reflects the uncertainty as 10^3 µm³. We indicate only an order of magnitude in cases when the expected imprecision is so large (practically, larger than threefold) that we cannot expect to have any sense of even one digit and have an estimate only of the number of digits in the accurate value. The digit 1 is special in the sense that it doesn't mean necessarily a value of 1, but rather signifies the order of magnitude. So, in such a case the number can be thought of as reported with less than one significant digit. If you write 100, how do you know if this is merely an order of magnitude, or should be actually interpreted as precise to within twofold or maybe even 10% (that is, also the following zero is precise)? In one convention, this ambiguity can be solved by putting an underline for the last significant digit. So 10̲0̲ shows the zero (and the 1) are significant digits, 1̲00 shows the 1 is a significant digit, whereas plain 100 is only to within an order of magnitude. We try to follow this convention in this book. Trailing zeros are by custom used as a replacement for the scientific notation (as in 3×10^3). The scientific notation is more precise in its usage of digits but less intuitive in many cases. The trailing zeros should not be interpreted as indicating a value of exactly zero for those digits, unless specifically noted (for example, with an underline).

We often will not write the uncertainty, because in many cases it was not reported in the original paper the value came from, and thus we do not really know what it is. Yet, from the way we write the property value, the reader can infer something about our ballpark estimate based on the norms above. Such an implicit indication of the expected precision should be familiar, as in the following example borrowed from the excellent book *Guesstimation* by Lawrence Weinstein and John A. Adam. A friend gives you driving directions and states you should be taking a left turn after 20 km. Probably when you reach 22 km and did not see a turn, you would start to get worried. But if the direction had been to take the turn after 20.1 km, you would probably have become suspicious before you reached even 21 km.

When aiming only to find orders of magnitude, we perform the rounding in log space—that is, 3000 would be rounded to 1000, while 4000 would be rounded to 10,000 because $\log_{10}(4) > 0.5$. We follow this procedure because our perception of the world, as well as many error models of measurement methods, are logarithmic (that is, we perceive fold changes rather than absolute values). Thus, the log scale is where the errors are expected to be normally distributed, and the closest round number should be found. When performing a series of calculations (multiplying, subtracting, etc.), it is often prudent to keep more significant digits than will be kept to report final results and perform the rounding only at the end result stage. This is most relevant when subtraction cancels out the leading digits making the following digits critical. We are under the impression that following such guidelines can improve the quantitative hygiene essential for properly using and interpreting numbers in cell biology.

THE GEOGRAPHY OF THE CELL

The vignettes that take center stage in the remainder of the book characterize many aspects of the lives of cells. There is no single path through the mass of data that we have assembled here, but nearly all of it refers to cells, their structures, the molecules that populate them, and how they vary over time. As we navigate the numerical landscape of the cell, it is important to bear in mind that many of our vignettes are intimately connected. For example, when thinking about the rate of rotation of the flagellar motor that propels bacteria forward, as discussed in the rates chapter, we will do well to remember that the energy source that drives this rotation is the transmembrane potential discussed in the energy and forces chapter. Further, the rotation of the motor is what governs the motility speed of those cells, a topic with quantitative enticements of its own. Though we will attempt to call out the reticular attachments between many of our different bionumbers, we ask the reader to be on constant alert for the ways in which our different vignettes can be linked up, many of which may harbor some new insights.

To set the cellular stage for the remainder of the book, in this brief section, we highlight three specific model cell types that will form the basis for the coming chapters. Our argument is that by developing intuition for the "typical" bacterium, the "typical" yeast cell, and the "typical" mammalian cell, we will have a working guide for launching into more specialized cell types. For example, even when introducing the highly specialized photoreceptor cells, which are the beautiful outcome of the evolution of "organs of extreme perfection" that so puzzled Darwin, we will still have our "standard" bacterium, yeast, and mammalian cells in the back of our minds as a point of reference. This does not imply a naïveté on our side about the variability of these "typical" cells—indeed, we have several vignettes on these very issues. It is rather an appreciation of the value of a quantitative mental description of a few standard cells that can serve as a useful benchmark to begin the quantitative tinkering that adapts to the biological case at hand, much as a globe gives us an impression of the relative proportion of our planet that is covered by oceans and landmasses, and the key geographical features of those landmasses, such as mountain ranges, deserts, and rivers.

Figure 0-1 gives a pictorial representation of our three standard cell types and **Figure 0-2** complements it by showing the molecular census associated with each of those cell types. This figure goes hand in hand with **Table 0-1** and can be thought of as a compact visual way of capturing the various numbers housed there. In some sense, much of the remainder of our book focuses on asking the following questions: Where do the numbers in the following figures and table come from? Do they make sense? What do they imply about the functional lives of cells? In what sense are cells the "same" and in what sense are they "different"?

Figure 0-1A shows us the structure of the bacterium *E. coli*. Figure 0-2A shows its molecular census. The yeast cell shown in Figure 0-1B and Figure 0-2B reveals new layers of complexity beyond that seen in the standard bacterium, as we see that these cells feature a variety of internal

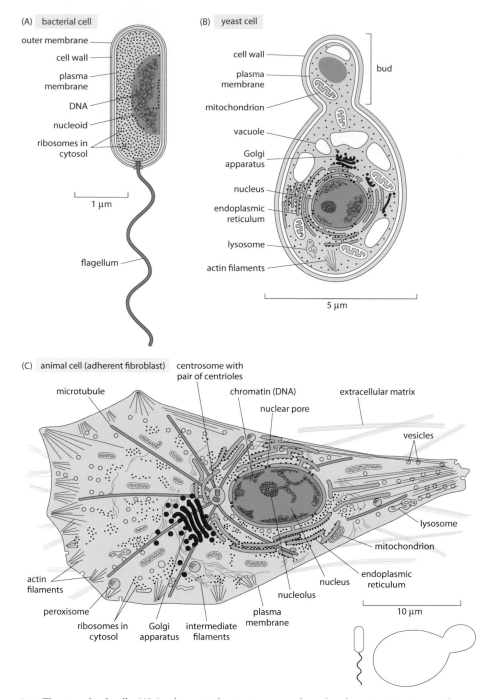

Figure 0-1 The standard cells. (A) A schematic bacterium revealing the characteristic size and components of *E. coli*. (B) A budding yeast cell showing its characteristic size, its organelles, and various classes of molecules present within it. (C) An adherent human cell. We note that these are very simplified schematics. For example, only a small fraction of ribosomes are drawn. Each cell is drawn to a different scale, as indicated by the distinct scale bars in each schematic. The relative sizes of the bacterial and yeast cells at the same scale as the mammalian cell are shown in the bottom right. (A, and C, adapted from Alberts B, Johnson A, Lewis J et al. [2015] Molecular Biology of the Cell, 6th ed. Garland Science.)

(A) bacterial cell (specifically, *E. coli*: $V \approx 1\ \mu m^3$; $L \approx 1\ \mu m$; $\tau \approx 1$ hour)

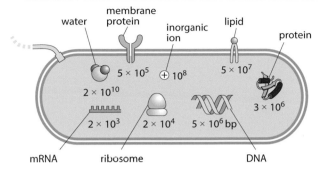

(B) yeast cell (specifically, *S. cerevisiae*: $V \approx 30\ \mu m^3$; $L \approx 5\ \mu m$; $\tau \approx 3$ hours)

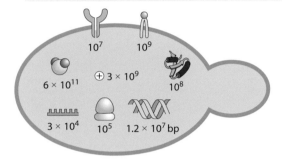

(C) mammalian cell (specifically, HeLa: $V \approx 3000\ \mu m^3$; $L \approx 20\ \mu m$; $\tau \approx 1$ day)

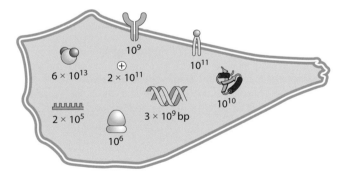

Figure 0-2 An order-of-magnitude census of the major components of the three model cells we employ often in the lab and in this book. A bacterial cell (*E. coli*), a unicellular eukaryote (the budding yeast *S. cerevisiae*), and a mammalian cell line (such as an adherent HeLa cell).

membrane-bound structures. One of the key reasons that yeast cells have served as representatives of eukaryotic biology is the way they are divided into various compartments, such as the nucleus, the endoplasmic reticulum, and the Golgi apparatus. Further, their genomes are packed tightly within the cell nucleus in nucleoprotein complexes known as

property	E. coli	budding yeast	mammalian (HeLa line)
cell volume	0.3–3 μm^3	30–100 μm^3	1000–10,000 μm^3
proteins per μm^3 cell volume	———————————	2–4 × 10^6	———————————
mRNA per cell	10^3–10^4	10^4–10^5	10^5–10^6
proteins per cell	~10^6	~10^8	~10^{10}
mean diameter of protein	———————————	4–5 nm	———————————
genome size	4.6 Mbp	12 Mbp	3.2 Gbp
number protein coding genes	4300	6600	21,000
regulator binding site length	10–20 bp	———— 5–10 bp	————
promoter length	~100 bp	~1000 bp	~10^4–10^5 bp
gene length	~1000 bp	~1000 bp	~10^4–10^6 bp (with introns)
concentration of one protein per cell	~1 nM	~10 pM	~0.1–1 pM
diffusion time of protein across cell ($D \approx 10$ μm^2/s)	~0.01 s	~0.2 s	~1–10 s
diffusion time of small molecule across cell ($D \approx 100$ μm^2/s)	~0.001 s	~0.03 s	~0.1–1 s
time to transcribe a gene	<1 min (80 nts/s)	~1 min	~30 min (incl. mRNA processing)
time to translate a protein	<1 min (20 aa/s)	~1 min	~30 min (incl. mRNA export)
typical mRNA lifetime	3 min	30 min	10 h
typical protein lifetime	1 h	0.3–3 h	10–100 h
minimal doubling time	20 min	1 h	20 h
ribosomes/cell	~10^4	~10^5	~10^6
transitions between protein states (active/inactive)	———————————	1–100 μs	———————————
time scale for equilibrium binding of small molecule to protein (diffusion limited)	———————	1–1000 ms (1 μM–1 nM affinity)	———————
time scale of transcription factor binding to DNA site	———————————	~1 s	———————————
mutation rate	———————	10^{-8}–10^{-10}/bp/replication	———————

Table 0-1 Typical parameter values for a bacterial *E. coli* cell, the single-celled eukaryote *S. cerevisiae* (budding yeast), and a mammalian HeLa cell line. These are crude characteristic values for happily dividing cells of the common lab strains. (Adapted from Alon U [2006] Introduction to Systems Biology. CRC Press. See full references at BNID 111494.)

nucleosomes, an architectural motif shared by all eukaryotes. Beyond its representative cellular structures, yeast has been celebrated because of the "awesome power of yeast genetics," meaning that in much the same way we can rewire the genomes of bacteria such as *E. coli*, we are now able to alter the yeast genome nearly at will. As seen in the table and figure, the key constituents of yeast cells can roughly be thought of as a scaled-up version of the same census results already sketched for bacteria in Figure 0-1A.

Figures 0-1C and 0-2C complete the trifecta by showing a "standard" mammalian cell. The schematic shows the rich and heterogeneous structure of such cells. The nucleus houses the billions of base pairs of the genome and is the site of the critical transcription processes taking place as genes are turned on and off in response to environmental stimuli and over the course of both the cell cycle and development. Organelles, such as the endoplasmic reticulum and the Golgi apparatus, are the critical site of key processes, such as protein processing and lipid biosynthesis. Mitochondria are the energy factories of cells where in humans, for example, about our body weight in ATP is synthesized each and every day. What can be said about the molecular players within these cells?

Given that there are several million proteins in a typical bacterium and these are the product of several thousand genes, we can expect the "average" protein to have about 10^3 copies. The distribution is actually very far from being homogenous in any such manner, as we will discuss in several vignettes in Chapter 2 on concentrations and absolute numbers. Given the rule of thumb from above that one molecule per *E. coli* corresponds to a concentration of roughly 1 nM, we can predict the "average" protein concentration to be roughly 1 mM. We will be sure to critically dissect the concept of the "average" protein, highlighting how most transcription factors are actually much less abundant than this hypothetical average protein and why components of the ribosome are needed in higher concentrations. We will also pay close attention to how to scale from bacteria to other cells. A crude and simplistic null model is to assume that the absolute numbers per cell tend to scale proportionally with the cell size. Under this null model, concentrations are independent of cell size.

Let's exemplify our thinking for a mammalian cell that has 1000 times the volume of a bacterial cell. Our first-order expectation will be that the absolute copy number will be about 1000 times higher and the concentration will stay about the same. The reader knows better than to take this as an immutable law. For example, some universal molecular players, such as ribosomes or the total amount of mRNA, also depend close to proportionally on the growth rate—that is, inversely with the doubling time. For such a case we should account for the fact that the mammalian cell divides, say, 20 times slower than the bacterial cell. So, for these cases we need a different null model. But in the alien world of molecular biology, where our intuition often fails, any guidance (that is, null model to rely on) can help. As a teaser example, consider the question of how many copies there are of your favorite transcription factor in some mammalian cell line—say, p53 in a HeLa cell. From the rules of thumb above, there are about 3 million proteins per μm^3 and a characteristic mammalian cell will be 3000 μm^3 in volume. We have no reason to think our protein is especially high in terms of copy number, so it is probably not taking one part in a hundred of the proteome (only the most abundant proteins will do that). So, an upper crude estimate would be 1 in 1000. This translates immediately into 3×10^6 proteins/μm^3 × 3000 μm^3/1000 proteins/our protein ~ 10 million copies of our protein. As we shall see, transcription factors are actually on the low end of the copy number range, and something between 10^5–10^6 copies would have been a more accurate estimate, but we suggest this is definitely better than being absolutely clueless. Such an estimate is the crudest example of an easily acquired "sixth sense." We find that those who master the simple rules of thumb discussed in this book have a significant edge in street-fighting cell biology (borrowing from Sanjoy Mahajan's gem of a book on "street-fighting mathematics").

The logical development of the remainder of the book can be seen through the prism of Figure 0-1. First, we begin by noting the structures and their sizes. This is followed in the second chapter by a careful analysis of the copy numbers of many of the key molecular species found within cells. Already at this point, the interconnectedness of these numbers can be seen, for example, in the relationship between the ribosome copy number and the cell size. In Chapter 3, we explore the energy and force scales that mediate the interactions between the structures and molecular species considered in the previous chapters. This is then followed in Chapter 4 by an analysis of how the molecular and cellular drama plays out over time. The various structures depicted in Figure 0-1 exhibit order on many different scales, an order that conveys critical information to the survival and replication of cells. Chapter 5 provides a quantitative picture of different ways of viewing genomic information and on the fidelity of information transfer in a variety of different cellular processes. Our final chapter punctuates the diversity of cells beyond what is shown in Figure 0-1 by considering a variety of other miscellany that defies being put into the simple conceptual boxes that characterize the other chapters.

SIZE

Chapter 1: Size and Geometry

In this chapter, all of our vignettes center in one way or another on the same simple question: "How big?" JBS Haldane, when he wasn't busy inventing population genetics or formulating the theory of enzyme kinetics (among many other things), wrote a delightful essay entitled "On Being the Right Size." There, he discusses how size is critical in understanding functional constraints on animals. For example, Haldane notes that when a human steps out of water, because of surface tension, he or she carries roughly a pound of water. On the other hand, an insect would carry comparatively much more, covered by about its own weight in water. The functional implications are often dire. In this same spirit, we aim to characterize the sizes of things in molecular and cellular biology with the hope of garnering insights into the kinds of functional implications explained by Haldane at larger scales.

Biological structures run the gamut in sizes from the nanometer scale of the individual macromolecules of life all the way up to the gigantic cyanobacterial blooms in the ocean that can be seen from satellites. As such, biologists can interest themselves in phenomena spanning more than 15 orders of magnitude in length scale. Though we find all of these scales fascinating (and important), in this book we primarily focus on those length scales that are smaller than individual organisms, as depicted in **Figure 1-1**.

In moving from the intuitive macroscopic world into the microscopic domain, a critical intellectual linkage will often be provided by Avogadro's number (see the Preface for historical efforts to determine its value). This important constant is defined as the number of hydrogen atoms contained in one gram of such atoms. With a value of about 6×10^{23}, this conversion

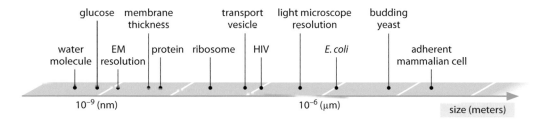

Figure 1-1 Range of characteristic sizes of the main biological entities relevant to cells. On a logarithmic scale we depict the range from single molecules, serving as the nuts and bolts of biochemistry, through molecular machines, to the ensembles that are cells.

factor reveals itself time and again, and the conversion was shown in the opening chapter.

One dilemma faced when trying to characterize biological systems is the extent to which we should focus on model systems. Often, the attempt to be comprehensive can lead to an inability to say anything concrete. As a result, we aim to give an intimate quantitative description of some common model cells and organisms, punctuated here and there by an attempt to remind the reader of the much larger diversity that lies beyond. We suggest that in those cases where we don't know better, it is very convenient to assume that all bacteria are similar to *E. coli*. We make this simplification for the sake of providing a general order-of-magnitude idea of the numbers that characterize most bacteria. In the same vein, our picture of a mammalian cell is built around the intuition that comes from using HeLa cells as a model system. We can always refine this crude picture when more information becomes available on, say, the volume or geometry of a specific cell line of interest. The key point is to have an order of magnitude to start with. A similar issue arises when we think about the changes in the properties of cells when they are subjected to different external conditions. Here again, we often focus on the simplified picture of happily dividing, exponentially growing cells, while recognizing that other conditions can change our picture of the "average" cell considerably. The final issue along this progression of challenges having to do with how to handle the diversity of biological systems is how we should deal with cell-to-cell variation—that is, how much do individual cells that have the same genetic composition and face the same external conditions vary? This chapter addresses these issues through a quantitative treatment both for cell size and protein abundance.

The geometries of cells come in a dazzling variety of shapes and sizes. Even the seemingly homogeneous world of prokaryotes is represented by a surprising variety of shapes and sizes. But this diversity of size and shape is not restricted only to cells. Within eukaryotic cells are found organelles with a similar diversity of form and a range of different sizes. In some cases, such as the mitochondria and chloroplasts (and perhaps the nucleus), the sizes of these organelles are similar to bacteria, which are their evolutionary ancestors through major endosymbiotic events. At smaller scales still, the macromolecules of the cell come into relief and, yet again, it is found that there are all sorts of different shapes and sizes, with examples ranging from small peptides, such as toxins, to the machines of the central dogma to the assemblies of proteins that make up the icosahedral capsids of viruses.

In thinking of geometrical structures, one of the tenets of many branches of science is the structure–function paradigm—the simple idea that form follows function. In biology, this idea has been a part of a long "structural" tradition that includes the development of microscopy and the emergence of structural biology. We are often tempted to figure out the *relative* scales of the various participants in some process of interest. In many of the vignettes, we attempt to draw a linkage between the size and the biological function.

Interestingly, even from the relatively simple knowledge of the sizes of biological structures, we can make subtle functional deductions. For example, what governs the burst size of viruses (that is, the number of viruses that are produced when an infected cell releases newly synthesized viruses)? Some viruses infect bacteria, whereas others infect mammalian cells, but the sizes of both groups of viruses are relatively similar, whereas the hosts differ in size by a characteristic volume ratio of 1000. This helps explain the fact that burst sizes from bacteria are about 100, whereas in the case of mammalian cells the characteristic burst size is in the thousands. Throughout the chapter, we return to this basic theme of reflecting on the biological significance of the many length scales we consider.

CELLS AND VIRUSES

How big are viruses?

In terms of their absolute numbers, viruses appear to be the most abundant biological entities on planet Earth. The best current estimate is that there are a whopping 10^{31} virus particles in the biosphere. We can begin to come to terms with these astronomical numbers by realizing that this implies that for every human on the planet, there is nearly an Avogadro's number worth of viruses. This corresponds to roughly 10^8 viruses to match every cell in our bodies. The number of viruses can also be contrasted with an estimate of 4–6×10^{30} for the number of prokaryotes on Earth (BNID 104960). However, because of their extremely small size, the mass tied up in these viruses is only approximately 5% of the prokaryotic biomass. The assertion about the total number of viruses is supported by measurements using both electron and fluorescence microscopy. For example, if a sample is taken from the soil or the ocean, electron microscopy observations reveal an order of magnitude more viruses than bacteria ($\approx$10:1 ratio, BNID

104962). These electron microscopy measurements are independently confirmed by light microscopy measurements. By staining viruses with fluorescent molecules, they can be counted directly under a microscope and their corresponding concentrations determined (for example, 10^7 viruses/mL).

Organisms from all domains of life are subject to viral infection, whether tobacco plants, flying tropical insects, or archaea in the hot springs of Yellowstone National Park. However, it appears that those viruses that attack bacteria (that is, so called bacteriophages—literally, bacteria eater—see **Figure 1-2**) are the most abundant of all, with these viruses present in huge numbers (BNID 104839, 104962, 104960) in a host of different environments ranging from soils to the open ocean.

As a result of their enormous presence on the biological scene, viruses play a role not only in the health of their hosts, but in global geochemical cycles affecting the availability of nutrients across the planet. For example, it has been estimated that as much as 20% of the bacterial mass in the ocean is subject to viral infection every day (BNID 106625). This can strongly decrease the flow of biomass to higher trophic levels that feed on prokaryotes (BNID 104965).

Viruses are much smaller than the cells they infect. Indeed, it was their remarkable smallness that led to their discovery in the first place. Researchers

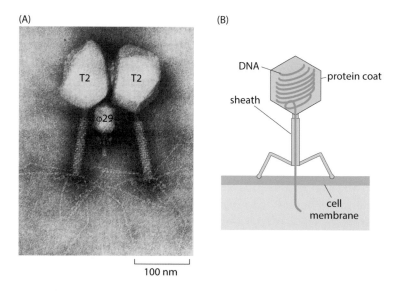

Figure 1-2 Geometry of bacteriophages. (A) Electron microscopy image of phi29 and T2 bacteriophages as revealed by electron microscopy. (B) Schematic of the structure of a bacteriophage. (A, adapted from Grimes S, Jardine PJ & Anderson D [2002] *Adv Virus Res* 58:255–280.)

were puzzled by remnant infectious elements that could pass through filters small enough to remove pathogenic bacterial cells. This led to the hypothesis of a new form of biological entity. These entities were subsequently identified as viruses.

Viruses are among the most symmetric and beautiful of biological objects, as shown in **Figure 1-3**. The figure shows that many viruses are characterized by an icosahedral shape with all of its characteristic symmetries (that is, twofold symmetries along the edges, threefold symmetries on the faces, and fivefold rotational symmetries on the vertices). The outer protein shell, known as the capsid, is often relatively simple since it consists of many repeats of the same protein unit. The genomic material is contained

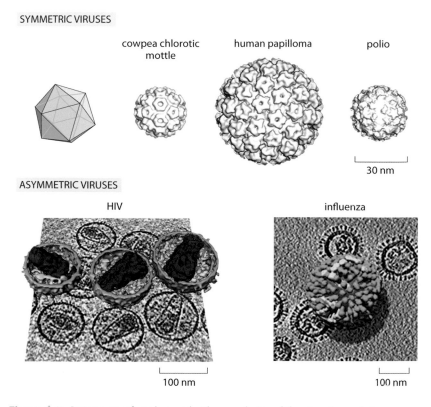

Figure 1-3 Structures of viral capsids. The regularity of the structure of viruses has enabled detailed, atomic-level analysis of their construction patterns. This gallery shows a variety of the different geometries explored by the class of nearly spherical viruses. HIV and influenza figures are 3D renderings of virions from the tomogram. (Symmetric virus structures adapted from Baker TS, Olson NH & Fuller SD [1999] *Microbiol Mol Biol Rev* 63:862–922. HIV structure adapted from Briggs JAG, Grünewald K, Glass B et al. [2006] *Structure* 14:15–20. Influenza virus structure adapted from Harris A, Cardone G, Winkler DC et al. [2006] *Proc Natl Acad Sci USA* 103:19123–19127.)

within the capsid. These genomes can be DNA or RNA, single-stranded or double-stranded (that is, ssDNA, dsDNA, ssRNA, or dsRNA) with characteristic sizes ranging from 10^3–10^6 bases (BNID 103246, 104073). With some interesting exceptions, a useful rule of thumb is that the radii of viral capsids themselves are all within a factor of 10 of each other, with the smaller viruses having a diameter of several tens of nanometers and the larger ones reaching diameters of several hundreds of nanometers, which is on par with the smallest bacteria (BNID 103114, 103115, 104073). Representative examples of the sizes of viruses are listed in **Table 1-1**. The structures of many viruses, such as HIV, have an external envelope (resulting in the label "enveloped virus") made up of a lipid bilayer. The interplay between the virus size and the genome length can be captured via the packing ratio, which is the percent fraction of the capsid volume taken by viral DNA. For phage lambda it can be calculated to be about 40%, whereas for HIV it is more than 10 times lower (BNID 111591).

virus	size (nm)	genome size (nucleotides)	genome type, capsid structure	BNID
porcine circovirus (PCV)	17	1760	circular ssDNA, icosahedral	106467, 106468
cowpea mosaic virus (CPMV)	28	9400	2 ssRNA molecules, icosahedral	106454, 106455
cowpea chlorotic mottle virus (CCMV)	28	7900	3 ssRNA molecules, icosahedral	106456, 106457
φX174 (*E. coli* bacteriophage)	32	5400	ssDNA, icosahedral	103246, 106442
tobacco mosaic virus (TMV)	40 × 300	6400	ssRNA, rod shaped	104376, 104375, 106453
polio virus	30	7500	ssRNA, icosahedral	103114, 111324
φ29 (*Bacillus* phage)	45 × 54	19,000	dsDNA, icosahedral (T3)	109734
lambda phage	58	49,000	dsDNA, icosahedral (with tail)	103122, 105770
T7 bacteriophage	58	40,000	dsDNA, 55 genes, icosahedral (T7)	109732, 109733
adenovirus (linear DNA)	88–110	36,000	dsDNA, icosahedral	103114, 103115, 106441
influenza A	80–120	14,000	ssRNA, roughly spherical	104073, 105768
HIV-1	120–150	9700	ssRNA, roughly spherical	101849, 105769
herpes simplex virus 1	125	153,000	dsDNA, icosahedral	103114, 106458
Epstein-Barr virus (EBV)	140	170,000	dsDNA, icosahedral	103246, 111424
mimivirus	500	1,200,000	dsDNA, icosahedral	105142, 105143
pandora virus	500 × 1000	2,800,000	dsDNA, icosahedral	109554, 109556

Table 1-1 Sizes of key representative viruses. The viruses in the table are organized according to their size, with the smallest viruses shown first and the largest viruses shown last. The organization by size gives a different perspective than typical biological classifications that use features such as the nature of the genome—RNA or DNA, single-stranded (ss) or double-stranded (ds)—and the nature of the host. Values are diameters rounded to one or two significant digits.

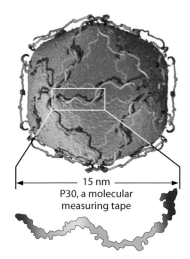

Figure 1-4 The P30 protein dimer serves as a measuring tape to help create the bacteriophage PRD1 capsid.

Some of the most interesting viruses have structures with less symmetry than those described above. Indeed, two of the biggest viral newsmakers, HIV and influenza, sometimes have irregular shapes, and even the structure from one influenza or HIV virus particle to the next can be different. Examples of these structures are shown in Figure 1-3. Why should so many viruses have a characteristic length scale of roughly 100 nm? If we consider the density of genetic material inside the capsid, a useful exercise for the motivated reader, it is found that the genomic material in bacterial viruses can take up nearly as much as 50% of the volume. Further, the viral DNA often adopts a structure that is close-packed and nearly crystalline to enable such high densities. Thus, in these cases, if we take as a given the length of DNA, which is tied in turn to the number of genes that viruses must harbor, the viruses show strong economy of size, minimizing the required volume to carry their genetic material.

To make a virus, the monomers making up the capsid can self-assemble; one mechanism is to start from some vertex and extend in a symmetric manner. But what governs the length of a facet? That is, what is the distance between two adjacent vertices that dictates the overall size of a virion? In one case, a nearly linear 83-residue protein serves as a molecular tape measure, helping the virus to build itself to the right size. The molecular players making this mechanism possible are shown in **Figure 1-4**. A dimer of two 15-nm-long proteins defines distances in a bacteriophage, which has a diameter of about 70 nm.

The recently discovered gigantic mimivirus and pandoravirus are about an order of magnitude larger (BNID 109554, 111143). The mechanism that serves to set the size of these viruses remains an open question. These viruses are larger than some bacteria and even rival some eukaryotes. They also contain genomes larger than 2 Mbp long (BNID 109556) and challenge our understanding of both viral evolution and diversity.

How big is an *E. coli* cell and what is its mass?

The size of a typical bacterium such as *E. coli* serves as a convenient standard ruler for characterizing length scales in molecular and cell biology. A rule of thumb, based upon generations of light and electron microscopy meas-urements, for the dimensions of an *E. coli* cell is to assign it a diameter of about ≈1µm, a length of ≈2µm, and a volume of ≈1µm³ (1 fL) (BNID 101788). The shape can be approximated as a spherocylinder—that is, a cylinder with hemispherical caps. Given the quoted diameter and length, we can compute a more refined estimate for the volume of ≈1.3 µm³ ($5\pi/12$, to be accurate). The difference between this value and the rule-of-thumb value quoted above shows the level of inconsistency we live with comfortably when using rules of thumb. One of the simplest routes to an estimate of the mass of a bacterium is to exploit the ≈1 µm³ volume of an *E. coli* cell and to assume it has the same density as water. This naïve estimate results in another standard value—namely, that a bacterium such as *E. coli* has a mass of ≈1 pg (pico = 10^{-12}). Because most cells are about two-thirds water (BNID 100044, 105482), and the other components, like proteins, have a characteristic density of about 1.3 times the density of water (BNID 101502, 104272), the conversion from cellular volume to mass is accurate to about 10%.

One of the classic results of bacterial physiology emphasizes that the plas-ticity in properties of cells derives from the dependence of the cell mass upon growth rate. Stated simply, faster growth rates are associated with larger cells. This observation refers to physiological changes where media that increase the growth rate also yield larger cells, as shown in **Figure 1-5**. This was also found to hold true genetically where long-term experimental evolution stud-ies that led to faster growth rates showed larger cell volumes (BNID 110462).

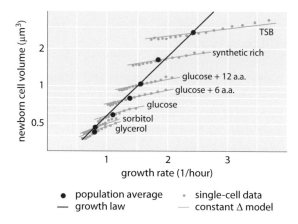

Figure 1-5 Relation between cell volume and growth rate. Using microscopy and microfluidic devices, cell volume can be measured at the single-cell level under various conditions, confirming that the average cell volume grows exponentially with growth rate. In contrast, variation among cells for a given condition scales differently. The variation in single-cell behavior is used to test models of cell size regulation. (Adapted from Taheri-Araghi S, Bradde S, Sauls JT et al. [2015] *Curr Biol* 25:385.)

Such observations help us dispel the myth of "the cell"—where people, often unwarily, use measurements about one cell to make inferences about other cell types or the same cell type under different conditions. Classic studies by Dennis and Bremer systematized these measurements and found that dry mass varies, as shown in **Table 1-2**, from an average value of 148 fg for cells dividing every 100 minutes to 865 fg for those with a 24-minute division time, indicating over a fivefold difference, depending upon the growth rate. A similar trend has been seen in other organisms (for example, for budding yeast; BNID 105103). At about 70% water, these values correspond to a range between about 0.4 to 2.5 μm^3 in terms of volume. How can we rationalize the larger sizes for cells growing at faster rates? This question is under debate to this day. Explanations vary from suggesting it has an advantage in the way resource allocation is done[*] to claiming that it is actually only a side effect of having a built-in period of about 60 minutes from the time a cell decides it has accumulated enough mass to begin the preparations for division and until it finishes DNA replication and the act of division. This roughly constant "delay" period leads to an exponential dependence of the average cell mass on the growth rate in this line of reasoning.[†]

generation time (min)	dry mass per cell (fg = 10^{-15} g)
100	150
60	260
40	430
30	640
24	870

Table 1-2 Relation between bacterial mass and division time. The dry mass per cell is given as a function of the generation (doubling) time. Mass is suggested to increase roughly exponentially with growth rate, as originally observed by Schaechter M, Maaloe O & Kjeldgaard NO (1958) *J Gen Microbiol* 19:592–606. The cell dry weight was calculated using a value of 173 µg per OD460 unit of 1 mL (BNID 106437). The strain used is B/r, a strain commonly used in early bacterial physiology studies. (Values taken from Neidhardt FC [1996] *Escherichia coli* and *Salmonella*: Cellular and Molecular Biology, Vol 1. ASM Press.)

Methods to measure cell volume range from the use of a Coulter Counter (BNID 100004), which infers volume based on changes in resistance of a small orifice as a cell passes in it, to more direct measurements using fluorescence microscopy that gauge cell lengths and diameters under different conditions (BNID 111480, 106577; see Figure 1-5). Surprisingly, the fact that different laboratories do not always converge on the same values may be due to differences in calibration methods or exact strains and growth conditions. An unprecedented ability to measure cell mass is achieved by effectively weighing cells on a microscopic cantilever. As illustrated in **Figure 1-6A**, fluid flow is used to force a cell back and forth in the hollowed-out cantilever. The measurement exploits the fact that the cell mass affects the oscillation frequency of the cantilever. This frequency can be measured to a phenomenal accuracy and used to infer masses with femtogram precision. By changing the liquid flow direction, the cell is trapped for minutes or more, and its mass accumulation rate is measured continuously at the single-cell level. In the initial application of this

---

[*] Molenaar D, van Berlo R, de Ridder D, & Teusink B (2009) Shifts in growth strategies reflect tradeoffs in cellular economics. *Mol Syst Biol* 5:323.
[†] Amir A (2014) Cell size regulation in bacteria. *Phys Rev Lett* 112:208102.

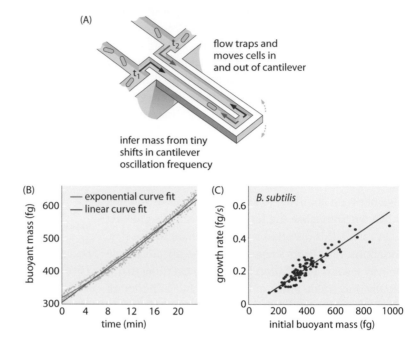

Figure 1-6 Using buoyant mass to measure the growth of single cells. (A) A micronscale cantilever oscillates at high frequency, and the mass of cells can be determined from changes in the oscillation frequency. (B) Measured over time, this results in a single-cell mass accumulation curve as shown. (C) Relation between growth rate and buoyant mass for *B. subtilis* cells. A comparison between the predictions of linear and exponential growth models are shown as best fits. The similarity demonstrates how close the two models are over a range of only a twofold increase over the course of the cell cycle. Cell dry weight is about four times the buoyant mass. (Adapted from Godin M, Delgado FF, Son S et al. [2010] *Nat Methods* 7:387–390. With permission from Macmillan Publishers Ltd.)

technique, it was shown that single cells that are larger also accumulate mass faster, shedding light on a long-standing question: Is cell growth linear with time or more appropriately described by an approximately exponential trend? The differences can be minute, but with these revolutionary capabilities it was clearly seen that the latter scenario better represents the situation in several cell types tested, as shown in **Figure 1-6B**.

How big is a budding yeast cell?

The budding yeast, *Saccharomyces cerevisiae* has served as the model eukaryote in much the same way that *E. coli* has served as the representative prokaryote. Due to its importance in making beer and baking bread (thus also called brewer's or baker's yeast), this easily accessible and simply cultured organism was also an early favorite of scientists, as interestingly recalled by James A. Barnett in his paper "Beginnings of microbiology and biochemistry: the contribution of yeast research." These cells are significantly larger than common bacteria and, as such, are a convenient single-celled organism to study under the microscope. In large part due to the ease with which its genome can be manipulated, yeast has remained at the forefront of biological research and, in 1996, was the first eukaryotic organism to have its genome completely sequenced. Another feature that makes yeast handy for geneticists is their dual lifestyle as either haploids or diploids. Haploid cells have only one copy of each chromosome, just like a human female egg cell, whereas diploid cells have two copies of each chromosome, just like somatic cells in our bodies. Haploids are analogous to our gametes—the egg cell and sperm cells. The haploid/diploid coexistence in budding yeast enables scientists to easily transfer mutations and study their effects.

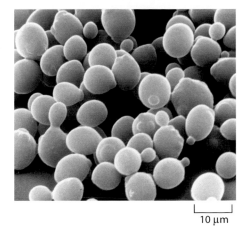

10 μm

Figure 1-7 Electron micrograph of budding yeast cells. (Courtesy of Ira Herskowitz and E. Schabtach.)

A simple rule of thumb for the dimensions of yeast cells is to think of them as spheres with a diameter of roughly 4 μm for haploids and roughly 6 μm for diploids, as shown in **Figure 1-7** (BNID 101796). To put the relative sizes of yeast and bacteria in everyday terms, if we think of a world in which *E. coli* is the size of a human, then yeast is about the size of an elephant. Prominent components of the cell volume include the nucleus, which takes up about 10% of the total cell volume (BNID 100491, 103952), the cell wall, often ignored but making up 10–25% (BNID 104593, 104592) of the total dry mass, and the endoplasmic reticulum and vacuole, which are usually the largest organelles.

One of the ideas that we repeatedly emphasize in a quantitative way is cell-to-cell variability and its role in establishing the different behaviors of cells in response to different environmental cues. As yeast replicate by budding off small daughter cells from a larger mother, any population has a large range of cell sizes spread around the median, as shown in **Figure 1-8**. The haploid strain shown has a median cell volume of 42 ± 2 μm³ (BNID 100427). Another common metric is the 25th–75th percentile range, which here is ≈30–60 fL. The median cell size itself is highly dependent on genetic

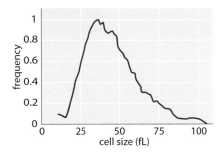

Figure 1-8 Histogram of distribution of cell sizes for wild-type budding yeast cells. (Adapted from Jorgensen P, Nishikawa JL, Breitkreutz BJ & Tyers M [2002] *Science* 297:395–400.)

and environmental factors. A diploid cell is almost twice as big as its haploid progenitors at ≈ 82 μm^3 (BNID 100490). This reflects the more general observation from cell biology that median cell size tends to grow proportionally to ploidy (DNA content). Yeasts, where ploidy can be manipulated to higher than two, serve as useful test cases for illuminating this phenomenon.

Beyond the bulk DNA content, the median cell volume can differ by more than twofold in different strains of *S. cerevisiae* that evolved in different parts of the world, or more recently, in different industries utilizing them. Finally, like *E. coli*, median cell size in yeast is correlated with growth rate—the better the environmental conditions and growth rate, the larger the cells (BNID 101747). An intriguing open question is whether there is an evolutionary advantage of shifting cell size in response to environmental conditions. Recent measurements (BNID 100490) have probed how sensitive yeast cell size is to single-gene deletions. In some of these deletion mutants, the median volume was only 40% of the wild-type size, whereas in others it was larger than wild type by >70%. These observations reveal strong coupling between size regulation and the expression of critical genes. It still remains largely unknown how genetic and environmental changes shift the median cell size in yeast.

How big is a human cell?

A human is, according to the most recent estimates, an assortment of $3.7 \pm 0.8 \times 10^{13}$ cells (BNID 109716), plus a similar complement of allied microbes. The identities of the human cells are distributed amongst more than 200 different cell types (BNID 103626, 106155) that perform a staggering variety of functions. The shapes and sizes of cells span a large range, as shown in **Table 1-3**. Size and shape, in turn, are intimately tied to the function of each type of cell. Red blood cells need to squeeze through narrow capillaries, and their small size and biconcave disk shape achieve that while also attaining a high surface-area-to-volume ratio. Neurons need to transport signals and, when connecting our brains to our legs, can reach lengths of over a meter (BNID 104901), but with a width of only about 10 μm. Cells that serve for storage, like fat cells and oocytes, have very large volumes.

The different shapes also enable us to recognize the cell types. For example, the leukocytes of the immune system are approximately spherical in shape, while adherent tissue cells on a microscope slide resemble a fried egg, with the nucleus analogous to the yolk. In some cases, such as the

cell type	average volume (µm³)	BNID
sperm cell	30	109891, 109892
red blood cell	100	107600
lymphocyte	200	111788
neutrophil	300	108241
beta cell	1000	109227
enterocyte	1400	111216
fibroblast	2000	108244
HeLa, cervix	3000	103725, 105879
hair cell (ear)	4000	108242
osteoblast	4000	108088
alveolar macrophage	5000	103566
cardiomyocyte	15,000	108243
megakaryocyte	30,000	110129
fat cell	600,000	107668
oocyte	4,000,000	101664

Table 1-3 Characteristic average volumes of human cells of different types. Large cell–cell variation of up to an order of magnitude or more can exist for some cell types, such as neurons or fat cells, whereas for others the volume varies by much less (for example, red blood cells). The value for a beta cell comes from a rat, but we still present it because average cell sizes usually change relatively little among mammals.

different types of white blood cells, the distinctions are much more subtle and are only reflected in molecular signatures.

Mature female egg cells are among the largest cell types, with a ≈120 µm diameter. Other large cell types include muscle fiber cells (which merge together to form syncytia, where multiple nuclei reside in one cell) and megakaryocytes (the bone marrow cells responsible for the production of blood platelets). Both of these cell types can reach 100 µm in diameter (BNID 106130). Red blood cells, also known as erythrocytes, are some of the smallest and most abundant of human cells. These cells have a characteristic biconcave disk shape with a depression where the nucleus was lost in maturation and have a corresponding diameter of 7–8 µm (BNID 100509) and a volume of ≈100 µm³ (BNID 101711, 101713). Sperm cells are even smaller, with a volume of about 20–40 µm³ (BNID 109892, 109891).

Certain human cell lines have been domesticated as laboratory workhorses. Perhaps the most familiar of all are the so-called HeLa cells, an example of which is shown dividing in **Figure 1-9**. Such immortal cancer cell lines divide indefinitely, alleviating the need to sacrifice primary animal tissue for experiments. These cell lines have been used for studies on topics such as the molecular basis of signal transduction and the cell cycle. In these cell types, the cell volumes are captured by a rule-of-thumb

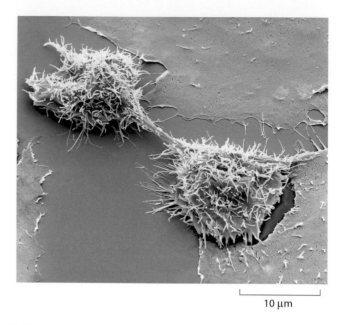

10 μm

Figure 1-9 Dividing HeLa cells as seen by a scanning electron micrograph (colored). The image was taken during cell division (cytokinesis). The transient connecting midbody formed by microtubules can be seen. (Courtesy of Steve Gschmeissner/Photo Researchers, Inc.)

value of 2000 μm³, with a range of 500–4000 μm³ (BNID 100434). HeLa cells adhere to the extracellular matrix and, like many other cell types on a microscope slide, spread to a diameter of ≈40 μm (BNID 103718, 105877, 105878), but only a few thin μm in height. When grown to confluence, they press on each other to compact the diameter to ≈20 μm such that in one of the wells of a 96-well plate, they create a monolayer of ≈100,000 cells. As in bacteria and yeast, average cell size can change with growth conditions. In the case of HeLa cells, a greater than twofold decrease in volume was observed when comparing cells three days and seven days after splitting and re-plating (BNID 108870, 108872). A snapshot of the variability of mammalian cells was achieved by a careful microscopic analysis of a mouse lymphocyte cell line, as shown in **Figure 1-10**. The distribution is

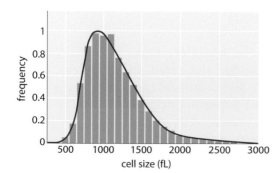

Figure 1-10 Distribution of cell sizes for L1210, a mouse lymphoblast cell line. The cell volumes are reported in units of fL (1 fL = 1 μm³). (Adapted from Tzur A, Kafri R, LeBleu VS et al. [2009] *Science* 325:167–171.)

centered at about 1000 μm³ with a variance of about 300 μm³. To put these cellular sizes in perspective, if we think of *E. coli* as having the size of a human being, then a HeLa cell is about the size of a blue whale.

Our examination of the sizes of different cell types will serve as a jumping-off point for developing intuition for a variety of other biological numbers we will encounter throughout the book. For example, when thinking about diffusion, we will interest ourselves in the time scale for particular molecules to traverse a given cell type, and this result depends in turn upon the size of those cells.

How big is a photoreceptor?

One of the greatest charms of biology is the overwhelming diversity of living organisms. This diversity is reflected, in turn, by the staggering array of different types of cells found in both single-celled and multicellular organisms. Earlier, we celebrated some of the most important "model" cells, such as our standard bacterium, *E. coli* and our single-celled eukaryote, the yeast *S. cerevisiae*. Studies of these model systems have to be tempered by a realization of both the great diversity of single-celled organisms themselves, as shown in the vignette on cell size diversity (pg. 21), as well as of the stunning specializations in different cell types that have arisen in multicellular organisms. The cells that make possible the sense of vision discussed in this vignette are a beautiful and deeply studied example of such specializations.

There is perhaps no sense that we each take more personally than our vision. Sight is our predominant means of taking in information about the world around us, a capacity made possible as a result of one of evolution's greatest inventions—namely, the eye—as shown in **Figure 1-11**. Eyes and the cells that make them up have been a central preoccupation of scientists of all kinds for centuries, whether in the hands of those like Hermann von Helmholtz, who designed instruments such as the opthalmoscope to study eyes of living humans, or those like Charles Darwin and his successors, who have mused on how evolution could have given rise to such specialized organs. Chapter VI of *The Origin of Species* is entitled "Difficulties on Theory" and is used by Darwin as a forum to explain what he referred to as a "crowd of difficulties" that "will have occurred to the reader." Darwin says that some of these difficulties are "so grave that to this day I can never reflect on them without being staggered; but, to the best of my judgment, the greater number are only apparent, and those that are real are not, I think, fatal to my theory." One of the most significant of those difficulties was what Darwin

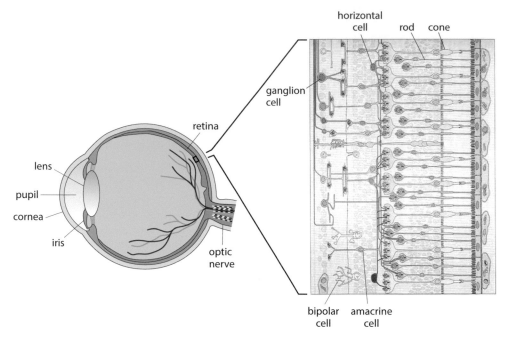

Figure 1-11 A multi-scale view of the retina. The schematic on the left shows the entire eye. The magnified view on the right illustrates the organization of the different cell types in the retina ranging from the photoreceptors that receive light to the ganglion cells that communicate electrical impulses as a result of stimulation of photoreceptors by light. (Adapted from Rodieck RW [1998]. The First Steps of Seeing. Sinauer Associates.)

thought of as "organs of extreme perfection," such as our eye. He goes on to say, "To suppose that the eye, with all its inimitable contrivances for adjusting the focus to different distances, for admitting different amounts of light, and for the correction of spherical and chromatic aberration, could have been formed by natural selection, seems, I freely confess, absurd in the highest possible degree. Yet reason tells me, that if numerous gradations from a perfect and complex eye to one very imperfect and simple, each grade being useful to its possessor, can be shown to exist; if further, the eye does vary ever so slightly, and the variations be inherited, which is certainly the case; and if any variation or modification in the organ be ever useful to an animal under changing conditions of life, then the difficulty of believing that a perfect and complex eye could be formed by natural selection, though insuperable by our imagination, can hardly be considered real." Our understanding of the long evolutionary history of eyes continues to evolve itself, and a current snapshot can be attained by reading a recent review.[‡]

What are these organs of extreme perfection like at the cellular level? Figure 1-11 provides a multi-scale view of the human eye and the cells that

[‡] Lamb T, Collin SP, & Pugh EN Jr. (2007) Evolution of the vertebrate eye: opsins, photoreceptors, retina and eye cup. *Nat Rev Neurosci* 8:960–976.

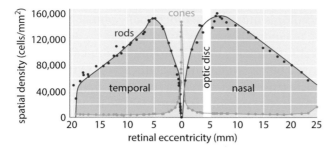

Figure 1-12 Distribution of rods and cones in the vertebrate retina. Note that if we consider 100,000 rods/mm^2 as the typical areal density, this corresponds to 10 μm^2 per rod cell, which jibes nicely with our simple estimate made above. (Adapted from Rodieck, RW [1988] The primate retina. In Comparative Primate Biology, Vol 4 [HD Steklis and J Erwin eds], pp. 203-278. Alan R. Liss Inc.)

make it work, giving a sense of the complexity and specialization that so staggered Darwin. Our focus here is on the retina, the 100–300-μm thick (BNID 109683) structure at the back of the eye. The mammalian retina harbors two types of photoreceptor cells, called rods and cones. Rods are mostly used for night vision, whereas cones enable color vision using three types of pigments. As seen in Figure 1-11, in addition to the rods and cones, the retina is also populated by layers of cells, such as horizontal cells, bipolar cells, amacrine cells, and the ganglion cells that convey the information derived from the visual field to the brain itself. One of the surprising features of the human eye is that the photoreceptors are actually located at the back of the retina, whereas the other cells responsible for processing the data and the optic nerve that conveys that information to the brain are at the front of the retina, thus blocking some of the photons in our visual field. This seems a strange feature for an organ considered a glaring example of optimality in nature. Indeed, in cephalopods, like the squid and octopus, the situation is reversed, with the nerve fibers routing behind rather than in front of the retina. Further, the human eye structure is not optimal not only in this respect, but also in the aberrations it features, many of which are corrected by the cells downstream of the photoreceptors.[§]

The distribution of rods and cones throughout the retina is not uniform. As shown in **Figure 1-12**, cones have the highest density at a central part of the retina called the fovea and thus enable extremely high resolution. To get a feeling for the optical properties of this collection of photoreceptors, it is perhaps useful to consider a comparison with digital cameras. We are used to cameras with 10 million pixels per image. Though

[§] Liang J & Williams DR (1997) Aberrations and retinal image quality of the normal human eye. *J Opt Soc Am A* 14:2873-2883.

a photoreceptor is much more functionally potent than a pixel, it is still interesting to contemplate how many photoreceptor cells we have and how this value compares to what we find in our cameras. To produce a naïve estimate of the number of photoreceptors in the human retina, we need a rough sense of how much area is taken up by each such cell. A human rod cell is ≈2 microns in diameter (BNID 107894), which is a few times the wavelength of light. If we maximally stacked them, we could get 500 by 500 such cells in a square millimeter—that is, ≈250,000 rods/mm^2. Figure 1-12 reports experimental values that confirm this is close to reality. To finish the estimate of the total number of receptors decorating the back surface of the eye, we consider the eyeball to be a hemisphere of 2–3 cm in diameter, as shown in Figure 1-11 (BNID 109680), implying an area of roughly 10^9 μm^2. The number of photoreceptors can be estimated as $(10^9$ μm^2/retina)/(4 μm^2/photoreceptor), which yields ≈200 million photoreceptors in each of our eyes and is of the same order of magnitude as estimates based on current knowledge and visualization techniques (BNID 105347, 108321). Digital cameras still have a long way to go before they reach this number, not to speak of the special adaptation and processing that each cell in our eye can perform and a digital pixel cannot.

The anatomy of these individual photoreceptors is remarkable. As seen in **Figure 1-13**, a typical photoreceptor cell such as a rod is roughly 100 μm

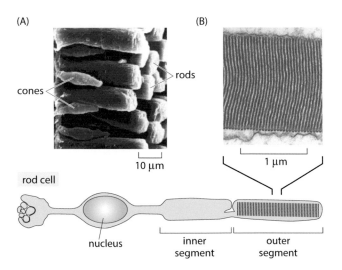

Figure 1-13 Anatomy of rods and cones. The schematic shows some of the key anatomical features of a photoreceptor cell. (A) A scanning electron micrograph illustrating the organization of rods and cones in the retina of a salamander. (B) Electron micrograph of the membrane disks of the outer segment of the photoreceptor. In both rods and cones, the proteins holding the light-absorbing retina are homologous opsins—that is, rhodopsins in rods and three types of spectrally distinct photopsins in cones. (A, courtesy of Scott Mittman and David R. Copenhagen. B, adapted from Dowling J [2012] The Retina. Belknap Press.)

long (BNID 108246, 109684) and is characterized by a number of special-ized features, such as the roughly 25-micron-long "outer segment" (BNID 107894, 107895) shown in Figure 1-13B that is filled with the rhodopsin molecules that absorb light. At the opposite extremity of these cells are the synapses—the structures used to communicate with adjacent cells. Synapses are crucial to the signal cascade that takes place following the detection of a photon by a photoreceptor cell. As seen in Figure 1-13, the outer segments of a photoreceptor rod cell are characterized by stacks of membrane discs. These discs are roughly 10 nm thick and are stacked in a periodic fashion with a spacing of roughly 25 nm. Given that the outer segment is ≈25,000 nm long, this means that there are roughly 1000 such discs in each of the ≈10^8 rod cells in the vertebrate retina (with about 10^8 rhodopsin molecules per rod cell, as discussed in the vignette entitled "How many rhodopsin molecules are in a rod cell?"; pg. 142). These 1000 effective layers increase the cross section available for intercepting pho-tons, thus making our eyes such "organs of extreme perfection."

What is the range of cell sizes and shapes?

Cells come in a dazzling variety of shapes and sizes. As we have already seen, deep insights into the workings of life have come from focused studies on key "model" types such as *E. coli*, budding (baker's) yeast, and certain human cancer cell lines. These model systems have helped develop a precise feel for the size, shape, and contents of cells. However, undue focus on model organisms can give a deeply warped view of the diversity of life. Stated simply, there is no easier way to dispel the myth of "the cell"—that is, the idea that what we say about one cell type is true for all others—than to show examples of the bizarre gallery of different cell types found both in unicellular and multicellular organisms.

In this vignette, we are interested in the broad question of the diversity of cell size and shape. Some representative examples summarizing the diver-sity of shapes and sizes in the microbial world are shown in **Figure 1-14**. Though this figure largely confirms our intuitive sense that microbial cells are usually several microns in size, the existence of the giant *Thiomarga-rita namibiensis* belies such simple claims in much the same way that the Star-of-David shape of *Stella humosa* is at odds with a picture of bacteria as nothing more than tiny rods and spheres.

Some of the most dramatic examples of cellular diversity include the beau-tiful and symmetrical coccolithophore, *Emiliana huxleyi* (**Figure 1-15**),

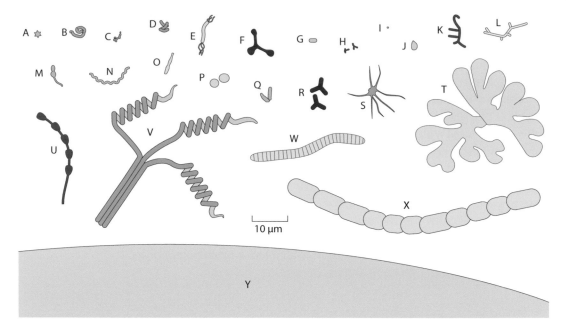

Figure 1-14 A gallery of microbial cell shapes. These drawings are based upon microscopy images from the original literature. (A) *Stella strain* IFAM1312 (380); (B) *Microcyclus* (a genus since renamed Ancylobacter) *flavus* (367); (C) *Bifidobacterium bifidum*; (D) *Clostridium cocleatum*; (E) *Aquaspirillum autotrophicum*; (F) *Pyroditium abyssi* (380); (G) *Escherichia coli*; (H) *Bifidobacterium sp.*; (I) transverse section of ratoon stunt-associated bacterium; (J) *Planctomyces sp.* (133); (K) *Nocardia opaca*; (L) chain of ratoon stunt-associated bacteria; (M) *Caulobacter sp.* (380); (N) *Spirochaeta halophila*; (O) *Prosthecobacter fusiformis*; (P) *Methanogenium cariaci*; (Q) *Arthrobacter globiformis* growth cycle; (R) gram-negative *Alphaproteobacteria* from marine sponges (240); (S) *Ancalomicrobium* sp. (380); (T) *Nevskia ramosa* (133); (U) *Rhodomicrobium vanniellii*; (V) *Streptomyces sp.*; (W) *Caryophanon latum*; (X) *Calothrix sp.*; (Y) A schematic of part of the giant bacterium, *Thiomargarita namibiensis* (290). All images are drawn to the same scale. (Adapted from Young KD [2006] *Microbiol Mol Biol Rev* 70:660–703.)

whose exoskeleton shield is very prominent and makes up the chalk rocks we tread on and much of the ocean floor, though its functional role is still not clear; the richly decorated protozoan, *Oxytricha* (for more single-cell

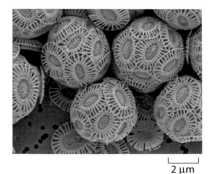

Figure 1-15 Scanning electron microscopy image of a collection of *E. huxleyi* cells, which illustrates their solid exterior. Each of these structures contains a single eukaryotic cell on its interior. (Courtesy of Jeremy R. Young.)

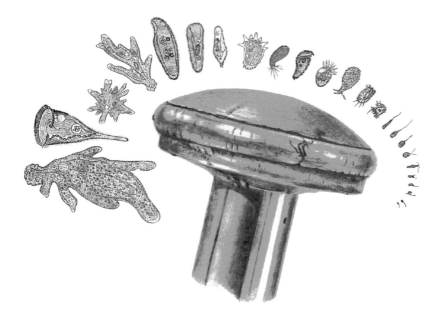

Figure 1-16 Protist diversity. This figure illustrates the morphological diversity of free-living protists. The various organisms are drawn to scale relative to the head of a pin about 1.5 mm in diameter. (Adapted from Finlay BJ [2002] *Science* 296:1061–1063.)

protists, see the diversity and relative scale depicted in **Figure 1-16**); and the sprawling geometry of neurons, which can have sizes of over a meter (even in our own bodies). One of the most interesting classes of questions left in the wake of these different examples concerns the mechanisms for the establishment and maintenance of shape and the functional consequences of different sizes and shapes.

Perhaps the most elementary measure of shape is cell size, with sizes running from sub micron to meters, exhibiting roughly a seven order-of-magnitude variability in cell sizes across the different domains of life, as shown in **Figure 1-17**. Though prokaryotes are typically several microns in size, sometimes they can be much larger. Similarly, even though eukaryotes typically span the range from 5 to 50 microns, they too have a much wider range of sizes, with the eggs of eukaryotes and the cells of the nervous system providing a measure of just how large individual cells can be. One of the most interesting challenges that remains in understanding the diversity of all of these sizes and shapes is gathering a sense of their functional implications and the evolutionary trajectories that gave rise to them.

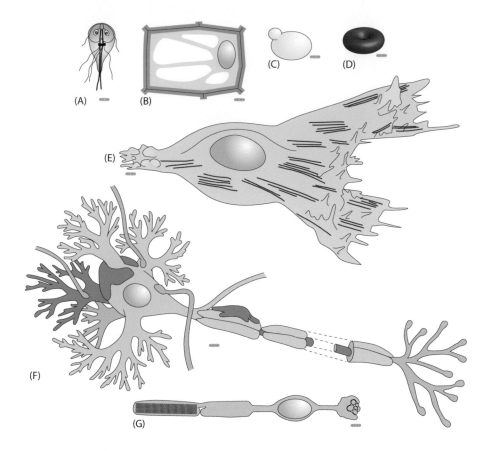

Figure 1-17 Cartoons of several different types of cells all referenced to a standard *E. coli* ruler of 1 micron width drawn in gray. (A) The protist, *Giardia lamblia*, (B) a plant cell, (C) a budding yeast cell, (D) a red blood cell, (E) a fibroblast cell, (F) a eukaryotic nerve cell, and (G) a rod cell from the retina.

ORGANELLES

How big are nuclei?

One of the most intriguing structural features of eukaryotic cells is that they are separated into many distinct compartments, each characterized by differences in molecular composition, ionic concentrations, membrane potential, and pH. In particular, these compartments are separated from each other and the surrounding medium (that is, the cytoplasm or extra-cellular solution) by membranes that themselves exhibit a great diversity of lipid and protein molecules, with the membranes of different compartments also characterized by different molecular compositions. Given the central role of genomes in living matter, there are few organelles as important as

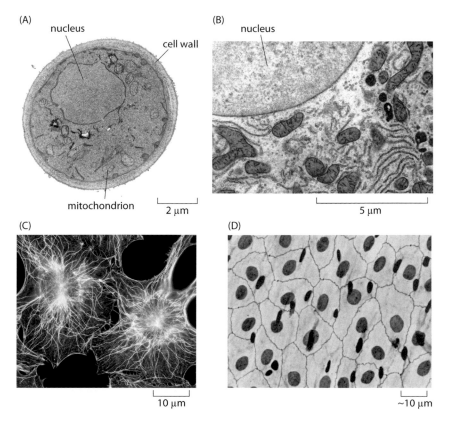

Figure 1-18 Nuclear size. (A) Electron microscopy image of a yeast cell, which reveals the roughly 2-micron-sized nucleus. (B) A portion of a rat liver cell that shows part of the nucleus and a variety of surrounding organelles, such as the endoplasmic reticulum, the mitochondria, and the Golgi apparatus. (C) Fluorescence image of a human fibroblast cell with the roughly 10-micron nucleus labeled in green. (D) Light microscopy image of a human epithelial sheet. The dark ovals are the cell nuclei stained with silver. (A, adapted from Alberts B, Johnson A, Lewis J et al. [2015] Molecular Biology of the Cell, 6th ed. Garland Science. B, adapted from an electron micrograph in Fawcett DW [1966] The Cell, Its Organelles and Inclusions: An Atlas of Fine Structure. W. B. Saunders. With permission from Elsevier Inc. C, and D, adapted from Phillips R, Kondev J, Theriot J & Garcia H [2013] Physical Biology of the Cell, 2nd ed. Garland Science.)

the eukaryotic nucleus, home to the chromosomal DNA that distinguishes one organism from the next. As seen in **Figure 1-18**, using both electron and light microscopy, it is possible to determine nuclear size variation with typical diameters ranging between 2 and 10 microns, though in exceptional cases such as oocytes, the nuclear dimensions are substantially larger.

One feature of organellar dimensions is their variability. We have already seen the range of sizes exhibited by yeast cells in an earlier vignette. **Figure 1-19** takes up this issue again by revealing the typical sizes and variability for the nuclei in haploid and diploid yeast cells, complementing the data presented earlier on cell size. For haploid yeast cells, the mean nuclear volume is 3 μm^3 (BNID 104709). With a genome length of

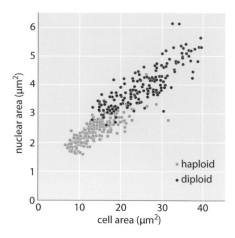

Figure 1-19 Nuclear size for haploid and diploid yeast cells. The cross-sectional areas of the nuclei are plotted as a function of the cross-sectional areas of the cells themselves. (Adapted from Jorgensen P, Edgington NP, Schneider BL et al. [2007] *Mol Biol Cell* 18:3523–3532. With permission from the American Society for Cell Biology.)

12 Mbp (BNID 100459), the DNA takes up roughly 0.3% of the nuclear volume. We can arrive at this estimate based on the rule of thumb that a base pair has a volume of ≈ 1 nm^3 (BNID 103778), in which case the DNA occupies roughly 0.01 μm^3. A similar value is found for diploid yeast. In contrast, for the yeast spore, the nuclear volume is an order of magnitude smaller—namely, 0.3 μm^3 (BNID 107660) or about 3% of the nuclear volume—indicating a much more dense packing of the genomic DNA.

These estimates for nuclear fraction are agnostic about the higher-level chromatin structure induced by nucleosome formation. In nucleosomes, 147 base pairs of DNA are wrapped roughly 1.75 times around an octamer of histone proteins, making a snub disk roughly 10 nm across (BNID 102979, 102985). In **Figure 1-20**, we show the so-called 30-nm fiber. When we travel 10 nm along the fiber, about six nucleosomes are packed in a staggered manner, and thus we have included on the order of 1000 bp. We can estimate the total volume taken up by yeast genomic DNA when in this structure by multiplying the area of the effective circular cross section by the height of the structure, resulting in

$$V = \pi (15 \text{ nm})^2 \times (10 \text{ nm}/1000 \text{ bp}) \times (10^7 \text{ bp}) \approx 10^8 \text{ nm}^3 = 0.1 \text{ μm}^3. \quad (1.1)$$

Given the volume of the yeast nucleus of roughly 4 μm^3, this implies a packing fraction of $\approx 2\%$, and is consistent with our earlier estimate, which was based on the volume of a base pair.

Questions about nuclear size in eukaryotes have been systematically investigated in other organisms besides yeast. It has been hypothesized

(A)

histone octamer

DNA

(B)

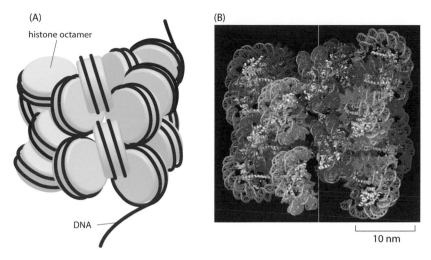

10 nm

Figure 1-20 DNA packing into higher-level compact structures. (A) Schematic illustrating how multiple nucleosomes can be arranged into a solenoidal structure. The histone octamer shown in yellow and the DNA as a red strand. (B) Models of nucleosome packing based upon high-resolution cryo-electron microscopy images of arrays of nucleosomes. In these *in vitro* experiments, nucleosome arrays were generated by using purified histones and specific DNA molecules of known sequence. (B, adapted from Robinson P, Fairall L, Huynh V & Rhodes D [2006] *Proc Natl Acad Sci USA* 103:6506–6511.)

that there is a simple linear relationship between the mean diameter of a plant meristematic cell (the plant tissue consisting of undifferentiated cells from which growth takes place) and the diameter of its nucleus. Such ideas have been tested in a variety of different plant cells, as shown in **Figure 1-21**, for example. In the experiments summarized there, the nuclear and cell volumes of 14 distinct species of herbaceous angiosperms, including some commonly known plants such as chickpeas and lily of the valley, were measured, resulting in a simple relationship of the following form (BNID 107802)

$$V_{nuc} \approx 0.2 \, V_{cell} . \tag{1.2}$$

The observations reported here raise the question of how the relative size of the nucleus compared to the whole cell is controlled. This is especially compelling since the nucleus undergoes massive rearrangements during each and every cell cycle as the chromosomes are separated into the daughter cells. We remind the reader that a relatively stable ratio is a common observation rather than a general law. In mammalian cells, this ratio can be very different between cell types. For example, in resting lymphocytes, the nucleus occupies almost the whole cell, while in macrophages or fat cells, the ratio of nucleus to cell volume is much smaller.

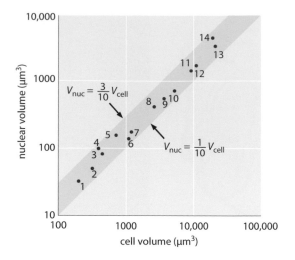

Figure 1-21 The relationship between the nuclear volume and cell volume in apical meristems of 14 herbaceous angiosperms. 1 *Arabidopsis thaliana*, 2 *Lobularia maritime* (Sweet Alison), 3 *Hypericum virginicum* (Marsh St. John's wort), 4 *Cicer arietinum* (chickpea), 5 *Nelumbo lutea*, 6 *Spinacia oleracea* (spinach), 7 *Cyanotis pilosa*, 8 *Anemone pulsatilla* (Meadow Anemone), 9 *Tradescania navicularis* (day flower), 10 *Convallaria majalis* (lily of the valley), 11 *Fritillaria laneeolata* (chocolate lily), 12 *Fritillaria camtschatcensis*, 13 *Lilium longiflorum* (Easter lily)(4×), 14 *Sprekelia formosissima* (Aztec lily). (Adapted from Price HJ, Sparrow AH & Nauman AF [1973] *Experientia* 29:1028–1029.)

How big is the endoplasmic reticulum of cells?

The endoplasmic reticulum, known to its friends as the ER, is often the largest organelle in eukaryotic cells. As shown in **Figure 1-22**, the structure of the ER is made up of a single, continuous membrane system, often spreading its cisternae and tubules across the entire cytoplasm. In addition to its exquisite and beautiful structure, it serves as a vast processing unit for proteins, with ≈20–30% of all cellular proteins passing through it as part of their maturation process (BNID 109219). As another indication of the demands made on the ER, we note that a mature secreting B cell can secrete up to its own weight in antibody each day (BNID 110220), all of which must first be processed in the ER. The ER is also noted for producing most of the lipids that make up the cell's membranes. Finally, the ER is the main calcium deposit site in the cell, thus functioning as the crossroads for various intracellular signaling pathways. Serving as the equivalent of a corporate mailroom, the ER activity, and thus its size, depends on the state of the cell.

When talking about the "size" of organelles such as the ER, there are several different ways we can characterize their spatial extent. One perspective is

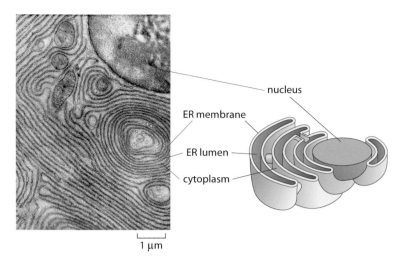

Figure 1-22 Structure of the endoplasmic reticulum. The left panel shows a thin-section electron micrograph of the region surrounding the nucleus in an acinar cell that comes from the pancreas of a bat. The schematic illustrates the connected membrane morphology of the ER, which is contiguous with the nuclear membrane. (Adapted from Fawcett DW [1966] The Cell, Its Organelles and Inclusions: An Atlas of Fine Structure. W. B. Saunders.)

to compare the total membrane area tied up in the organelle of interest relative to that of the plasma membrane, for example. A second way of characterizing the spatial extent of the organelle is by appealing to the volume enclosed within the organelle of interest and comparing it to the total cell volume. As can be inferred from the electron micrograph image of an acinar cell from the pancreas in charge of secretion (Figure 1-22), the undulating shape of the convoluted ER membrane ensures that its surface area is actually 10–20 times larger than the outer surface area of the cell itself (the plasma membrane). The distribution of membrane surface area among different organelles in liver and pancreatic cells is quantified in **Table 1-4**. The table shows that the membrane area allocation is dominated by the ER (as much as 60%), followed by the Golgi and mitochondria. The cell plasma membrane in these mammalian cells tends to be a small fraction of less than 10%. In terms of volume, the ER can comprise >10% of the cellular volume, as shown in **Table 1-5**.

In recent years, the advent of both fluorescence microscopy and tomographic methods in electron microscopy have made it possible to construct a much more faithful view of the full three-dimensional structure of these organelles. One of the insights to emerge from these studies is the recognition that they are made up from a few fundamental structural units—namely, tubules, which are 30–100 nm in diameter (BNID 105175, 111388), and sheets, which bound an internal space known as the ER lumen, as shown in Figure 1-22. As with studies of other organelles, such

membrane type	percentage of total cell membrane	
	liver hepatocyte	pancreatic exocrine cell
plasma	2	5
rough ER	35	60
smooth ER	16	<1
Golgi apparatus	7	10
mitochondria outer	7	4
mitochondria inner	32	17
nucleus inner	0.2	0.7
secretory vesicle	—	3
lysosome	0.4	—
peroxisome	0.4	—
endosome	0.4	—

Table 1-4 The percentage of the total cell membrane of each membrane type in two model cells. The symbol "—" indicates that the value was not determined. (Adapted from Alberts B, Johnson A, Lewis J et al. [2015] Molecular Biology of the Cell, 6th ed. Garland Science.)

as the mitochondria, early electron microscopy images were ambiguous since in cross section, even planar cisternae have a tubular appearance. The more recent three-dimensional membrane reconstructions have clarified such issues by making it possible to actually see tubular structures unequivocally and to avoid mistaking them for cuts through planar structures. These more detailed studies have revealed that the ER's fundamental structures are spatially organized, with the sheets being predominant in the perinuclear ER and tubules found primarily at the peripheral ER. Thus, it appears that the various parts of the cell "see" different ER architecture. The ER is in contact with most organelles through membrane contact sites. For example, the mitochondria–ER contact site is composed of a complex of membrane proteins that span either

intracellular compartment	percentage of total cell volume
cytosol	50–60
mitochondria	20
rough ER cisternae	10
smooth ER cisternae plus Golgi cisternae	6
nucleus	6
peroxisomes	1
lysosomes	1
endosomes	1

Table 1-5 The volume fraction occupied by different intracellular compartments in a liver hepatocyte cell. (Adapted from Alberts B, Johnson A, Lewis J et al. [2015] Molecular Biology of the Cell, 6th ed. Garland Science.)

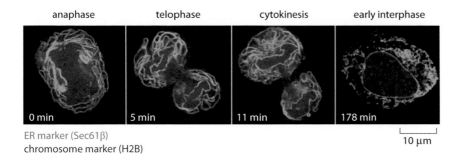

anaphase   telophase   cytokinesis   early interphase

0 min   5 min   11 min   178 min

ER marker (Sec61β)
chromosome marker (H2B)

10 μm

Figure 1-23 Structural dynamics of the endoplasmic reticulum during the cell cycle. Confocal images of HeLa cells. The chromosomes are labeled in red using a fusion of a fluorescent protein with histone H2B. The ER is labeled in green by virtue of a fusion of a fluorescent protein to a molecular member of the ER segregation apparatus (Sec61β-GFP). The sequence of images shows the changes in ER morphology as a function of time during the cell cycle. (Adapted from Lu L, Ladinsky MS & Kirchhausen T [2009] *Mol Biol Cell* 20:3471–3480.)

organelle. Similar contacts are found between the ER and the vacuole, peroxisome, and cell membrane.

One of the deceiving aspects of images like those shown in Figure 1-22 is that they give the illusion that these structures are static. However, given the cell's imperative to reproduce itself, it is clear that during the process of cell division, when the nuclear envelope dissolves away, the ER must undergo substantial rearrangement as well, cutting it in two parts to later re-engulf the two nuclei to be. Beautiful recent studies have made it possible to watch the remodeling of the ER structure during the cell cycle in real time, as shown in **Figure 1-23**. By making a stack of closely spaced confocal images, it is possible to gain insights into the three-dimensional structure of the organelle over time. In these images, we see that during interphase the ER is reticular (netlike). To appreciate the tangled arrangement of organellar membranes even more deeply, **Figure 1-24** provides a reconstructed image using X-ray microscopy of the ER and other ubiquitous membrane systems in the cell. In this cell type and under

Figure 1-24 X-ray microscopy image of cellular ultrastructure highlighting the endoplasmic reticulum. This image is a volumetric rendering of images of a mouse adenocarcinoma cell. The numbers represent percent of the volume occupied by the different compartments. (Adapted from Schneider G, Guttmann P, Heim S et al. [2010] *Nature Meth* 7:985–987.)

■	lysosomes	13%
■	mitochondria	17%
■	endoplasmic reticulum	3%
▨	vesicles	2%
	external	65%

these growth conditions, the reconstruction reveals that the mitochondria and lysosomes are more dominant in terms of volume than the ER. The cytoplasm itself occupies more than one-half of the volume, even if it is deemed transparent in these reconstructions, which take a wide slice (depth of focus) and project it into a dense two-dimensional image. Structural images like these serve as a jumping-off point for tackling the utterly mysterious microscopic underpinnings for how the many complex membrane structures of the ER and other organelles are set up and change during the course of the cell cycle.

How big are mitochondria?

Mitochondria are famed as the energy factories of eukaryotic cells, the seat of an array of membrane-bound molecular machines synthesizing the ATP that powers many cellular processes. It is now thought that mitochondria in eukaryotic cells came from some ancestral cell that took up a prokaryote through a process such as endocytosis or phagocytosis and, rather than destroying it, opted for a peaceful coexistence in which these former bacteria eventually came to provide the energy currency of the cell. One of the remnants of this former life is the presence of a small mitochondrial genome that bears more sequence resemblance to its prokaryotic precursors than to its eukaryotic host.

Beyond their fascinating ancestry, mitochondria are also provocative due to their great diversity in both size and shape. Though probably familiar to many for the morphology depicted in **Figure 1-25**, with its characteristic micron-sized, bacterium-like shape and series of internal lamellae shown in magnified form in Figure 1-25, mitochondria have in fact a host of different structural phenotypes. These shapes range from onion-like morphologies to reticular structures such as those shown in Figures 1-25C and 1-25D, in which the mitochondrion is one extended object, to a host of other bizarre shapes that arise when cells are exposed to an oxygen-rich environment or that emerge in certain disease states. These reticular mitochondria can spread over tens of microns.

As shown in **Figure 1-26**, electron microscopy images of mitochondria encourage their textbook depiction as approximately spherocylindrical in shape (that is, cylinders with hemispherical caps), with a length of roughly two microns and a diameter of roughly one micron. These organelles are characterized by two membrane systems that separate the space into three distinct regions—namely, the space external to the mitochondrion, the intermembrane space between the mitochondrial inner and outer

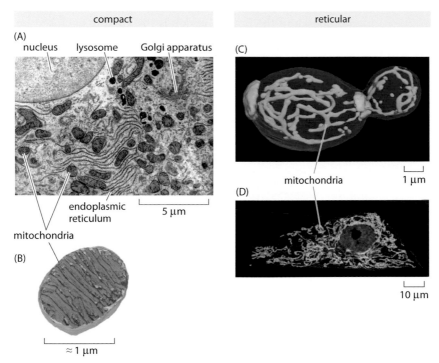

Figure 1-25 Shapes and sizes of mitochondria. (A) Electron microscopy image of a rat liver cell, which highlights many of the important organelles and illustrates the size and shape of mitochondria. (B) Cryo-electron microscopy reconstruction of the structure of a lamellar mitochondrion. (C) Reticular structure of mitochondria in a budding yeast cell. Bud scars are labeled separately in blue. (D) Reticular mitochondrial network in a PtK2 kangaroo rat cell. The mitochondria are visible in green and were labeled with an antibody against the proteins responsible for the transport of proteins across the mitochondrial membranes. The tubulin of the microtubules are labeled in red and the nucleus is shown in blue. (A, adapted from Fawcett DW [1966] The Cell, Its Organelles and Inclusions: An Atlas of Fine Structure. W. B. Saunders. With permission from Elsevier Inc. B, courtesy of Terry Frey and Guy Perkins. C, adapted from Egner A, Jakobs S & Hell SW [2002] *Proc Natl Acad Sci USA* 99:3370–3375. With permission from the National Academy of Sciences. D, adapted from Schmidt R, Wurm CA, Punge A et al. [2009] *Nano Lett* 9:2508–2510.)

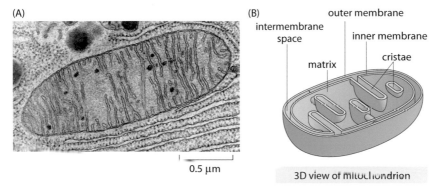

Figure 1-26 The structure of a mitochondrion. (A) Electron microscopy image of a mitochondrion from the pancreas of a bat. (B) Schematic illustrating the three membrane spaces relevant to mitochondria as well as the connectivity. (A, adapted from Fawcett DW [1966] The Cell, Its Organelles and Inclusions: An Atlas of Fine Structure. W. B. Saunders.)

membranes, and the matrix, which is the volume enclosed by the inner membrane. Different mitochondrial morphologies all respect this basic organizational connectivity.

How many mitochondria are in a cell? A characteristic order of magnitude for yeast would be 10^1 and for mammalian cells it would be 10^3–10^4, but beware that the very idea of "counting" mitochondria can be tricky in many cases, because the mitochondria are sometimes reticular and indistinct, peanut-shaped objects. For example, yeast cells grown on ethanol contain a larger number (20–30; BNID 103070) of small, discrete mitochondria, whereas when these same cells are grown on glucose, they contain a smaller number (~3; BNID 103068) of large, branched mitochondria. These distinct morphologies do not significantly affect the fraction of the cellular volume occupied by the mitochondria and probably relates to the different demands in a respiratory-versus-respiro-fermentative lifestyle.

How big are chloroplasts?

Chloroplasts play a key role in the energy economy of the cells that harbor them. Chloroplasts are less well known than their mitochondrial counterparts, though they are usually much larger and play a key role in producing the reduced compounds that store the energy that is then broken down in mitochondria. Chloroplasts have the pivotal role in the biosphere of carrying out the chemical transformations linking the inorganic world (CO_2) to the organic world (carbohydrates). This feat of chemical transformation enables the long-term storage of the fleeting sun's energy in carbohydrates and its controlled release in energy currencies such as ATP and NADPH. Those same carbon compounds also serve to build all of the biomass of cells as a result of downstream metabolic transformations.

Chloroplasts in vascular plants range from being football- to lens-shaped and, as shown in **Figure 1-27**, have a characteristic diameter of ≈4–6 microns (BNID 104982, 107012), with a mean volume of ≈20 μm^3 (for corn seedling; BNID 106536). In algae, they can also be cup-shaped, tubular, or even form elaborate networks, paralleling the morphological diversity found in mitochondria. Though chloroplasts are many times larger than most bacteria, in their composition they can be much more homogeneous, as required by their functional role, which centers on carbon fixation. The interior of a chloroplast is made up of stacks of membranes, in some ways analogous to the membranes seen in the rod cells found in the visual systems of mammals. The many membranes that make up a chloroplast are fully packed with the apparatus of light capture, photosystems, and

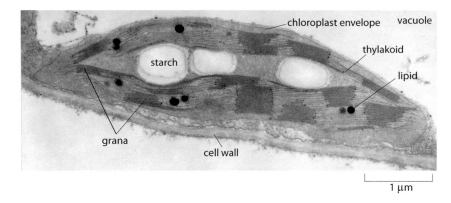

chloroplast envelope   vacuole

thylakoid

starch

lipid

grana

cell wall

1 µm

Figure 1-27 Electron micrograph of a chloroplast. The light reactions occur in the membrane-bound compartment called the thylakoid. There are usually about 40–60 stacks of disks termed grana per chloroplast (BNID 107013) covering 50–70% of the thylakoid membrane surface (BNID 107016). Each single stack has a diameter of 0.3–0.6 µm (BNID 107014). The sugar produced is stored in starch granules. (Adapted from Alberts B, Johnson A, Lewis J et al. [2015] Molecular Biology of the Cell, 6th ed. Garland Science.)

related complexes. The rest of the organelle is packed almost fully with one dominant protein species—namely, Rubisco, the protein serving to fix CO_2 in the carbon fixation cycle. The catalysis of this carbon fixation reaction is relatively slow, thus necessitating such high protein abundances.

The number of chloroplasts per cell varies significantly between organisms. Even within a given species, this number can change significantly, depending upon growth conditions. In the model algae, *Chlamydomonas reinhardtii*, there is only one prominent cup-shaped chloroplast per cell, whereas in a typical photosynthetic leaf cell (mesophyll) from plants such as *Arabidopsis* and wheat, there are about 100 chloroplasts per cell (BNID 107030, 107027, 107029). A vivid example from a moss is shown

Figure 1-28 Chloroplasts in the moss *Plagiomnium affine*, found in old-growth boreal forests in North America, Europe, and Asia, growing in moist woodland and turf. The lamina cells shown here are elongated, with a length of about 80 microns and a width of 40 microns. These cells, as with most plant cells, have their volume mostly occupied by large vacuoles, so the cytoplasm and chloroplasts are at the periphery. Chloroplasts also show avoidance movement, in which they move from the cell surface to the side walls of cells under high light conditions to avoid photodamage. (Courtesy of Ralf Wagner.)

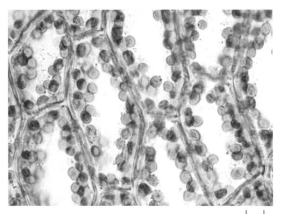

10 µm

in **Figure 1-28**. Each chloroplast has tens to hundreds of copies (BNID 107105, 107107, 107108) of the chloroplast genome, which is ≈100 kbp in length (BNID 105918). This creates a fascinating challenge of how to balance the expression of genes that are coded in the chloroplast genome at thousands of gene copies per cell with the expression of genes that have a single copy in the main nuclear genome. In some cases, such as the protein Rubisco, they form a complex at one-to-one stoichiometric ratios!

Much evidence points to the idea that chloroplasts originated in a process of endosymbiosis—that is, they were originally free-living cells (probably photosynthetic cyanobacteria) that were engulfed (or enslaved) a billion years ago (BNID 107041) by cells that became their new hosts. With time, these originally distinct cells forged a tight collaboration in which most genes transferred from the engulfed cell to the host nucleus, in much the same way that the mitochondrial genome obtained its tiny size. From genomes that probably originally contained over 3000 genes, only about 130 genes remain in the chloroplasts of contemporary plants (BNID 106553, 106554).

These processes of engulfment followed by adaptation can still be observed today. Through a process known as kleptoplasty, different organisms (ranging from dinoflagellates to sea slugs) are able to digest algae while keeping the chloroplasts of these algae intact. These captured plastids are kept functional for months and are used to "solar power" these organisms. Not only the act of engulfing, but also the slower process of adaptation between the host and the organelle, can be observed. In one study it was determined that in 1 out of ~10,000 pollen grains, a reporter gene is transferred from the chloroplast to the nuclear genome (BNID 103096). How can such a low value be assessed reliably? A drug-resistance gene that can only function in the nucleus was incorporated into the chloroplasts of tobacco plants. Pollen from these plants was used to pollinate normal plants. Next, 250,000 seeds were screened and 16 showed resistance to the drug. Now here is the catch: Chloroplast genomes are transferred only through the mother. The pollen has only nuclear genes. The only way for the resistance gene to arrive through the pollen was shown to be through infiltration from the chloroplast genome into the nuclear genome. Measuring the rate of this process gives some insight into how genomes of organelles can be so small. It leaves open the question of what is the selective advantage of transferring the genomic information from the organelle's DNA to the central cell repository in the nucleus.

All told, chloroplasts are organelles of great beauty and sophistication. Their intriguing evolutionary history is revealed in their compact genomes. Structurally, their stacked membrane systems provide a critical system for capturing light and using its energy for the synthesis of the carbohydrates that are at the center of food chains across Earth.

How big is a synapse?

So far in the book, we have mainly focused on the sizes of individual cells and the organelles within them. Multicellularity, however, is all about partnerships between cells. A beautiful example of our own multicellularity is provided by the cells in our nervous system. These cells are part of a vast and complex array of interactions that are only now beginning to be mapped. The seat of interactions between neighboring neurons are synapses, the interface between cells in which small protrusions adopt a kissing configuration, as seen in **Figure 1-29** and **Figure 1-30** for the cases of a neuromuscular junction and a synapse in the brain, respectively. These synapses are responsible for the propagation of information from one neuron to the next. Interestingly, information transmission in the nervous system is partly electrical and partly chemical. That is, when an action potential travels along a nerve, it does so by transiently changing the transmembrane potential from its highly negative resting value to a nearly equal positive potential. When the action potential reaches the synapse, this leads to vesicle fusion and subsequent release of chemical signals (neurotransmitters), which induce channel gating in the neighboring cell with which it has formed the synapse. This results, in turn, in an action potential in the neighboring cell. As seen in Figures 1-29 and 1-30, the synapse is composed of a presynaptic terminal on the axon of the transmitting neuron and a postsynaptic terminal with a so-called synaptic cleft between them. The total number of such synapses in the human brain has been vaguely stated to be in the range of 10^{13}–10^{15} (BNID 106138, 100693), with every cubic millimeter of cerebral cortex having about a billion such synapses (BNID 109245).

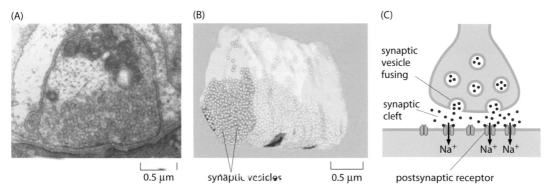

(A) (B) (C)

synaptic vesicle fusing

synaptic cleft

Na^+ Na^+ Na^+

0.5 µm   synaptic vesicles   0.5 µm   postsynaptic receptor

Figure 1-29 Structure of a neuromuscular junction. (A) Electron microscopy image of a nerve terminal and its synapse with a neighboring cell in a neuromuscular junction. (B) Cryo-electron microscopy reconstruction image of a fraction of the presynaptic neuron showing the synaptic vesicles it harbors for future release. (C) Schematic of a synapse. Note that the synaptic cleft, vesicles, etc., are not drawn to scale. (A, and B, adapted from Rizzoli SO & Betz WJ [2005] *Nat Rev Neurosci* 6:57–69.)

(A)

synapse   dendrite   axon

(B)

synapse

(C)

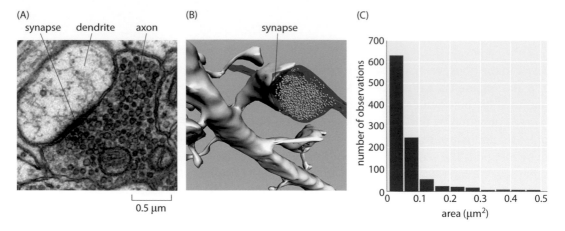

0.5 µm

Figure 1-30 Size of synapses in the brain. (A) Electron microscopy image of a synapse between an axon and a dendrite. (B) Reconstruction of a synapse like that shown in (A), but illustrating the synaptic vesicles. (C) Distribution of synapse sizes as measured using electron microscopy. (Courtesy of Linnaea Ostroff.)

(A)

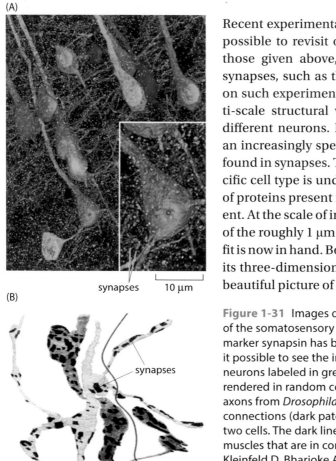

synapses   10 µm

(B)

synapses

4 µm

Recent experimental developments have now made it possible to revisit order-of-magnitude estimates, like those given above, based on volume renderings of synapses, such as those shown in **Figure 1-31**. Based on such experiments, we have begun to garner a multi-scale structural view of the connections between different neurons. Further, these maps are providing an increasingly specific view of the chemical diversity found in synapses. That is, depending upon which specific cell type is under consideration, the complement of proteins present in the synapse region will be different. At the scale of individual synapses, a close-up view of the roughly 1 µm-size box into which most synapses fit is now in hand. Both classic electron microscopy and its three-dimensional tomographic extensions paint a beautiful picture of synapses with their complement of

Figure 1-31 Images of the synapse. (A) Volume rendering of the somatosensory cortex of a mouse. The synaptic marker synapsin has been immunolabeled, thereby making it possible to see the individual synapsin puncta connecting neurons labeled in green. The individual synapses were rendered in random colors. (B) Reconstruction of two axons from *Drosophila* that show the location of synaptic connections (dark patches). Color is used to distinguish the two cells. The dark line is the boundary between the two muscles that are in contact with the axons. (A, adapted from Kleinfeld D, Bharioke A, Blinder P et al. [2011] *J Neurosci* 31:16125–16138. B, adapted from Rizzoli SO & Betz WJ [2005] *Nat Rev Neurosci* 6:57–69.)

synaptic vesicles. Using tomographic techniques and a combination of light and electron microscopy, scientists have mapped out the rich network of connections between neurons, including their complements of synaptic vesicles. Figures 1-29 and 1-30 show several different views of the distributions of these micron-sized synapses on individual neurons. Indeed, Figure 1-30C gives a precise quantitative picture of the range of synaptic sizes in the brain. To get a sense of scale, recall that a bacterium has a volume of ≈ 1 μm^3, so each of these synapses is roughly the size of a bacterium (BNID 111086).

To better understand the intimate connection between structure and function in the cells of the nervous system, consider the processes that take place when we read a book. First, photons reflected from the page are absorbed by rhodopsin in the photoreceptors of our eyes. This photon absorption results in a signal cascade in the photoreceptor of the retina, which culminates in the release of neurotransmitters at the synapse. Specifically, the synaptic vesicles fuse with the membrane of the presynaptic cell, as shown in Figures 1-29C and 1-30A (though the microscopy image in Figure 1-29 shows a neuromuscular junction and not the synapse of a photoreceptor), and release 10^3–10^4 neurotransmitter molecules from each vesicle (BNID 108622, 108623, 102777). Common neurotransmitters are glutamate, used in about 90% of synapses (BNID 108663), as well as acetylcholine and GABA, which are packed at a high concentration of 100–200 mM in the synaptic vesicles (BNID 102777). Each vesicle is about 10^{-5} μm^3 in volume (BNID 102776), so our rule of thumb that 1 nM concentration in 1 μm^3 is about 1 molecule enables us to verify that there are indeed about 1000 neurotransmitter molecules per vesicle.

These molecules then diffuse into the synaptic cleft and bind receptors on the postsynaptic cell surface. The signal propagating to our brain is carried by electric action potentials within neurons and relayed from one neuron to the next by similar synaptic fusion events. Vesicle release is triggered by 10^2–10^4 Ca^{2+} ions (BNID 103549). The energy expended per vesicle release has been estimated to be about 10^5 ATP/vesicle (BNID 108667). Synapses are cleared within about 1 ms, thus preparing the way for future communication. Rapid clearing is essential because neuronal firings can reach rates of over 100 times per second (BNID 107124), though the average firing rate is estimated to be 1–10 times per second in the cortex (BNID 108670). The delay created by the time it takes a neurotransmitter to diffuse across the synaptic cleft (not drawn to scale in the schematic of Figure 1-29) is part of the response time of humans to any reflex or neural action of any sort. Conveniently, it takes less than 1 ms to traverse the 20- to 40-nm synaptic divide (BNID 100721, 108451), as the reader can verify after reading the vignette entitled "What are the time scales for diffusion in cells?" (pg. 211). Interestingly, this can be compared to the time it takes for the action potential to propagate down a nerve, which is on the ms time scale, as discussed in the vignette entitled "How fast are electrical signals propagated in cells?" (pg. 249).

CELLULAR BUILDING BLOCKS

How big are biochemical nuts and bolts?

The textbook picture of the molecules of life is dominated by nucleic acids and proteins, in no small measure because of their fascinating linkage through the processes of the central dogma. On the other hand, this picture is terribly distorted biochemically because many of the key reactions, even in the central dogma, would not happen at all were it not for a host of biochemical allies, such as water and the many ions that are needed as co-factors for the enzymes that make these reactions go. Further, we cannot forget the substrates themselves—namely, the nucleotides and amino acids from which the nucleic acids and proteins are constructed. Energizing all of this busy activity are small sugar molecules, energy carriers such as ATP, and other metabolites. In this vignette, we take stock of the sizes of the many biochemical "nuts and bolts" that provide the molecular backdrop for the lives of cells, as shown in **Figure 1-32**.

Probably the single most important biochemical nut and bolt of them all is water. It is no accident that the search for life beyond Earth often begins with the question: Is there water? Though part of the reason for this might be a lack of imagination about what other life-supporting chemistries

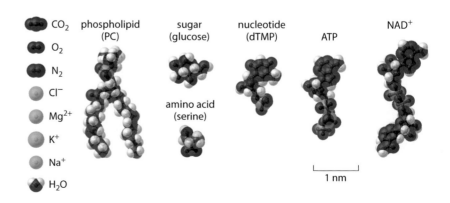

Figure 1-32 A structural view of some of the basic constituents of a cell.

might look like, the simplest reason for this obsession with water is that without it, life as we know it could not exist. One of the easiest ways for us to characterize the size of a water molecule, which is a convenient standard molecular ruler for biology, is by reference to the roughly 0.1-nm bonds (BNID 106548) between its hydrogen and oxygen atoms. Since water molecules are not spherically symmetric, it is hard to assign an effective radius to such a molecule. As another estimate for the size of a water molecule, we appeal to the mean spacing between such molecules by using the density of water. In particular, given that there are 55 moles of water per liter, the volume of a water molecule is 0.03 nm^3, and the mean spacing between molecules is roughly 0.3 nm (BNID 106548). We will also find it convenient to use the 18 Da mass of water as a way of comparing the sizes of these various molecular players.

We all come from the ocean. Despite our human dependence on fresh water for drinking and maintaining the many plants and animals that feed us, real biological water bears the signature of our watery origins in the ocean. Our first impression on hopping into the ocean (besides that it is cold!) is likely the salty taste it leaves on our tongues. A simple estimate of the saltiness of the ocean can be garnered from remembering that a kilogram of water has roughly 55 moles of water molecules (that is, 1000 g/18 g/mole). This same seawater has roughly 1 mole of salt (BNID 100802), meaning that 1 out of every 55 molecules is an ion. If we look within cells, we find a number of different ions, such as H^+, Na^+, K^+, Mg^{2+}, and Cl^-, that add up to about one-quarter of the concentration of seawater, as discussed in the vignette entitled "What are the concentrations of different ions in cells?" (pg. 91). The sizes of these ions can be captured by the so-called ionic radii, which are given by $Na^+ = 0.09$ nm, $K^+ = 0.13$ nm, $Mg^{2+} = 0.07$ nm, and $Cl^- = 0.18$ nm (BNID 108517, 104162, 109742, 109743, 103950). These ionic radii reveal the so-called "bare" ionic radius, whereas the hydrated ionic radius is usually much larger, and more similar among ions, at 0.3–0.4 nm (BNID 108517). These surrounding water molecules are exchanged on the micro- to nanosecond time scales (BNID 108517). The hydrated ions radii are shown to scale next to other nuts and bolts of the cell in Figure 1-32.

To build up the nucleic acids and proteins of the cell requires molecular building blocks. The nucleotides that are the building blocks of nucleic acids have a mass of ≈300 Da. Their physical size is compared to water in the gallery shown in Figure 1-32, though we can also get a feel for this size by remembering that the DNA double helix has a radius of roughly 1 nm and an average spacing between bases along the chain of 0.3 nm. This means that a plasmid of, say, 10 kbp will have a circumference of about 3000 nm—that is, a diameter of about 1 micron. The common depiction of plasmids as small circles inside a bacteria are easy to understand, but they do not do justice to the physical size of plasmids. Indeed, plasmids in cells must be curled up to fit in. The amino acids that make up proteins range in size from the tiny glycine, with a molecular mass of roughly 75 Da, to the 204 Da mass of tryptophan, the largest of the naturally occurring amino acids. Their respective lengths vary from 0.4 to 1 nm (BNID 106983). Adopting a mass of 100 Da per amino acid (aa) in a protein polymer serves as a very useful and calculationally convenient rule of thumb. Here, too, the sizes of the amino acids with respect to a water molecule are shown in Figure 1-32.

All of this emphasis on nucleic acids and proteins can lead us to forget the critical role played in the lives of cells by both lipids and sugars. The emerging field of lipidomics has shown that just as there is immense diversity in the character of the many proteins that inhabit cells, the membranes of the cells and their organelles are similarly characterized by widely different concentrations of an entire spectrum of different lipids (see the vignette entitled "What lipids are most abundant in membranes?"; pg. 99). In simplest terms, the lipids making up these membranes have a cross-sectional area of between 0.25 and 0.5 nm^2, and a length on the order of 2 nm, as shown in Figure 1-32. More generally, the lengths of the lipid chains (L) are dictated by the number of carbons they contain, with a rule of thumb that

$$L = a + b \times n, \qquad (1.3)$$

where n is the number of carbons in the tail and a and b are constants depicting, respectively, the terminal group size outside the carbon chain and the length extension per carbon atom. The masses of lipids are between 700 and 1000 Da as a general rule.

Cellular life is powered by a number of other key molecules besides those discussed so far. To grow new cells, biologists use various kinds of growth media, but some of the most standard ingredients in such media are sugars such as glucose. With a chemical formula of $C_6H_{12}O_6$, glucose has a molecular mass of 180 Da. Structurally, the glucose molecule is a six-membered ring, as shown in Figure 1-32, with typical carbon–carbon bond lengths of $\approx$0.15 nm and an overall molecular size of roughly 1 nm, as measured by the long axis of the cyclic form or the length of the open chain form (BNID 110368, 106979). Once sugars are present within a cell, they can be remodeled to build the carbon backbones of molecules such as the nucleotides and amino acids described above, and also for the synthesis of key energy carriers such as ATP. The size of ATP (effective diffusion diameter of $\approx$1.4 nm; BNID 106978) is compared to the rest of the biochemical nuts and bolts in Figure 1-32. ATP is a nucleotide adapted to piggyback energized phosphate groups, and it has a molecular mass of roughly 500 Da. The other major energy sources are electron donors, with NADH and NADPH being the prime shakers and movers with a mass of about 700 Da and a length of about 2.5 nm (BNID 106981).

In summary, if you have to remember one round number to utilize for thinking about the sizes of small building blocks such as amino acids, nucleotides, energy carriers, and so on, 1 nm is an excellent rule of thumb.

Which is bigger, mRNA or the protein it codes for?

The role of messenger RNA molecules (mRNAs), as epitomized in the central dogma, is one of fleeting messages for the creation of the main movers and shakers of the cell—namely, the proteins that drive cellular life. Words like these can conjure a mental picture in which an mRNA is thought of as a small blueprint for the creation of a much larger protein machine. In reality, the scales are exactly the opposite of what most people would guess. Nucleotides, the monomers making up an RNA molecule, have a mass of about 330 Da (BNID 103828). This is about three times heavier than the average amino acid mass, which weighs in at ≈110 Da (BNID 104877). Moreover, since it takes three nucleotides to code for a single amino acid, this implies an extra factor of three in favor of mRNA such that the mRNA coding a given protein will be almost an order of magnitude heavier when one compares codons to the residues they code for. A realistic depiction of a mature mRNA versus the protein it codes for, in this case the oxygen-binding protein myoglobin, is shown in **Figure 1-33**. As shown in the figure, in the microscopic world, our everyday intuition that the blueprint (mRNA) should be smaller than the object it describes (protein) does not hold. In eukaryotes, newly transcribed mRNA precursors are often richly decorated with introns that skew the mass imbalance even further.

What about the spatial extent of these mRNAs in comparison with the proteins they code for? The mass excess implies a larger spatial scale as well, though the class of shapes adopted by RNAs are quite different than their protein counterparts. Many proteins are known for their globular structures (see the vignette entitled "How big is the 'average' protein?"; pg. 45). By way of contrast, mRNA is more likely to have a linear structure punctuated by secondary structures in the form of hairpin stem-loops and pseudoknots, but is generally much more diffuse and extended. The "threadlike" mRNA backbone has a diameter of less

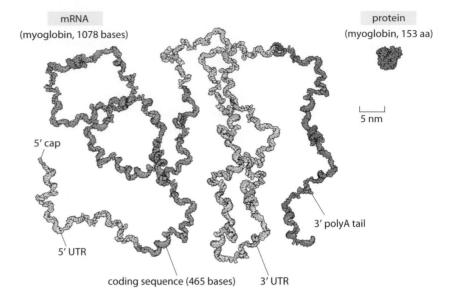

Figure 1-33 The relative sizes of a globular protein and the mRNA that codes for it. The myoglobin protein is drawn to scale next to the mRNA transcript that leads to it. The coding sequence of an mRNA alone is about an order of magnitude heavier by mass than the protein. The myoglobin protein is in blue. For the mRNA, the 5′ cap and 3′ polyA tail are in purple, the 5′ and 3′ untranslated regions (UTRs) are in yellow, and the coding sequence is in orange. (Courtesy of David Goodsell.)

than 2 nm, much smaller than the diameter of a characteristic globular protein, which is about 5 nm (BNID 100481). On the other hand, a characteristic 1000-nucleotide-long mRNA (BNID 100022) will have a linear length of about ≈300 nm (BNID 100023). The most naïve estimate of mRNA size is to simply assume that the structure is perfectly base paired into a double-stranded RNA molecule. For a 1000-base-long mRNA, this means that its double-stranded version will be 500 bp long, corresponding to a physical dimension of more than 150 nm, using the rule of thumb that a base pair is about 0.3 nm in length. This is an overestimate, however, because these structures are riddled with branches and internal loops, which shorten the overall linear dimension. Recent advances have made it possible to visualize large RNA molecules in solution using small-angle X-ray scattering and cryo-EM, as shown in **Figure 1-34**. One of the useful statistical measures of the spatial extent of such structures is the so-called radius of gyration, which can be thought of as the radius of a sphere of an equal effective size. For RNAs, this was found to be roughly ≈20 nm (BNID 107712), indicating a characteristic diameter of ≈40 nm. Hence, contrary to the expectation of our uncoached intuition, we note that like the mass ratio, the spatial extent of the characteristic mRNA is about 10-fold larger than the characteristic globular protein.

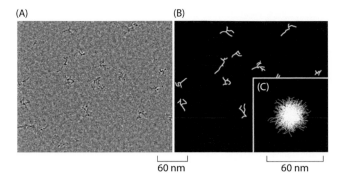

(A)   (B)

(C)

60 nm   60 nm

Figure 1-34 Cryo-electron microscopy images of RNAs in vitrified solution. (A) Fourier band-pass filtered images of a 975-nt RNA from chromosome XII of *S. cerevisiae*. Individual RNA molecules, suspended in random orientations, are seen as dark branched objects. (B) Traced skeletons of the molecules in panel A. (C) Depiction of 100 traced projections superimposed with their centers of mass in registry. (B, adapted from Gopal A, Zhou ZH, Knobler CM & Gelbart WM [2012] *RNA* 18:284–299.)

How big is the "average" protein?

Proteins are often referred to as the workhorses of the cell. An impression of the relative sizes of these different molecular machines can be garnered from the gallery shown in **Figure 1-35**. One favorite example is provided by the Rubisco protein shown in the figure, which is responsible for atmospheric carbon fixation, literally building the biosphere out of thin air. This molecule, one of the most abundant proteins on Earth, is responsible for extracting about a hundred Gigatons of carbon from the atmosphere each year. This is $\approx$10 times more than all the carbon dioxide emissions made by humanity from car tailpipes, jet engines, power plants, and all of our other fossil-fuel-driven technologies. Yet carbon levels keep on rising globally at alarming rates because this fixed carbon is subsequently re-emitted in processes such as respiration and the like. This chemical fixation is carried out by these Rubisco molecules with a monomeric mass of 55 kDa, fixating CO_2 one at a time, with each CO_2 having a mass of 0.044 kDa (just another way of writing 44 Da that clarifies the 1000:1 ratio in mass). For another dominant player in our biosphere, consider the ATP synthase (MW $\approx$ 500–600 kDa; BNID 106276), also shown in Figure 1-35, which decorates our mitochondrial membranes and is responsible for synthesizing the ATP molecules (MW = 507 Da) that power much of the chemistry of the cell. These molecular factories churn out so many ATP molecules that all the ATPs produced by the mitochondria in a human body in one day would

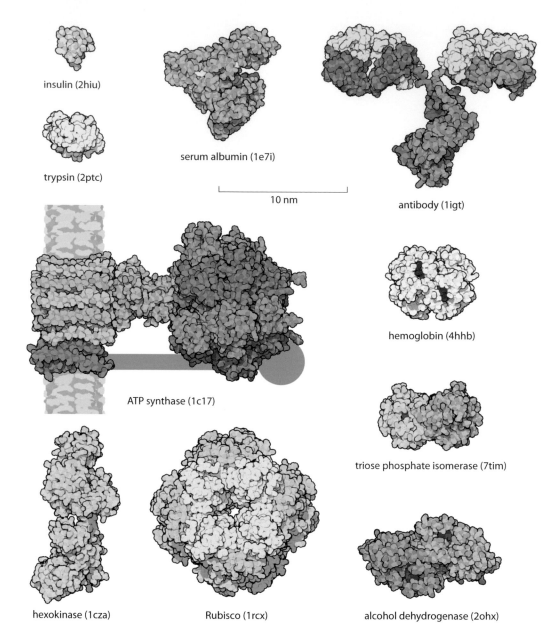

insulin (2hiu)

trypsin (2ptc)

serum albumin (1e7i)

10 nm

antibody (1igt)

ATP synthase (1c17)

hemoglobin (4hhb)

triose phosphate isomerase (7tim)

hexokinase (1cza)

Rubisco (1rcx)

alcohol dehydrogenase (2ohx)

Figure 1-35 Gallery of proteins. Representative examples of protein size are shown with examples drawn from across biology to illustrate some of their key functional roles. Examples range from the antibodies so important to the immune system to Rubisco and photosynthesis. All the proteins in the figure are shown on the same scale to give an impression of their relative sizes. The small red objects shown on some of the molecules are the substrates for the protein of interest. For example, in hexokinase, the substrate is glucose. The handle in ATP synthase is known to exist, but the exact structure was unavailable and thus only schematically drawn. Names in parentheses are the PDB database identifiers. (Courtesy of David Goodsell.)

have nearly as much mass as the body itself. As we discuss in the vignette entitled "What is the turnover time of metabolites?" (pg. 228), the rapid turnover makes this less improbable than it may sound.

The size of proteins, such as Rubisco, ATP synthase, and many others, can be measured both geometrically, in terms of how much space they take up, and in terms of their sequence size, as determined by the number of amino acids that are strung together to make the protein. Given that the average amino acid has a molecular mass of 100 Da, we can interconvert between mass and sequence length. For example, the 55 kDa Rubisco monomer has roughly 500 amino acids making up its polypeptide chain. The spatial extent of soluble proteins and their sequence size often exhibit an approximate scaling property where the volume scales linearly with sequence size, in which case the radii or diameters tend to scale as the sequence size to the one-third power. A simple rule of thumb for thinking about typical soluble proteins like the Rubisco monomer is that they are 3–6 nm in diameter, as illustrated in Figure 1-35, which shows not only Rubisco, but many other important proteins that make cells work. In roughly one-half the cases, it turns out that proteins function when several identical copies are symmetrically bound to each other, as shown in **Figure 1-36**. These are called homo-oligomers to differentiate them from the cases where different protein subunits are bound together, thus forming the so-called hetero-oligomers. The most common states are the dimer and tetramer (and the non-oligomeric monomers). Homo-oligomers are about twice as common as hetero-oligomers (BNID 109185).

There is an often-surprising size difference between an enzyme and the substrates it works on. For example, in metabolic pathways, the substrates are metabolites, which usually have a mass of less than 500 Da, while the corresponding enzymes are usually about 100 times heavier. In the glycolysis pathway, small sugar molecules are processed to extract both energy and building blocks for further biosynthesis. This pathway is characterized by a host of protein machines, all of which are much larger than their sugar substrates, with examples shown in the bottom right corner of Figure 1-35, where we see the relative size of the substrates denoted in red when interacting with their enzymes.

Concrete values for the median gene length can be calculated from genome sequences as a bioinformatic exercise. **Table 1-6** reports these values for various organisms, which show a trend towards longer protein coding sequences when moving from unicellular to multicellular organisms. In **Figure 1-37A**, we go beyond mean protein sizes to characterize the full distribution of coding sequence lengths on the genome, reporting values for three model organisms. If our goal was to learn about the spectrum of protein sizes, this definition based on the genomic length might

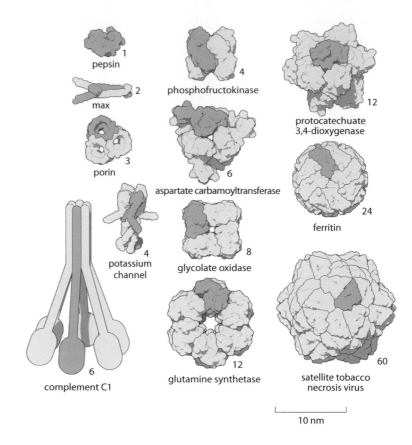

Figure 1-36 A gallery of homo-oligomers that shows the beautiful symmetry of these common protein complexes. The monomeric subunits making up each oligomer are highlighted in pink. (Courtesy of David Goodsell.)

organism	median protein length (amino acids)
H. sapiens	375
D. melanogaster	373
C. elegans	344
S. cerevisiae	379
A. thaliana	356
5 eukaryotes (above)	361
67 bacteria	267
15 archaea	247

Table 1-6 Median length of coding sequences of proteins based on genomes of different species. The entries in this table are based upon a bioinformatic analysis (BNID 106444). As discussed in the text, we propose an alternative metric that weights proteins by their abundance as revealed in recent proteome-wide censuses using mass spectrometry. (Adapted from Brocchieri L and Karlin S [2005] *Nuc Acids Res* 33:3390.)

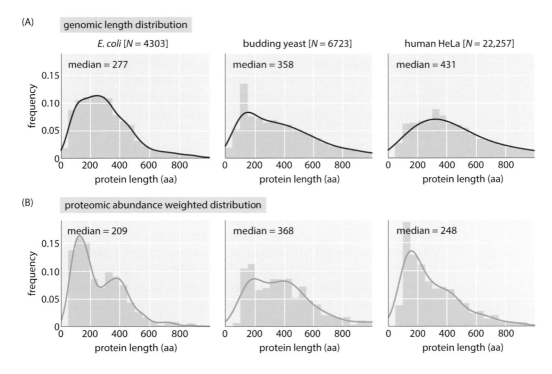

Figure 1-37 Distribution of protein lengths in *E. coli*, budding yeast, and human HeLa cells. (A) Protein length is calculated in amino acids (aa), based on the coding sequences in the genome. (B) Distributions are drawn after weighting each gene with the protein copy number inferred from mass spectrometry proteomic studies. Continuous lines are Gaussian kernel-density estimates for the distributions, and they serve as a guide to the eye. (*E. coli* copy number weighted results based on M9+glucose media data from Schmidt A, Kochanowski K, Vedelaar S et al. [2015] *Nat Biotechnol* (forthcoming). *S. cerevisiae* weighted results based on defined media data from de Godoy LMF, Olsen JV, Cox J et al. [2008] *Nature* 455:1251–1254. *H. sapiens* data from Geiger T, Wehner A, Schaab C et al. [2012] *Mol Cell Proteomics* 11:M111.014050.)

be enough. But when we want to understand the investment in cellular resources that goes into protein synthesis, or to predict the average length of a protein randomly chosen from the cell, we advocate an alternative definition, which has become possible thanks to recent proteome-wide censuses. For these kinds of questions, the most abundant proteins should be given a higher statistical weight in calculating the expected protein length. We thus calculate the weighted distribution of protein lengths shown in **Figure 1-37B**, giving each protein a weight proportional to its copy number. This distribution represents the expected length of a protein randomly fished out of the cell rather than randomly fished out of the genome. The distributions that emerge from this proteome-centered approach depend on the specific growth conditions of the cell. In this book, we chose to use ≈300 aa in prokaryotes and ≈400 aa in eukaryotes as the simple rule of thumb for the length of the "typical" protein. The distributions in Figure 1-37 show this is a reasonable estimate, though it might be an overestimate in some cases.

One of the charms of biology is that evolution necessitates very diverse functional elements, thus creating outliers in almost any property (which is also the reason we discussed medians and not averages above). When it comes to protein size, titin is a whopper of an exception. Titin is a multifunctional protein that behaves as a nonlinear spring in human muscles, with its many domains unfolding and refolding in the presence of forces and giving muscles their elasticity. Titin, with its 33,423 aa polypeptide chain (BNID 101653), is about 100 times longer than the average protein. Identifying the smallest proteins in the genome is still controversial, but short ribosomal proteins of about 100 aa are common.

It is very common to use GFP tagging of proteins in order to study everything from their localization to their interactions. Armed with the knowledge of the characteristic size of a protein, we are now prepared to revisit the seemingly innocuous act of labeling a protein. GFP is 238 aa long, and is composed of a beta barrel within which key amino acids form the fluorescent chromophore, as discussed in the vignette entitled "What is the maturation time for fluorescent proteins?" (pg. 237). As a result, for many proteins the act of labeling should really be thought of as the creation of a protein complex that is now twice as large as the original unperturbed protein.

How big are the molecular machines of the central dogma?

Molecular machines manage the journey from genomic information in DNA to active and functioning proteins in the processes of the central dogma. The idea of directional transfer of information through a linked series of processes, termed the central dogma, started out in 1956 as a fertile hypothesis in the hands of Francis Crick, as shown in **Figure 1-38**. In the time since its original suggestion, this hypothesis has been confirmed in exquisite detail, with the molecular anatomy of the machines that carry out these processes now coming into full relief.

The machines that mediate the processes of the central dogma include RNA polymerase, which is the machine that takes the information stored in DNA and puts it in a form suitable for protein synthesis by constructing messenger RNA molecules, and the ribosome, the universal translation machine that synthesizes proteins. Proteins do not survive indefinitely, and their fate is often determined by another molecular machine, the proteasome—the central disposal site that degrades the proteins so carefully assembled by the ribosome. Our understanding of these macromolecular

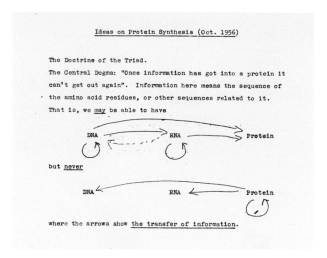

Figure 1-38 Notes of Francis Crick on the central dogma. This was an early draft for the article published as Crick FHC [1958] *Symp Soc Exp Biol* XII:139-163. The 1958 paper did not include this visual depiction, which later appeared in Crick FHC [1970] *Nature* 227, 561–563.

complexes has evolved from the point where three to four decades ago, it was only possible to infer their existence, to the present era in which it is possible to acquire atomic resolution images of their structures in different conformational states.

As seen in Figure 1-38, there is an arrow from DNA to itself, which signifies DNA replication. This process of replication is carried out by a macromolecular complex known as the replisome. The *E. coli* replisome is a collection of distinct protein machines that include helicase (52 kDa, each of six subunits; BNID 104931), primase (65 kDa; BNID 104932), and the DNA polymerase enzyme complex (791 kDa in several units of the complex; BNID 104931). To put the remarkable action of this machine in focus, an analogy has been suggested in which one thinks of the DNA molecule in human terms by imagining it to have a diameter of 1 m[¶] (to get a sense of the actual size of the replication complex relative to its DNA substrate, see **Figure 1-39**). At this scale, the replisome has the size of a FedEx truck, and it travels along the DNA at roughly 600 km/h. Genome replication is a 400-km journey in which a delivery error occurs only once every several hundred kilometers, this despite the fact that a delivery is being made roughly six times for every meter traveled. During the real replication process, the error rate is even lower as a result of accessory quality control steps (proofreading and mismatch correction) that ensure that a wrong delivery happens only once in about 100 trips.

[¶] Baker TA & Bell SP (1998) Polymerases and the replisome: machines within machines. *Cell* 92:295–305.

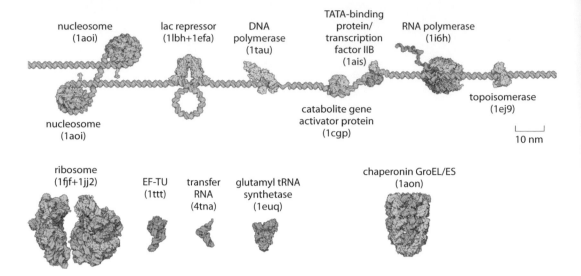

Figure 1-39 Structures of the machines of the central dogma. The machines responsible for replication, transcription, and translation are all shown drawn to a scale relative to the DNA substrate. The notations in parentheses are the PDB database names for the protein structures shown. (Courtesy of David Goodsell.)

Transcription is another key process in the central dogma and is intimately tied to the ability of cells to "make decisions" about which genes should be expressed and which should not at a given place within an organism at a given moment in time. The basal transcription apparatus is an assembly of a variety of factors surrounding the RNA polymerase holoenzyme. As shown in Figure 1-39, the core transcription machinery, like many oligomeric proteins, has a characteristic size of roughly 5 nm and a mass in *E. coli* of roughly 400 kDa (BNID 104927, 104925). Comparison of the machines of the central dogma between different organisms has been the most powerful example of what Linus Pauling referred to as using "molecules as documents of evolutionary history." Polymerases have served in that capacity and as such the prokaryotic and eukaryotic polymerases are contrasted in **Figure 1-40**.

The ribosome, a collection of three RNA chains (BNID 100112) and over 50 proteins (56 in bacteria, BNID 100111; and 78–79 in eukaryotes, BNID 102492), is arguably the most studied of all of the machines of the central dogma. Its importance can be seen from any of a number of different perspectives. For fast-growing microorganisms like *E. coli*, it can make up over one-third of the total protein inventory. From a biomedical perspective, it is the main point of attack of many of the most common and effective antibiotics, which utilize the intricate differences between the bacterial and eukaryotic ribosomes to specifically stop translation of the former and halt their growth. The ribosome has also served as the basis of a quiet revolution in biology that has entirely rewritten the tree of life. Because of its universality, the comparison of ribosomal sequences from

prokaryotic
RNA polymerase (4kmu)

prokaryotic
ribosome (2wdk+2wdl)

eukaryotic
RNA polymerase (1i6h)

eukaryotic ribosome
(3u5b+3u5c+3u5d+3u5e)

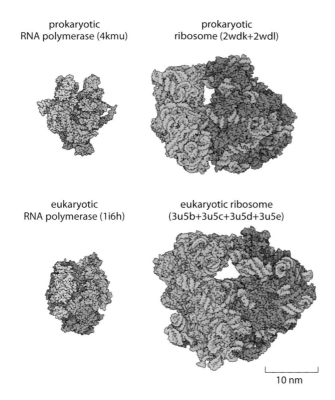

10 nm

Figure 1-40 Comparison of the structures of the RNA polymerase and ribosomes from prokaryotic and eukaryotic (in this case, yeast) organisms. The yeast ribosome at 3.3 MDa is intermediate between the bacterial ribosome at about 2.7 MDa and the mammalian ribosome at 4.3 MDa (BNID 111560, 100118). The notations in parentheses are the PDB database names for the protein structures shown. (Courtesy of David Goodsell.)

different organisms has served as the basis of a modern version of phylogeny, which tells a story of the history of life like no other.

Befitting its central role, the ribosome is also a relatively large molecular machine with a diameter of 20–30 nm (BNID 102320, 111542). In *E. coli*, it is composed of ≈7500 amino acids (BNID 101175, 110217, 110218) and ≈4600 nucleotides (BNID 101439) with a total mass of 2.7 MDa (BNID 106864, 100118, 111560) if the ribosome was made only of carbon atoms there would be about 200,000 of them. Given that the characteristic mass of an amino acid is ≈100 Da (BNID 104877) and that of an RNA nucleotide is ≈300 Da (BNID 104886), these numbers imply that the RNA makes up close to two-thirds of the mass of the ribosome and proteins make up only one-third. Indeed, crystal structures have made it clear that the function of the ribosome is performed mainly by the RNA fraction, exposing its origins as a ribozyme (that is, an enzyme based on catalytic RNA). The ribosome volume is ≈3000–4000 nm^3 (104919, 102473,

BNID 102474), implying that for rapidly dividing cells a large fraction of the cellular volume is taken up by ribosomes, a truth that is now seen routinely in cryo-electron microscopy images of bacteria.

Ending with a somewhat less dogmatic view of the central dogma, the diligent reader might have noticed the broken line in Crick's note from RNA back to DNA. This feat is achieved through reverse transcriptase, which in HIV is a heterodimer of 70- and 50-kDa subunits with a DNA polymerization rate of 10–100 nt/s (BNID 110136, 110137).

What is the thickness of the cell membrane?

One of the key defining characteristics of living organisms is that cells are separated from their external environment by a thin, but highly complex and heterogeneous, cell membrane. These membranes can come in all sorts of shapes and molecular compositions, though generally they are made up of a host of different lipid molecules and they are riddled with membrane proteins. Indeed, if we take the mass of all of the proteins that are present in such a membrane and compare it to the mass of all of the lipids in the same membrane, this so-called protein-to-lipid mass ratio is often greater than one (BNID 105818). This assertion applies not only to the plasma membranes that separate the cellular contents from the external world, but also to the many organellar membranes that are one of the defining characteristics of eukaryotic cells.

The thickness of this crucial but very thin layer in comparison to the diameter of the cell is similar to the thickness of an airplane fuselage in comparison with the plane's body diameter. The key point of this analogy is simply to convey a geometric impression of the thickness of the membrane relative to the dimensions of the cell using familiar everyday objects. In the case of an airplane, the thickness of the exterior shell is roughly 1 cm in comparison with the overall diameter of roughly 5 m, resulting in an aspect ratio of 1:500. How can we estimate the aspect ratio for the biological case? With a few exceptions, such as in Archaea, the lipid part of the cell membrane is a bilayer of lipids with the tails on opposite leaflets facing each other (see **Figure 1-41**). These membranes spontaneously form as a relatively impermeable and self-mending barrier at the cell's (or organelle's) periphery, as discussed in the section on the cell's membrane permeability. The length scale of such structures is given by the lipid molecules themselves, as shown in **Figure 1-42**. For example, the prototypical phospholipid, dipalmitoyl phosphatidylcholine, has a head-to-tail length of 2 nm (BNID 107241,

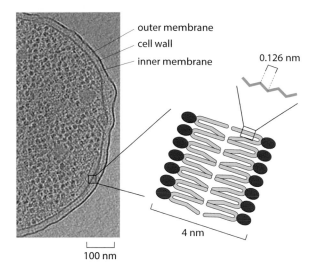

outer membrane
cell wall
inner membrane

0.126 nm

4 nm

100 nm

Figure 1-41 An electron micrograph of an *E. coli* cell that highlights the width of the cell inner and outer membranes and the cell wall. Zoom in: a schematic of the lipid bilayer. The red circle denotes the hydrophilic head, which consists of a polar phosphoglycerol group, and the pink lines represent the hydrocarbon chains that form a tight hydrophobic barrier, thereby excluding water as well as polar or charged compounds. Two tails are drawn per head, but there could also be three or four. (Adapted from Briegel A, Ortega DR, Tocheva E et al. [2009] *Proc Natl Acad Sci USA* 106:17181–17186.)

107242). This implies an overall bilayer membrane thickness of 4 nm, 3 nm of which are strongly hydrophobic and the rest of which are composed of the polar heads (BNID 107247). For a 2-micron cell diameter (a relatively large bacterium or a very small eukaryotic cell), the 4-nm thickness implies an aspect ratio of 1:500, similar to the case of an airplane. Larger numbers are sometimes quoted, probably resulting from the effective increase due to proteins and lipopolysaccharides sticking out of the membrane.

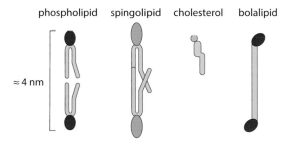

phospholipid   spingolipid   cholesterol   bolalipid

≈ 4 nm

Figure 1-42 Characteristic relative sizes and shapes of the lipid molecules making up biological membranes.

The story of how lipid size was initially estimated has a long and interesting history, as vividly described in Charles Tanford's little book, *Ben Franklin Stilled the Waves*. Specifically, the story begins with experiments of Benjamin Franklin, who explored the capacity of oils to still the waves. Franklin performed his experiments in a pond near London and said of them, "The oil, though not more than a teaspoonful, produced an instant calm over a space several yards square, which spread amazingly and extended itself gradually until it reached the leeside, making all that quarter of the pond, perhaps half an acre, as smooth as a looking glass." The calming of the waves is attributed to a monolayer of oil forming on the surface of the water and causing damping through energy dissipation. A similar approach to calming waves was taken by sailors at the time of the Romans, who dumped oil (such as whalers using blubber) in rough seas. Energy is dissipated as the oil film flows and gets compressed and dilated during the movement of the waves. Using Franklin's own dimensions for the size of his oil slick (that is, 0.5 acre $\approx$ 2000 m^2) and the knowledge of the initial teaspoon volume (that is, 1 teaspoon $\approx$ 5 cm^3), we see that his oil formed a single layer with a thickness of several nanometers. To be precise, using the numbers above, we find a thickness of roughly 2.5 nm. More precise measurements were undertaken by Agnes Pockels, who invented an experimental technique used to construct lipid monolayers that made it possible to settle the question of molecular dimensions precisely. Lord Rayleigh performed small-scale versions of the Franklin experiment in an apparatus similar to what is now known as the "Langmuir trough," which allowed him to spread a monolayer of molecules on a liquid surface and detect their presence with a small wire that squeezed this monolayer.

Each layer of the cell membrane is made up of molecules similar in character to those investigated by Franklin, Rayleigh, and others. In particular, the cell membrane is composed of phospholipids, which contain a head group and a fatty acid tail that is roughly 10–20 carbons long. An average carbon–carbon bond length projected on the chain, and thus accounting for the tail's zigzag shape arising from carbon's tetrahedral orbital shape, is $l_{cc} = 0.126$ nm (BNID 109594). The overall tail length is $n_c \times l_{cc}$, where n_c is the number of carbon atoms along the chain length. Overall, the two tails end-to-end plus the phosphoglycerol head groups have a length of $\approx$4 nm (BNID 105821, 100015, 105297, 105298).

Unsurprisingly, membrane proteins are roughly as thick as the membranes they occupy. Many membrane proteins like ion channels and pumps are characterized by transmembrane helices that are $\approx$4 nm long and have physicochemical properties like that of the lipids they are embedded in. Often these proteins also have regions that extend into the space on either side of the membrane. This added layer of protein and carbohydrate fuzz adds to the "thickness" of the membrane. This is evident in **Figure 1-43**, where some of the membrane-associated proteins are shown to scale in cross section. Due

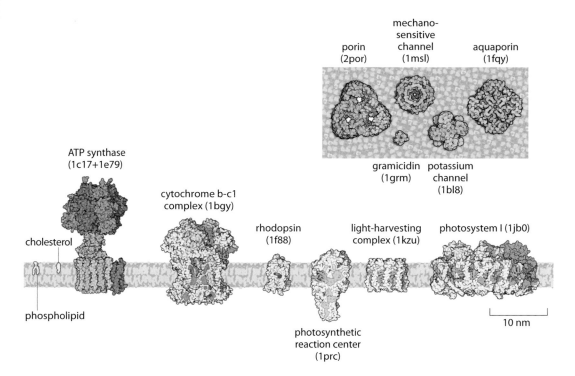

porin
(2por)

mechano-
sensitive
channel
(1msl)

aquaporin
(1fqy)

ATP synthase
(1c17+1e79)

gramicidin
(1grm)

potassium
channel
(1bl8)

cytochrome b-c1
complex (1bgy)

rhodopsin
(1f88)

light-harvesting
complex (1kzu)

photosystem I (1jb0)

cholesterol

phospholipid

10 nm

photosynthetic
reaction center
(1prc)

Figure 1-43 The membrane with some notable constituents. The extent of protrusion of proteins from the cell membrane is evident. The fraction of membrane surface occupied by proteins in this cross-section depiction is similar to that actually found in cells. (Courtesy of David Goodsell.)

to these extra constituents, which also include lipopolysaccharides, the overall membrane width is variably reported to be anywhere between 4 and 10 nm. The value of 4 nm is most representative of the membrane shaved off from its outer and inner protrusions. This value is quite constant across different organellar membranes, as shown recently for rat hepatocytes via X-ray scattering, where the ER, Golgi, basolateral, and apical plasma membranes were 3.75 ± 0.04 nm, 3.95 ± 0.04 nm, 3.56 ± 0.06 nm, and 4.25 ± 0.03 nm, respectively (BNID 105819, 105820, 105822, 105821). We conclude by noting that the cell membrane area is about one-half protein (BNID 106255), and the biology and physics of the dynamics taking place there is still intensively studied and possibly holds the key to the action of many future drugs.

How big are the cell's filaments?

Cell biology is a subject of great visual beauty. Indeed, magazines such as *National Geographic* and microscope manufacturers exploit this beauty with contests to see who can come up with the most stunning microscopy images

(A)   (B)

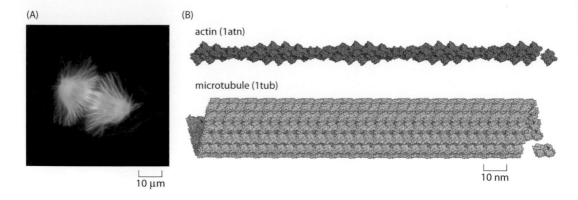

Figure 1-44 The cytoskeleton. (A) Fluorescence microscopy image of an epithelial cell (PtK1) during furrow onset. The cell was fixed roughly five minutes after onset of anaphase, with filamentous actin labeled in red, microtubules labeled in green, and DNA labeled in blue. (B) Structural models of actin and microtubules. (A, Courtesy of Julie C. Canman and E.D. (Ted) Salmon, University of North Carolina. B, courtesy of David Goodsell.)

of cellular structures. An example of such an image is shown in **Figure 1-44A**. One of the mainstays of these images is colorful depictions of the many cytoskeletal filaments (actin and microtubules, shown in **Figure 1-44B**) that crisscross these cells. These filaments serve in roles ranging from helping cells move around to providing a molecular superhighway for cell traffic to pulling chromosomes apart during the process of cell division. How should we characterize this molecular network structurally? How long is the typical filament and how many of them span the "typical" eukaryotic cell?

To consider the "size" of the cytoskeleton, we take a hierarchical view starting at the level of the individual monomers that make up the filaments of the cytoskeleon, and then pass to the properties of individual filaments, followed finally by the structural properties and extent of the cytoskeletal networks found in cells. Cytoskeletal filaments are built up of individual monomeric units, which are the building blocks of filaments, as seen in Figure 1-44B. Their properties are summarized in **Table 1-7**. For example, each of the monomeric units making up an actin filament is roughly 5 nm in size with a molecular mass of about 40 kDa. Tubulin subunits that make up microtubules have comparable dimensions. To be more precise, tubulin dimers made up of alpha and beta tubulin subunits (each of mass roughly 50 kDa) form protofilaments with a periodicity of 8 nm. Like with many other proteins, the structural features of these proteins have been determined using X-ray crystallography, and their sizes are quite typical for globular proteins.

Figure 1-45 provides an opportunity to delve more deeply into these structures and to develop intuition about the length scales of this protein by showing the mRNA, protein monomer, and a fragment of an actin filament showing 1% of the persistence length at correct proportion. If we

	actin	microtubules
functions	cell motility, cytokinesis, muscle contraction, hearing	cell division, intracellular transport
subunit	actin monomer	α-tubulin + β-tubulin
subunit weight	≈40 kDa	≈50 kDa
subunit size	5 nm	4 nm (dimer)
protofilaments number	2	13 (variable)
cross section area	20 nm^2	200 nm^2
filament diameter	6 nm	25 nm
helical repeat period	36 nm	8 nm
persistence length	10 µm	1–10 mm
filament length	from 35 nm in erythrocyte cortex to 10–100 µm in ear hair cells	from <1 µm in S. pombe through 100 µm in rat neurons to >1 mm in insect sperm

Table 1-7 Properties of the main cytoskeleton components: actin and microtubules (BNID 107897).

take individual monomeric units of actin and mix them in solution, they will spontaneously polymerize over time into filamentous structures like that shown in Figure 1-44B. For actin, these filaments are microns in length with a corresponding diameter of only 6 nm, meaning that they have an extremely large aspect ratio. Similarly, tubulin monomers come together to

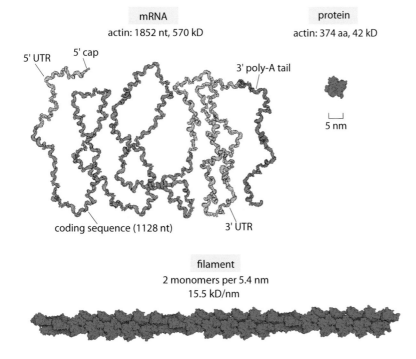

mRNA
actin: 1852 nt, 570 kD

protein
actin: 374 aa, 42 kD

5' UTR   5' cap

3' poly-A tail

5 nm

coding sequence (1128 nt)   3' UTR

filament
2 monomers per 5.4 nm
15.5 kD/nm

Figure 1-45 Sizes of actin mRNA, protein, and filament. The mRNA molecule is shown next to the corresponding protein monomer that it codes for (based on human actin A). The monomers assemble into actin filaments such as the one indicated schematically at the bottom. For reference, this filament fragment is only 1% of the measured persistence length of these structures. (Courtesy of David Goodsell.)

form hollow cylindrical filaments usually made up of 13 separate protofilaments. In this case, the hollow cylindrical structure has a diameter of roughly 25 nm. To get a sense of these aspect ratios when applied to everyday objects, we consider a microtubule with a typical length of 10 μm. To compare this to a human hair, note that a human hair is roughly 50 μm in diameter, meaning that for such a hair to have a comparable aspect ratio, it would be 400 times longer, with a length of 2 cm. Because of their slender geometries, these filaments have fascinating mechanical properties, which permit them to apply forces in key cellular processes, as we will see in the vignette entitled, "How much force is applied by cytoskeletal filaments?" (pg. 172). A useful parameter that characterizes the elastic behavior of these filaments is the persistence length, which is a measure of the length scale over which a filament is "stiff" or "straight"—that is, how far you have to proceed along a thermally fluctuating filament before the two ends have uncorrelated orientations.

A number of clever methods have been worked out for measuring the mechanical properties of individual filaments, ranging from measuring the spontaneous thermal fluctuations of the filaments in solution to working out the force at which they buckle when subject to an applied load at their end. Classic early experiments exploited the ability to fluorescently label filaments, which permitted the visual inspection of their dynamics under a microscope. The spectrum of vibrations is related, in turn, to the persistence length, and such measurements yield a persistence length of roughly 10 μm for actin (BNID 106830) and a whopping 1–10 mm for microtubules (BNID 105534).

Cytoskeletal filaments are generally not found in isolation. In most biological settings, it is the behavior of collections of filaments that is of interest. One of the most beautiful and mysterious examples is the orchestrated segregation of chromosomes during the process of cell division, the physical basis for which is mediated by a collection of microtubules known as the mitotic spindle (see **Figure 1-46A**). This figure shows a key stage in the cell cycle known as metaphase. The chromosomes of the daughter cells that are about to be formed are aligned in a structure that is surrounded by oriented microtubules, which pull those chromosomes apart during the subsequent stage of anaphase. What is the distribution of microtubule lengths within such spindles? The answer to this question is a mechanistic prerequisite to understanding both the nature of the spatial organization of the mitotic spindle, as well as how force is generated during the process of chromosome segregation.

Several approaches have shed light on such questions. Imaging with electron microscopy, it is possible to resolve individual microtubules within the spindle and to measure their lengths. An example of the kind of images used to perform such measurements, as well as the resulting distributions, is shown in **Figure 1-46B** for the relatively small spindles of yeast cells. These

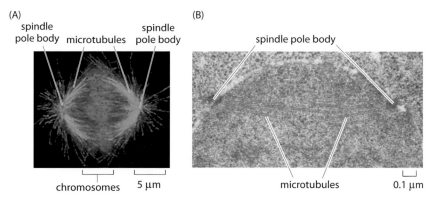

Figure 1-46 Microtubules and the mitotic spindle. (A) Fluorescence image showing the microtubule distribution in a dividing human HeLa cell. DNA is labeled in blue, microtubules are labeled in green, and the kinetochores are labeled in red. (B) Electron microscopy image of a mitotic spindle. (A, courtesy of Iain Cheeseman. B, adapted from Winey M, Mamay CL, O'Toole ET et al. [1995] *J Cell Biol* 129: 1601–1615.)

spindles have a size of roughly 2 μm and involve on the order of 50 different filaments connected in parallel, each one composed of several microtubules connected together by molecular motors (BNID 111478, 111479). However, such studies become much more difficult for characterizing the entire distribution of microtubules in the spindle of animal cells, for example. In this case, approaches using fluorescence microscopy and exploiting microtubule depolymerization dynamics make it possible to characterize the length distribution for much larger spindles, such as those in the egg of the frog *Xenopus laevis*, as shown in **Figure 1-47**.

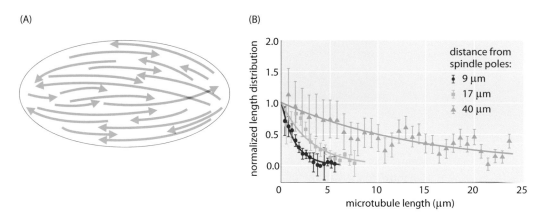

Figure 1-47 Microtubule length distribution in *Xenopus* egg extract. (A) Schematic of the microtubule distribution in the mitotic spindle of a *Xenopus* egg extract. The schematic illustrates both the differing polarities of the microtubules, as indicated by the arrows, as well as the variation in microtubule lengths as a function of their position relative to the spindle pole. (B) Distribution of microtubule lengths as a function of distance from the spindle pole. (Adapted from Brugués J, Nuzzo V, Mazur E & Needleman DJ [2012] *Cell* 149:554–564.)

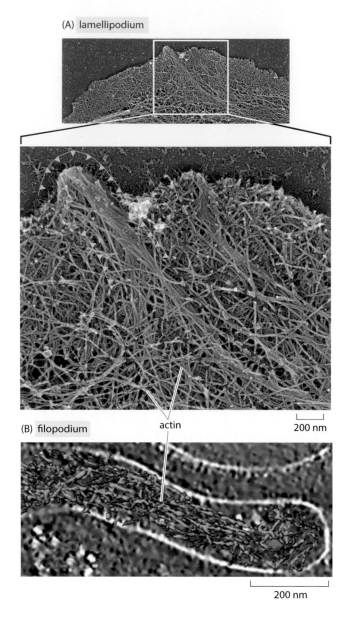

(A) lamellipodium

(B) filopodium

actin

200 nm

200 nm

Figure 1-48 Actin in the leading edge of cells. (A) Leading edge of a mouse skin cell showing the distribution of actin. (B) Cryo-electron microscopy image of the actin distribution in a filopodium of *Dictyostelium*. (A, adapted from Svitkina TM, Bulanova EA, Chaga OY et al. [2003] *J Cell Biol* 160: 409–421. B, adapted from Medalia O, Beck M, Ecke M et al. [2007] *Curr Biol* 17: 79–84.)

Just as microtubules serve in many different roles, actin too is one of the central players in a diverse array of processes in biology. One example that will serve as a useful entry into the collective properties of filaments is that of cell motility. Motion of crawling cells, such as keratocytes, are driven by two distinct actin-related processes, one through the relatively thin protrusions known as filopodia and the other through the much broader lamellipodium protrusions. The speed of keratocyte migration is about one-quarter of a micron per second. Given that the addition of an actin monomer increases the length of the filament by about 3 nm, we infer a net incorporation rate of roughly 100 monomers per second in each actin filament. The act of crawling can be broken down to its microscopic components by appealing to electron microscopy images like those shown in **Figure 1-48**. Figure 1-48A shows the leading edge of a fibroblast that has had its membrane peeled away and the actin filaments decorated with metals, thus rendering the filaments of the lamellipodium visible. In the second example, these same kinds of filaments are viewed without metal staining by using cryo-electron microscopy to reveal the filopodium in a *Dictyostelium* cell. As a look at older cell biology textbooks reveals, it was once thought that cytoskeletal filaments were the exclusive domain of eukaryotes. However, a series of compelling discoveries over the last several decades rewrote the textbooks by showing that bacteria have cytoskeletal analogs of actin, microtubules, and intermediate filaments. Like their eukaryotic counterparts, these filaments are engaged in a sweeping array of cellular activities, including the segregation of plasmids, the determination and maintenance of cellular shape, and the cell division process. The drama plays out on a much smaller stage, however, so the cytoskeletal filaments of bacteria have sizes that are constrained by the sizes of the cells themselves.

NUMBER

Chapter 2: Concentrations and Absolute Numbers

In this chapter, all of our vignettes center in one way or another on a simple question: How many? We challenge ourselves to think about critical issues, such as how many mRNAs or ribosomes are in a cell, measured in units of absolute number of copies per cell, rather than in relative amounts. These questions are tied in turn to other concerns, such as the concentrations of ions, metabolites, signaling proteins, and other key components (for example, the molecules of the cytoskeleton), all of which join the molecules of the central dogma as key players in the overall molecular inventory of the cell. In our view, knowing the concentrations and absolute numbers of the biological movers and shakers inside the cell, as depicted in **Figure 2-1**, is a prerequisite to being able to progress from qualitative depictions of mechanisms to constructing models with quantitative predictive power. Before we detail the results of the most advanced molecular techniques for surveying the molecular census of cells, we start with the major energy and matter transformations that sustain life. Today, many high school biology students can recite the stoichiometric equation for the carbon fixation that takes place during photosynthesis. But, in fact, the ability to account for the molecular census in this problem required the invention by the early "pneumochemists"—literally meaning chemists of the air—of new ways of accurately measuring the quantities of different gases taken up and liberated during photosynthesis. **Figure 2-2** shows how these early experimenters positioned leaves under water and painstakingly measured the volume of so-called "pure air" (oxygen) released. Such careful and accurate quantitation was at the heart of both revealing and proving the photochemical basis for this secret of life that had earlier garnered metaphysical vitalistic explanations.

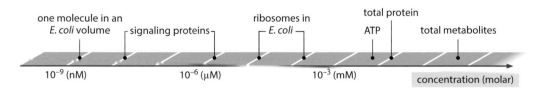

Figure 2-1 Range of characteristic concentrations of main biological entities from one molecule in a cell to the entire metabolite pool. Wherever an organism is not specified, the concentrations are characteristic for most cells in general.

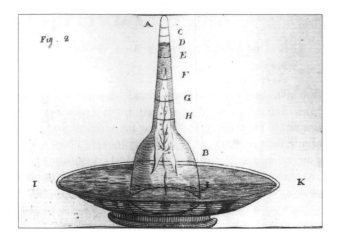

Figure 2-2 Experimental apparatus developed in order to learn about the gases taken up and liberated by plants. This setup made it possible to quantify the volume of oxygen gas produced by a leaf submerged in water. Many substances were added to the water to investigate their effect on oxygen production. This research effort in the second half of the 18th century culminated in the discovery that carbon dioxide, available at very low concentrations, is the substrate that plants feed on in the process of photosynthesis. (Adapted from Farmer EE [2010] *Arch Sci* 63:185–192.)

Chemistry is all about interactions between atoms and molecules of different types. Paul Ehrlich noted, "*Corpora non agunt nisi ligata*," meaning "a substance is not effective unless it is linked to another." One of the tenets of this chapter is the assertion that the propensity to form such linkages depends critically on the concentrations (and affinities) of the binding partners. The familiar case of hemoglobin illustrates the sensitive dependence of the binding of the essential oxygen to this protein. A similar story plays out in the context of DNA and the transcription factors that are in charge of the regulation of expression. Probably the most well-known example of such a regulator is the repressor protein LacI of *lac* operon fame. Less well known is the fact that in an *E. coli* cell this transcription factor has a copy number on the order of ≈ 10 tetramers per cell. Such a low copy number is interesting because it immediately raises questions about small-number effects that lead to cell-to-cell variability.

Not only do we have to think about the contents within the cell, but similar questions abound in the context of the cell surface, whose "real estate" is in limited supply. The cell surface is riddled with a dense population of different membrane proteins, many of which serve as conduits for communicating information about the external environment to the cell. Here, too, the binding between these surface receptors and their ligand partners is a sensitive function of the concentrations of both species. Real estate on the cell surface can limit the absolute number of transporter proteins, and we speculate on how it can play a part in putting a speed limit on maximal growth rates. In this chapter, we aim to convince the reader that the same basic approach, which pays careful attention to the quantitative abundance of different molecular species, can be repeated again and again for nearly all of the different provinces of molecular and cell biology, with great rewards for our intuition of what it means to function as a cell.

Over the course of this chapter, we go beyond taking stock only of molecular quantities by asking other census questions, such as, "What is the concentration of bacterial cells in a saturated culture?" One of the reasons this number is interesting and useful is that it tells us something about that most elemental of microbiological processes—namely, the growth of cells in a culture tube. If we are to place a single bacterium in a 5-mL culture tube containing growth media with some carbon source, a few hours later, that one cell will have become more than 10^9 such cells. How have the molecular constituents present in the culture tube been turned into complex living matter and what are the relative concentrations of the proteins, nucleic acids, sugars, and lipids that make up these cells? This simple growth experiment serves as one gateway from which to examine the chemical and macromolecular census of various types of cells. Though each cell is different, some handy and general rules of thumb can still be derived. For example, at $\approx 30\%$ dry mass, of which $\approx 50\%$ is carbon, a 1-μm^3 cell volume will contain 100–200×10^{-15} g of carbon. Because the molecular mass of carbon is 12 Da, this is equivalent to just over 10^{-14} mol, or by remembering Avogadro's number, 10^{10} carbon atoms, a fact already introduced in the opening chapter discussion of rules of thumb.

Our choice of topics is idiosyncratic rather than comprehensive. One of the motivations for our choices is a desire to see if we can figure out which molecular players are in some sense dominant. That is, which proteins are the most abundant in some cell type and why? The protein Rubisco, the key for turning inorganic carbon into organic matter in the form of sugars, has sometimes been called the "most abundant protein on Earth," and as such and due to its impact on agricultural productivity, has garnered much interest. We show, however, that this claim about Rubisco is likely exaggerated, with other proteins such as collagen, a protein present in the extracellular matrix and that makes up about one-third of the protein content in humans and livestock, coming in at even higher numbers. Similarly, as evidenced by their role in mapping out the diversity of life on Earth, ribosomes are a nearly universal feature of the living world and in rapidly dividing cells are major cellular components. In different vignettes, we interest ourselves in the molecular census of these and other dominant fractions of the biomass of cells in an attempt to sharpen our intuition about what cells are all about.

MAKING A CELL

What is the elemental composition of a cell?

One of the most interesting chemical asymmetries associated with life on Earth is the mismatch between the composition of cells and of inanimate matter. As a result of the rich and diverse metabolic processes that make cells work, living chemistry is largely built around carbon, oxygen, nitrogen, and hydrogen, with these elemental components serving as the key building blocks making up the cell's dry weight.

The dry weight of *E. coli* contains, for every nitrogen atom, about two oxygen atoms, seven hydrogen atoms, and four carbon atoms. Hence, the empirical composition can be approximated as $C_4:H_7:O_2:N_1$. The empirical composition on a per-carbon basis yields the equivalent empirical composition of $C_{1.00}:H_{1.77}:O_{0.49}:N_{0.24}$ (BNID 101800). In absolute terms, there are about 10^{10} atoms of carbon in a medium-sized *E. coli* cell (BNID 103010), on the order of the number of humans on Earth and, interestingly, less than the number of transistors in a state-of-the-art computer chip. For budding yeast, the proportional composition is similar—namely, $C_{1.00}:H_{1.61}:O_{0.56}:N_{0.16}$ (BNID 101801). How many atoms are in the human body? It depends (for example, on the weight). But we much prefer to estimate the order of magnitude, as shown in **Estimate 2-1**. Think of an adult as having a mass of about 100 kg and an atom in the human body having an average mass of 10 Da, thus arriving at about 1000 mol or somewhere between 10^{27}–10^{28} atoms. Those interested in a more detailed breakdown of the so-called "human empirical formula" may enjoy seeing our detailed

How many atoms are there in the human body?

main elements:

hydrogen	oxygen	carbon
(1 Da)	(16 Da)	(12 Da)

average MW ≈ 10 Da = 10 g/mol

$mass_{human}$ ≈ 100 kg

$\Longrightarrow$ N_{human} ≈ $\dfrac{10^5 \, g}{10 \, g/mol}$ = 10^4 mol ≈ 6×10^{27} atoms

Avogadro's number

1 mol ≈ 6×10^{23} atoms

10^{27}–10^{28} atoms in the human body

Estimate 2-1

(A) elemental composition of Earth vs. human

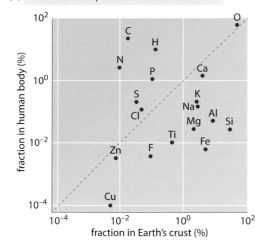

(B) elemental composition of ocean vs. human

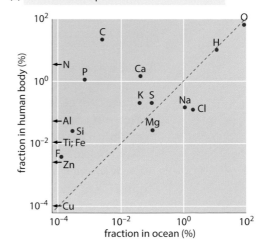

Figure 2-3 Comparing the elemental composition by weight in percent for the most abundant elements in the human body (A) to the Earth's crust and (B) to the oceans. Only elements that are at a concentration of at least 1 part per million in the human body are depicted. Some elements whose concentration is lower than the minimal value on the *x*-axis range are denoted with an arrow. (Data from BNIDs 110362, 107256, 107257, 107258, and 103490.)

stoichiometry, which (per atom of vanadium) can be written as (BNID 111243)

$$H_{2.5E9}O_{9.7E8}C_{4.9E8}N_{4.7E7}P_{9.0E6}Ca_{8.9E6}K_{2.0E6}Na_{1.9E6}S_{1.6E6}Cl_{1.3E6}$$
$$Mg_{3.0E5}Fe_{5.5E4}F_{5.4E4}Zn_{1.2E4}Si_{9.1E3}Cu_{1.2E3}B_{7.1E2}Cr_{98}Mn_{93}Ni_{87}Se_{65}Sn_{64}I_{60}Mo_{19}Co_{17}V_1.$$

As noted above, it is interesting to compare the composition of cells to that of the Earth's crust or the oceans, as shown in **Figure 2-3** (BNID 110362). Strikingly, carbon and hydrogen, majority players in living matter, are relatively rare in the Earth's crust. Carbon comes in as only the 17th most abundant element, whereas hydrogen comes in 10th place, still way behind the major constituents oxygen (60.5%), silicon (20.5%), and aluminum (6.2%). Similarly, in the atmosphere, the main carbon-containing compound, CO_2, makes up merely ≈ 400 parts per million (at the time of this writing, although this is one of the most dynamical atmospheric numbers as a result of human activity). Extracting this dilute resource is the main reason for the need to water plants. Plants lose water when opening their stomata, the small pores on leaves that are the channels for importing carbon dioxide molecules. This mundane process accounts for a staggering two-thirds of humanity's water consumption (BNID 105887). Hydrogen, which was prevalent in the early Earth's atmosphere, was lost to space during Earth's history. This process of loss is a result of hydrogen's low mass, because the thermal velocities it attains at the high temperatures prevailing in the atmosphere's upper layers provide enough kinetic energy to overcome Earth's gravitational pull. This trickling continues today at a rate of $\approx$ 3 kg/s from Earth's atmosphere (BNID 111477).

Can we say something more about the elemental compositions of living matter by thinking about the makeup of the key macromolecular constituents of cells? In particular, how might we infer the elemental ratios from our acquaintance with the cell's components without consulting the empirical measurement? A bacterial cell is made up of about 55% protein, 20% nucleic acids, 10% lipids, and another 15% of various other components (by weight, BNID 101436). Exploiting the fact that the mass ratio of proteins to nucleic acids is about 3:1, we explore in **Estimate 2-2** how far a few simple facts about these two dominant components can take us in estimating the elemental composition of a cell.

A nucleotide is composed of a phosphate (PO_4) and ribose ($C_5H_8O_2$) backbone and a base ($\sim N_5C_5O_1H_6$, using guanine as our representative example). Thus, the total chemical composition is $P_1N_5O_7C_{10}H_{14}$, with a total mass of about 350 Da (BNID 104886). An amino acid consists of a backbone with a peptide bond–RC(O)NH–, where the first group (R) is a carbon harboring a residue that on average is crudely about three carbons, one oxygen, and six hydrogens, leading to a total elemental composition of $N_1C_5O_2H_8$ and a mass of about 110 Da (BNID 104877). If we focus our attention only on the protein and nucleic acid content of cells, we are now prepared to estimate

Main constituents of a cell

amino acids

$\approx C_3O_1H_6$

$\approx N_1C_5O_2H_8$
$\approx 110\,Da$

nucleotides

e.g., guanine $N_5C_5O_1H_6$

base

HO

$\approx PO_3H$

OH (RNA)
or
H (DNA)

$\approx C_5H_8O_2$

$\approx P_1N_5O_7C_{10}H_{14}$
$\approx 350\,Da$

total protein mass in cell $\approx$ 3 times larger than RNA + DNA

$\implies$ composition $\approx 10 \times (N_1C_5O_2H_8) + 1 \times (P_1N_5O_7C_{10}H_{14}) \approx P_1N_{15}O_{27}C_{60}H_{94})$

$3 \times 350\,Da\,/\,110\,Da \approx 10$

close to empirical formula $1N : 20 : 4C : 7H$
and plankton Redfield ratio $1P : 16N : 106C$

Estimate 2-2

the overall composition of a cell. To reason this out, recall that the mass of protein in a typical bacterium is roughly threefold larger than the mass of nucleic acids. Further, since nucleic acids have roughly three times the mass of amino acids, this implies that for every nucleotide there are roughly 10 amino acids. We need to evaluate the chemical composition of a mix of 10 amino acids and 1 nucleic acid, resulting in the stoichiometric relation $10 \times (N_1C_5O_2H_8) + 1 \times (P_1N_5O_7C_{10}H_{14}) = (C_{60}O_{27}H_{94}N_{15}P_1)$. Normalizing by the number of nitrogen atoms, we can rewrite this composition as $C_4H_{6.3}O_{1.8}N_1$, which is pretty close to the empirical value of $C_4H_7O_2N_1$ (or $C_{1.00}H_{1.77}O_{0.49}N_{0.24}$). This estimate can be refined further if we include the next largest contributor to the cell mass—namely, the lipids that account for ~10% of that mass. These molecules are mostly composed of fatty acids that have about twice as many hydrogens as carbons and very little oxygen. Including lipids in our estimate will thus increase the proportion of H and decrease that of O, which will bring our crude estimate closer to the measured elemental formula of cell biomass. Estimating something that is already known through an empirical formula serves as a critical sanity check of our understanding of the main biological components that determine the cell's composition.

Why were these particular elements chosen to fulfill biological roles? Why is carbon the basis of life as we know it? These questions are discussed in detailed books on the subject.[*] We end by noting that there could still be surprises lurking in the field of the elemental stoichiometry of life. For example, a recent high-profile publication claimed to reveal the existence of bacteria that replace the use of phosphate by the element arsenic that is one line lower in the periodic table and highly abundant in Mono Lake, California. However, more rigorous studies showed these organisms to be highly resistant to arsenic poisoning but still in need of phosphate. The vigorous discussion refuting the original claims led to renewed interest in how elemental properties constrain evolution.

What is the mass density of cells?

The density of biological material is responsible for the settling of cells to the bottom of our laboratory tubes and multi-well plates and serves as the basis of the routine centrifuging that is part of the daily life of so many biologists. These very same differences in density between cells and their watery exterior are also the basis of the contrast we observe in phase microscopy

[*]Williams RJP & Fraústo da silva JJR (2001) The Biological Chemistry of the Elements. Oxford University Press; Sterner RW & Elser JJ (2002) Ecological Stoichiometry. Princeton University Press.

object	density	BNID
DNA (unhydrated)	2.0	107858, 111208
RNA	2.0	111208
DNA (in solution with 7 M CsCl)	1.7–1.8	107857
chromatin	1.4	106492
proteins	1.2–1.4	104272, 111208
chloroplasts	1.1–1.2	106492, 109442
mammalian viruses	1.1–1.2	106492, 106494, 109442
mitochondria	1.05–1.2	106492, 106494, 109442
hepatocyte	1.05–1.15	106494, 109441
erythrocyte	1.1	101502, 109441
E. coli	1.08–1.10	103875, 102239, 110096
budding yeast	1.08–1.10	106439
skeletal muscle	1.06	111214
synaptic vesicle	1.05	101502
HeLa	1.04–1.08	109441
fibroblast	1.03–1.05	101502, 106494, 109441
membrane (including proteins)	1.02–1.18	106492, 106494, 109442
phospholipid (+ cholesterol)	1.01	108142
adipocyte tissue (fat cells)	0.92	111213

Table 2-1 Densities of biological objects relative to water. The densities shown here can be approximately thought of as appearing in units of g/mL or 1000 kg/m³. Values are sorted in descending order. Unless otherwise stated, values were measured in sucrose or ficoll solution.

images. These differences are also important outside the lab setting. For example, plankton have to contend with this density difference to remain at a depth in the ocean where sunlight is plentiful, rather than sinking to the blackened depths. Given that most biologists and biochemists make use of separation based on density on a daily basis, it seems surprising how rarely densities such as those collected in **Table 2-1** are actually discussed.

What is the underlying basis for the varying densities of different organelles and cell types? To a large extent, these differences can be attributed to the ratio between water content and dry mass. Proteins have a density of ≈ 1.3–1.4 (BNID 104272, 101502) relative to water (or almost equivalently in units of g/mL or 1000 kg/m³). Given the benchmark value of 1 for the density of water, a spectrum of intermediate values for the cell density between 1 and 1.3 are obtained based on the relative abundance of proteins and water. Lipids are at the low end of the spectrum next to water at a density of about 1 (BNID 108142). At the other extreme, starch granules with a density of ≈ 1.5 (BNID 103206) and nucleotides at ≈ 1.7 can shift the overall mass balance in the opposite direction.

Density is often known based on the location at which a given biological component settles when spun in a centrifuge containing a gradient of concentrations, often produced by sucrose, or in the case of DNA, cesium chloride. The density reflects the mass divided by the volume, but for charged compounds in solution, the density is also affected by shells of so-called bound water. The density in this case becomes an effective density, reduced by the bound water, and thus somewhat dependent on the salt concentration (BNID 107858).

The rate of sedimentation in a centrifuge is measured in units of svedberg (S), which is the origin of the names 70S, 23S, and so on, for the ribosome and its rRNA subunits. A 23S rRNA will sediment at a velocity of 23×10^{-12} m/s under normal gravity. In an ultracentrifuge producing an acceleration of one million g, the velocity will proportionally scale to 23×10^{-6} m/s or about 1 mm/min. The rate of sedimentation depends on the density, size, and shape of the molecule. For similar shapes and densities, the sedimentation rate scales as the square root of the molecular mass. For such cases, the molecular mass goes as the square of the sedimentation rate, such that the 23S and 16S subunits of the ribosome have a molecular mass with a ratio of roughly $(23/16)^2$, or about 2, which is closely in line with measurements of 0.9 and 0.5 MDa, respectively (BNID 110972, 110967). In the clinic, the sedimentation rate of erythrocytes (red blood cells) is routinely used to measure inflammation. Rates much higher than 10 mm/h usually indicate the presence of the pro-sedimentation factor fibrinogen, which is a general indicator of an inflammatory condition.

Water is the most abundant molecular fraction of cells, but just how abundant is it exactly? If we examine tissues from multicellular organisms, finding the water content is a simple task of measuring the mass of the tissue before and after drying. But how can we perform such measurements for cells? When we weigh a mass of cells before and after drying, how do we measure only the cells without any water around them? Even after centrifugation, there is water left in the cell pellet, resulting in ambiguity about the dry mass itself.

Once again, radioisotopic labeling comes to the rescue.[†] First, labeled water (using tritium, ^{3}H) is measured in a cell pellet. This indicates the sum of water inside and outside the cells. Then, another soluble compound that is labeled but that cannot enter the cell, such as ^{14}C-inulin or ^{3}H-PEG, measures the volume of water outside the cells in a centrifuged pellet (for example, in *E. coli*, it is about 25–35% of the pellet volume). The difference indicates the water content inside cells. Such methods lead to typical values ranging from ≈60–65% by mass for budding yeast and red blood

---

[†] Cayley S, Lewis BA, Guttman HJ & Record MT Jr (1991) Characterization of the cytoplasm of *Escherichia coli* K-12 as a function of external osmolarity. Implications for protein-DNA interactions in vivo. *J Mol Biol* 222:281–300.

cells, to ≈70% for *E. coli* and the amoeba *D. discoideum*, and up to ≈80% for rat muscle and pig heart tissues (BNID 105938, 103689). Since the dry matter contribution is dominated by constituents of density ≈1.3 (that is, proteins), this leads to the characteristic overall density of ≈1.1 (BNID 103875, 106439, 101502). From these characteristic fractions, the dry mass per volume can be inferred to be about 300–500 mg/mL (BNID 108131, 108135, 108136), but during slow growth, values can be higher. Low densities are common in dry seeds and underwater plants that have buoyant parts with densities of less than the surrounding water, thus allowing them to float. Densities lower than that of water can be achieved either by gas, as in kelp and some bacteria, or by using solutes of molecular weight (MW) lower than the surrounding media [for example, replacing sodium (MW ≈ 23) with ammonium (MW ≈ 18)], as in the small crustaceans, Antarctic copepods.

Humans are made of about 60% water (40% in cells, 15% in interstitial fluid, and 5% in blood plasma; BNID 110743), and most of us have experienced the strong effects of dehydration after forgetting to drink even just a few glasses. Yet, some cells can be surprisingly robust to a decrease in their water content. For example, the rate of glucose metabolism in rat liver cells was not affected by a 25% loss of intracellular water. Such a decrease can be attained by osmosis—that is, by changing the tonicity (solute concentration) of the extracellular fluid. An extreme example is that of the remarkable brine shrimp. Living in environments where the outside salt concentration can fluctuate and be very high, brine shrimp have cysts that can be desiccated to only 2% water without irreversible damage, and at hydration levels of higher than 37% (only about half of its fully hydrated state), their physiology behaves as normal. This robustness in the face of water loss might be related to a distinction sometimes made between two forms of water in the cell interior. Normal "bulk water," which is more dispensable, and "bound water," which is associated with the cellular components and serves as a solvent that is essential for proper functioning.

What are environmental O_2 and CO_2 concentrations?

The air we breathe is made up of 20% oxygen. The concentration of carbon dioxide in that same air has recently surpassed levels of 400 parts per million, the highest in millions of years, pumped up by human activities. These atmospheric gases are critical to the life styles of plants and animals alike. However, biological reactions take place in liquid media and thus should depend upon the solubility of these key inorganic constituents. What concentrations of oxygen and carbon dioxide do cells see in the watery media within which they live?

Living organisms are built out of four main types of atoms: carbon, oxygen, nitrogen, and hydrogen. In the human body, together they amount to $\approx$96% of the wet weight and $\approx$87% of the dry weight, as shown in the vignette entitled "What is the elemental composition of a cell?" (pg. 68). However, the pool of these constituents in the cellular milieu is often in limited supply. For example, as we will discuss below, oxygen is soluble in water to only about 10 parts per million. In the case of carbon and nitrogen, these atoms are tied up in a relatively inert inorganic form sequestered in CO_2 and N_2, respectively. As a result, cells must find ways to draw these molecules out of these otherwise inaccessible reservoirs and convert them into some usable form. Though "water" and "air" are known to all in the same way that anyone who lives in northern climes has a visceral response to the word "snow," it is often forgotten that these words from the common vernacular mask a rich molecular reality.

Carbon enters the biosphere when it is transformed from its oxidized form in CO_2 to a reduced form mostly in the carbohydrate repeating motif $(CHOH)n$. This motif makes up sugars in general, and is the prime component of the cell walls present in both the microbes and plants that make up most of the organic matter in the biosphere. This transformation occurs in a process known as carbon fixation, which is performed by plants, algae, and a range of bacteria known as autotrophs. The concentration of dissolved CO_2 in water at equilibrium with the atmosphere is $\approx$10 μM (BNID 108697), as shown in **Figure 2-4**. This means there are only about 10^4 CO_2 molecules in a water volume the size of a bacterium. This should be compared to the 10^{10} carbon atoms that are required to constitute a bacterium. The concentration of O_2 is similarly quite low, at $\approx$100–300 μM (BNID 109182; see Figure 2-4 to appreciate how this solubility changes with temperature). The solubility of oxygen in water is about 50 times smaller than that of CO_2. As a result, even though oxygen in air is about 500 times more abundant than CO_2, the concentration ratio between O_2 and CO_2 in solution is about 10 rather than 500. By definition, each mg/L in Figure 2-4 is one part per million in terms of mass, so the rarity of oxygen and carbon dioxide can be directly appreciated by noting that the concentration of these gases are in the single-digit regime in terms of mg/L and are thus only a very few parts per million. CO_2 has the added feature that it reacts with water to give, at physiologically relevant pH values, mostly bicarbonate (HCO_3^-). At pH 7, there is about 10-fold more inorganic carbon in the form of bicarbonate than dissolved CO_2. At pH 8, characteristic of ocean water, there is 100-fold more bicarbonate than dissolved CO_2. These values are of importance to anyone who aims to gauge the pools of inorganic carbon

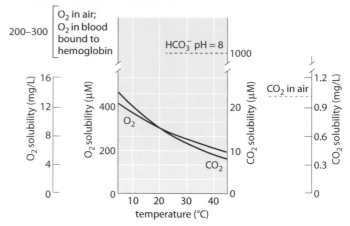

$[O_2]/[CO_2] \approx 20$ but axis scale chosen to show relative slope

Figure 2-4 The oxygen and carbon dioxide solubility in water and their dependence on temperature under normal air composition. The *y*-axis values for the two gases were chosen to enable comparison of the change with temperature, but note that the oxygen concentration scale is 10 times larger. The concentration of oxygen in air is about 500 times higher than CO_2, but oxygen is about 50 times less soluble. For both gases, the concentration is lower at higher temperatures. As the temperature increases, the availability of CO_2 decreases faster than that of oxygen. Bicarbonate (HCO_3^-) is the most abundant inorganic form of carbon in the pH range 6–10. Oxygen in blood is carried mostly bound to hemoglobin at a concentration similar to that of oxygen in air. This concentration is about 50 times higher than would be carried by the blood liquid without hemoglobin. Plot refers to fresh water; solubility is about 20–30% lower in ocean salt water. The data in the curves was calculated by the authors based on Henry's law.

available to cells. Specifically, the census of these molecular reservoirs is important for understanding the carbon sequestration in the oceans or the transport of inorganic carbon in our blood from tissues to the lungs. The transition from CO_2 to bicarbonate and vice versa is enhanced by the action of the enzyme carbonic anhydrase. This transition allows the cell to replenish the quickly depleted small pool of CO_2 from the much bigger pool of inorganic carbon in the form of bicarbonate.

In many aqueous environments, the low solubility and slow diffusion of O_2 is a major limitation for the aerobic metabolism of organisms. For example, consider the acute environmental problem of eutrophication, the process whereby oxygen gets depleted when excessive amounts of fertilizers containing nitrogen and phosphorus are washed to a water basin, leading to plankton blooms. Limited oxygen supply translates into enormous dead zones such as in the Gulf of Mexico, some as large as the area taken up by

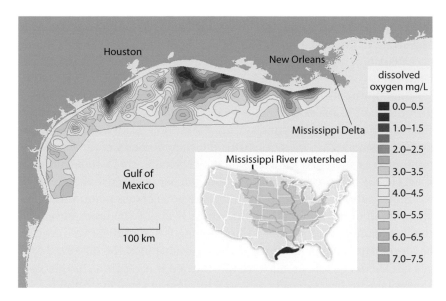

Figure 2-5 Dead zone in the Gulf of Mexico due to agricultural fertilizers borne by the Mississippi River. The long-term average for the area of midsummer bottom water hypoxia (colored red), where dissolved oxygen levels are <2 mg/L (also known as the dead zone, where oxygen depletion leads to fish suffocation) is 13,000 km², about the area of Connecticut. Values report the oxygen as measured at sea bottom stations. Without the eutrophication effect caused by agricultural fertilizers, oxygen levels would be close to equilibrium with the atmosphere—that is, about 7–9 mg/L, as shown in Figure 2-4. Note: 1 mg/L is about 1 part per million. Values for other parts of the Gulf of Mexico are not shown because there are no measurement stations located there. (Courtesy of the National Oceanic and Atmospheric Administration/Department of Commerce.)

the state of Connecticut, as shown in **Figure 2-5**. While the concentration of oxygen can be limiting for respiration in some organisms, it can actually be too high for those that perform carbon fixation. As noted in Figure 2-4, solubility depends on temperature, such that there is relatively less CO_2 with respect to O_2 at higher temperatures. This is suggested to drive the selective pressure leading to C4 plants (for example, maize and sugar cane), which employ metabolic pumps to locally increase CO_2 concentrations for carbon fixation.

To illustrate the meaning of the low oxygen concentrations found in marine environments in a familiar lab context, think of an overnight culture of bacteria. The cells grow from a small number up to saturation at an OD_{600} of about 1 (corresponding to about 100–1000 million bacterial cells per mL, as discussed in the vignette entitled "What is the concentration of bacterial cells in a saturated culture?"; pg. 84), under conditions that can largely be described as aerobic. The growth is facilitated by a sugar, such as glucose, in the media (say 0.2% by mass, equivalent to ≈ 10 mM). A simple calculation regarding the oxygen requirements of such growth is schematically depicted in **Estimate 2-3**. As a reasonable benchmark scenario, consider that about half of this sugar will be used for building biomass and the other half to make energy (as evidenced in the observation that the yield of carbon stored as biomass from the carbon taken from the growth media is usually ≈ 0.5; BNID 105318). The stoichiometry of the process of respiration is such that for each glucose molecule, six O_2 molecules are used. Hence, in a closed system, 5 mM of glucose respired to make energy will require about 30 mM of oxygen. The oxygen concentration was noted above to be in the hundreds of μM, which is about 100 times lower. We can thus conclude, as calculated in Estimate 2-3, that there will need to be more than 100 replenishment cycles (turnovers) of the dissolved oxygen pool in the growth media to supply the needs of respiring the glucose. The replenishment is usually achieved by vigorous shaking, bubbling, or by special impellers. The growth media is surrounded by air that has an oxygen fraction of 20%, equivalent to about 10 mmol per liter (of air). As analyzed in Estimate 2-3, a headspace of a few times the culture volume contains enough oxygen for the culture growth, as long as the aeration is vigorous enough to dissolve the oxygen from the headspace into the liquid media. As an alternative way to think about this estimate, consider the rule of thumb that the conversion of glucose to bacterial biomass requires about 1 g of O_2 per 1 g of cell dry weight produced (most of it emitted upon

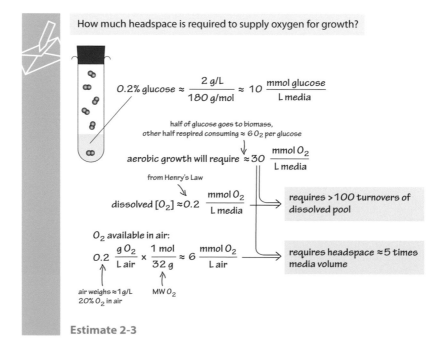

Estimate 2-3

respiration as CO_2). An OD_{600} of 1 has about 1 g cell dry weight per liter, which will require 1 g of oxygen, or 30 mmol, in accordance with the above derivation.

Oxygen is not the only critical cellular component that is in limited supply. Nitrogen, which comprises about 80% of the Earth's atmosphere, is highly inert, because it is almost exclusively tied up in the form of N_2. This nitrogen arrived in the atmosphere through the action of bacteria that utilize nitrogen as an electron acceptor in a process known as denitrification (another example of how biology helps shape the Earth). To make the atmospheric nitrogen available again for biochemistry, there is a challenging process—namely, turning nitrogen into ammonium (NH_4^+), nitrates (NO_3^-), or nitrites (NO_2^-). The organisms able to perform this nitrogen fixation process are single-celled organisms, such as the microbial symbiotic partners found at the roots of legumes. Only one enzyme, called nitrogenase, is able to carry out this process. Nitrogenase is oxygen sensitive, so it requires a local environment that is devoid of oxygen, a fact that leads some microbial systems to develop specialized cells known as heterocysts, as shown in **Figure 2-6**, which are the site of these nitrogen transactions. On a global scale, the natural cycle of nitrogen fixation is increased by humanity through a comparable amount of reduced nitrogen achieved in

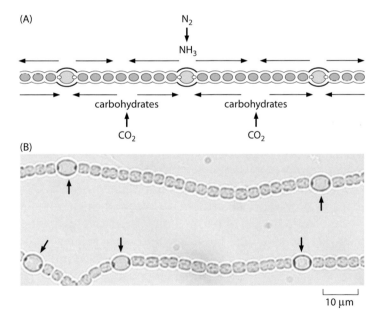

Figure 2-6 Heterocysts in *Anabaena*. (A) Schematic showing the regular positioning of the heterocysts in *Anabaena* that convert dinitrogen into ammonia. (B) Microscopy images showing both vegetative cells and heterocysts (labeled with arrows). (Adapted from Risser DD & Callahan SM [2009] *Proc Natl Acad Sci USA* 106: 19884–19888.)

the industrial Haber–Bosch process, resulting in fertilizers that are essential for feeding a large part of humanity but that also result in the ecological eutrophication mentioned previously. The fact that humans are making changes to major biogeochemical cycles involving the pools of these key inorganic substances alerts us to think about what is effectively a giant, human-run experiment engaged in altering the biosphere.

What quantities of nutrients need to be supplied in growth media?

Explanations that are widely accepted often turn out to be wrong. An everyday scientific example concerns the most commonly used media for growing bacteria across the globe—namely, LB media. Inquisitive students are usually told that this acronym originates from the names of its developers, Luria and Bertani. This story seems to make sense, and the explanation is widely "known." Yet, Giuseppe Bertani himself states that it was actually lysogeny broth, which led to the coining of the famed acronym, in reference to the experiments in which this media was used to study the lysogenic phase of bacteriophage in *E. coli*. Standard lore being off the mark regarding even such well-known recent human inventions suggests caution when considering seemingly beautiful explanations for the origins and purpose of ancient evolutionary inventions.

The LB medium contains mostly yeast extract and tryptone (as well as other trace constituents), which supply the building blocks needed for fast growth. Using substances such as yeast extract automatically implies significant differences in composition between batches, making it an ill-defined medium whose use is discouraged for physiological and quantitative studies. Originally LB contained glucose, but when formalized as a common lab media, it was defined without glucose, and whenever glucose is added—commonly, 1–4 g/L (0.1–0.4%)—that is indicated separately. For fast growth, LB is very useful, but when better concoctions were developed, they adopted names like super optimal broth (SOB), and when supplemented with glucose, super optimal broth with catabolite repression (SOC). The battle for impressive names to indicate potent media did not end there, but continued the hyperbole with names such as super broth (SB), terrific broth (TB), and so on.

When repeatability and accuracy is of importance in characterizing bacterial physiology, defined media is used, most commonly M9 minimal media. How much biomass can be expected as yield from such media?

Let's start with the question of carbon supply in such media. In minimal media, the only carbon present comes from the sugar added, often 2 g/L for bacteria, which is 0.2% by weight, recalling that the mass of the water used to make the medium is 1000 g/L (other organisms like yeast are usually grown at higher carbon-source concentrations, typically 2%). For aerobic growth, a characteristic yield factor from sugar to biomass is about one-half—that is, ≈ 0.5 g cell dry weight per 1 g of sugar (BNID 105318). The rest of the mass is often released as CO_2 through respiration or decarboxylation reactions, or alternatively, it is emitted as acids such as acetate as part of overflow metabolism. Interestingly, the evolutionary motivation for overflow metabolism, which excretes much of the imported carbon atoms back to the media, is still under lively discussion and is the subject of intensive research. In light of these numbers, the 2 g/L of sugar present in the media can be converted into 1 g/L of cell dry weight.

We are now in a convenient position to connect the amount of sugar we put in the media to the resulting optical density—that is, to the number of cells and number of atoms per cell. Converting from cells to optical density at 600 nm (OD_{600}) can be performed by using the rule of thumb that 1 OD_{600} unit corresponds to ≈ 0.5 g dry cell weight per liter (BNID 107924). We thus expect a final OD_{600} from 0.2% glucose of ≈ 2. One should take care not to be confused by the fact that many measurements today are performed in plate readers on multi-well plates, where the path length is usually about one-half or one-third of the 1 cm used in standard cuvettes, and thus the expected OD reading will be smaller by that factor. If one is interested in the number of cells, a useful rule of thumb states that an OD_{600} of 1 corresponds to about 10^9 E. coli cells/mL (BNID 106028, 100985; for budding yeast, the conversion factor between cell number and OD_{600} is roughly 10^7; BNID 100986, 106301). At about 10^{10} carbon atoms per cell of 1 μm^3 volume (as derived in the introduction to this chapter), this approximation implies 10^{19} carbon atoms per mL of our $OD_{600} = 1$ medium, or 10^{22} carbon atoms per liter. This is rewardingly consistent with our starting point of 2 g/L of sugar, which is 1/100th of a mole and thus at a carbon yield of about one-half, so the numbers are consistent with our quick sanity check.

Several points are worth noting about the rule of thumb regarding the conversion of OD readings to number of cells. One is that the accuracy of this value is relatively low, because under different growth conditions the cell size can vary about fivefold, and thus the number could be correspondingly higher or lower. This is in contrast to the rule of thumb regarding the conversion from OD to dry mass, which is much more robust, and thus preferable whenever the number of cells is not a must. A more trivial point is that most spectrophotometers are not linear at the range of OD of 1, so it is more accurate to work around OD_{600} of 0.1, which is equivalent to about

10^8 cells/mL (again noting the several-fold possible variation with growth conditions and strain).

Are we entitled to focus on carbon when estimating yield? For comparison, let's look at the oxygen requirements needed to synthesize cells and how this relates to the available oxygen. The needs of respiration in the form of oxygen are consumed at a ratio of about 1 g O_2 per 1 g cell dry weight (BNID 105317). This oxygen will come from the headspace in the vessel used, because the amount of oxygen soluble in the media is negligible, as shown in Estimate 2-3 of the vignette entitled "What are environmental O_2 and CO_2 concentrations?" (pg. 74). Beyond ensuring that there is enough shaking to achieve aeration, there must be enough headspace volume if the growth chamber is closed to oxygen replenishment. How much headspace volume? To achieve 1 g cell dry weight per liter, we need 1 g O_2. One liter of air weighs about 1 g, but it is only one-fifth oxygen, so about 5 L of headspace air volume will be needed per 1 L of media, or a ratio of fivefold. This is in line with common practices in microbiology that call for a headspace about 5–10-fold larger than the media space used. Further analysis and calculations on the oxygen requirements and availability are given in the vignette entitled "What are environmental O_2 and CO_2 concentrations?" (p. 74).

We can similarly analyze nitrogen, phosphate, and other macronutrients. Media are designed to make sure these are in excess and will not become growth limiting unless it is specifically intended to do so. In the back-of-the-envelope calculation shown in **Estimate 2-4**, we illustrate how this

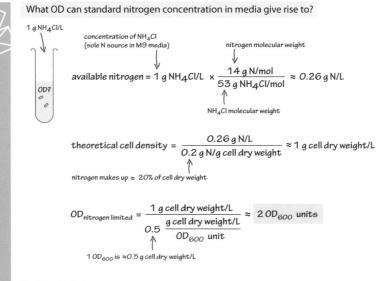

What OD can standard nitrogen concentration in media give rise to?

1 g NH_4Cl/L

OD?

concentration of NH_4Cl
(sole N source in M9 media)

nitrogen molecular weight

$$\text{available nitrogen} = 1 \text{ g } NH_4\text{Cl/L} \times \frac{14 \text{ g N/mol}}{53 \text{ g } NH_4\text{Cl/mol}} \approx 0.26 \text{ g N/L}$$

NH_4Cl molecular weight

$$\text{theoretical cell density} = \frac{0.26 \text{ g N/L}}{0.2 \text{ g N/g cell dry weight}} \approx 1 \text{ g cell dry weight/L}$$

nitrogen makes up ≈ 20% of cell dry weight

$$OD_{\text{nitrogen limited}} = \frac{1 \text{ g cell dry weight/L}}{0.5 \dfrac{\text{g cell dry weight/L}}{OD_{600} \text{ unit}}} \approx 2 \ OD_{600} \text{ units}$$

1 OD_{600} is ≈0.5 g cell dry weight/L

Estimate 2-4

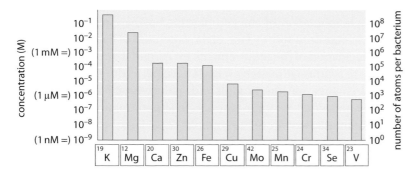

Figure 2-7 Metal content of *E. coli* cells grown in LB and glucose minimal medium as determined by mass spectrometry. The *E. coli* metallome—that is, the total metal content of the cell—is represented in terms of both concentrations and atoms per cell (grown in minimal medium) for each metal ion. The values shown here are the mean of three independent measurements; error bars are small on this log scale and are not shown. (Adapted from Outten CE & O'Halloran TV [2001] *Science* 292:2488–2492.)

works out for the case of nitrogen, a cellular building block usually supplied in the form of ammonium. Growth media constructions are usually strict and pedantic about all major elements, but in many cases trace elements like iron, copper, and so on are not explicitly mentioned or added and somehow life in the lab seems to go on. Yet small amounts of these trace elements are essential, as shown in **Figure 2-7** (about 10^5 Fe, Zn, and Ca atoms are required per *E. coli* cell and 10^3–10^4 Cu, Mn, Mo, and Se atoms; BNID 108825, 108826). This puzzle is resolved by the fact that trace elements are often contained as impurities in the distilled water used to make the media or even exist in the plastic and glassware used for growth. How much of this contamination needs to exist? One hundred thousand (10^5) iron atoms per bacterial cell volume correspond to a concentration of about 100 μM, and given that cells at saturation usually occupy about 1/1000th of the media volume, an initial concentration of 0.1 μM in the water will suffice. Indeed, tap water is allowed by standard to contain about 100 times more iron than this requirement and often contains 1 μM. Yet, if the water used is purified enough, one might find a lower yield that can be overcome by adding a concoction of trace elements.

The discussion above focused on bacteria and reflects the critical role played by prokaryotes in the development of modern molecular biology. However, interest in the medical applications of biology engendered parallel efforts aimed at figuring out the growth requirements of eukaryotic cells in culture. One of the pioneering efforts in this regard was spearheaded by Harry Eagle. Just as with the LB media described above, early efforts to grow mammalian cells in tissue culture involved undefined media derived from serum and embryo extracts, with HeLa cells, for example, originally grown in a combination of chicken plasma, bovine embryo extract, and human placental cord serum. One of the outcomes of Eagle's experiments was the elucidation of the requirements for essential amino acids that we are unable to synthesize ourselves. Eagle's recipe, a common staple in labs to this day, ensures that amino acids and vitamins that bacteria synthesize themselves are added in ample amounts to support mammalian cells that have evolutionarily lost those biosynthetic capacities.

What is the concentration of bacterial cells in a saturated culture?

Once we overcome our amazement at the exponential phase of cell growth in liquid media and the questions it engenders, the next mystery centers on when and why growth abates in what is known as stationary phase. In most labs, the use of overnight cultures is standard fare. The scheme is that an inoculum of several thousand cells is pipetted into a 5-mL tube and then grown overnight. During those 8–12 hours, the transparent and vacant media transforms into a saturated culture, as shown in **Figure 2-8**, with a characteristic density of cells measured via the optical density at 600 nm (OD_{600}) with a value of ≈ 2. With a calibration curve, or using the collection of characteristic conversion factors shown in **Table 2-2**, we can

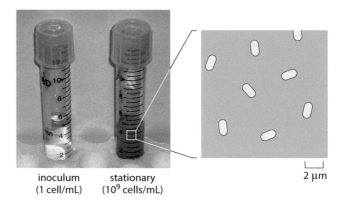

inoculum   stationary
(1 cell/mL)   (10^9 cells/mL)   2 μm

Figure 2-8 Depiction of the density of *E. coli* cells in saturation. A saturated cell culture contains about 10^9 cells per mL. The average spacing is about 10 μm between cells. The blowup is drawn to show a characteristic density at such conditions. In order to represent a three-dimensional situation in two dimensions, the figure shows all cells in a layer about 10 μm thick and the cells rotated to be seen sidewise. When viewed under a microscope, the layer thickness that is in focus is called the optical depth, and it is usually around one to several microns, depending on magnification.

organism	(cells/mL)/OD_{600}	CDW/OD [(g/L)/OD_{600}]	yield (g CDW/g sugar)	specific cell volume (mL/g CDW)	BNID
E. coli	$0.6\text{–}2 \times 10^9$	0.3–0.5	0.17–0.55	1.3–2.8	106578, 107919, 107924, 109835, 108126-8, 108135
S. cerevisiae	$0.8\text{–}3 \times 10^7$	0.5–0.6	0.45	2	106301, 100986, 100987, 102324, 107923, 108131

Table 2-2 Conversion between optical density (OD) and cell concentrations. CDW is cell dry weight. Yield is the ratio of CDW to mass of sugar consumed. The overall mass balance is that the total sugar mass plus oxygen consumed is equal to the biomass produced plus CO_2 emitted and by-products excreted. Note that values vary with growth rate (based on carbon source, etc.).

transform the OD value into a cell count of 10^9 cells/mL (BNID 104831). Under these conditions, the cells occupy about 0.1% of the total medium volume. The mean spacing between the cells is roughly 10 microns, a high density, but still not nearly as high as the cell densities in environments such as the guts of animals, which are typically a factor of 10 higher (BNID 104951, 104952, 104948, 102396). Examples of the extreme crowding in such environments are shown in **Figure 2-9A**, illustrating the crowded cellular environment in the termite gut, and **Figure 2-9B**, showing a trophosome—an organ in deep-sea tube worms packed with bacterial symbionts supplying its energy from sulfur oxidation in a biological process completely independent of the sun's energy. In fact, many biologists make use of such dense environments on a daily basis by growing colonies of bacteria on agar plates. Even in the sediments of the ocean floor, the bacterial densities are sometimes as high as those found in a saturated culture, as shown in **Figure 2-10**.

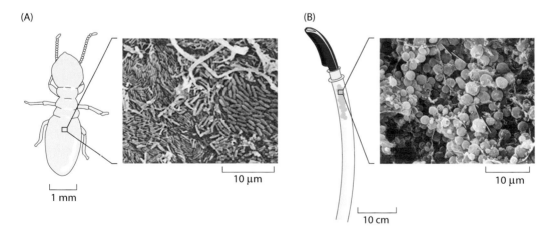

(A)

(B)

10 µm

1 mm

10 µm

10 cm

Figure 2-9 Bacterial crowding in multicellular organisms. (A) Scanning electron micrograph of the paunch section of a termite gut. (B) *Riftia pachyptila*. Scanning electron micrograph showing Gram-negative bacterial symbionts within trophosome. (A, adapted from Breznak JA & Pankratz HS [1977] *Appl Environ Microbiol* 33:406–426. B, adapted from Cavanaugh CM, Gardiner SL, Jones ML et al. [1981] *Science* 213:340–342.)

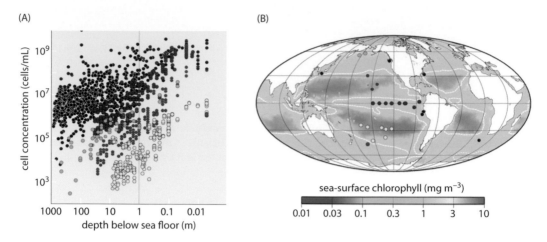

Figure 2-10 Cell counts from sub-seafloor sediments. (A) Cell concentration as a function of depth below sea floor. (B) The locations used for sampling in the study overlain on a map of time-averaged sea surface chlorophyll-a, indicating the level of photosynthetic primary productivity. (Adapted from Kallmeyer J, Pockalny R, Adhikari RR et al. [2012] *Proc Natl Acad Sci USA* 109:16213–16216.)

Given that bacteria have become the workhorses of many different parts of the biotechnology industry, it comes as no surprise that massive efforts have been undertaken to push the limits and efficiency of cell cultures. The concentration of cells at the final stages of growth, also known as the yield, is a dominant factor in the overall economic viability of many biotechnology applications. Thus, in industrial fermenters, effort is being made to optimize the conditions to reach as high a yield as possible. Strikingly, yields of about 200 g of dry weight per liter have been reported (BNID 104943), equivalent to several hundreds in OD units. Remembering that *E. coli* are usually about 70% water (BNID 100044), this leaves very little except cell mass in these extremely saturated cultures. Indeed, the cell density corresponds to a mean spacing between cells of just over a micron. At these densities, the cells are literally on top of each other. To achieve such high concentrations, methods such as dialysis fermentation have been developed, where the cells are separated by a semipermeable membrane such that low-molecular-weight excreted products are removed from the growth medium and a fresh supply of nutrients, including oxygen, is maintained.

How can such a multitude of bacteria be helpful? There are many circumstances in which we are interested in generating many copies of some DNA of interest. Preparing to transform a cell or checking the presence of a gene by running the corresponding DNA on a gel are two everyday examples from the lab. Consider again a 5-mL tube of LB media saturated overnight with a batch of bacteria. It will consist of about 10^{10} cells. If it expresses a very high copy number plasmid [that is, ≈ 100–1000 plasmids per cell (BNID 103857, 103860)], then there are $\approx 10^{13}$ copies of that gene in the culture. This is roughly the same number of copies as if that gene were present on the genome in each of the cells of your body (BNID 102390). If you need many copies of the gene, then extracting the gene from the bacterial culture will give you as many copies of that gene as would be obtained by extracting them from a whole human body.

CELL CENSUS

What is the pH of a cell?

Hydrogen ions play a central role in the lives of cells. For example, changes in hydrogen ion concentration are intimately tied to the charge of the side chains in proteins. This charge state, in turn, affects the activity of enzymes, as well as their folding and localization. Further, the ATP synthases that churn out the ATPs that power many cellular processes are driven by gradients in hydrogen ions across membranes.

The abundance of these ions and, as a result, the charge state of many compounds is encapsulated in the pH, which is defined as

$$pH = -\log_{10}([H^+]/1M), \tag{2.1}$$

where the brackets, [ ], denote the concentration (or more formally the activity) of the charged hydrogen ions (H^+, or more accurately, the sum of the hydronium, H_3O^+, as well as the functionally important but often overlooked Zundel, $H_5O_2^+$, and Eigen, $H_7O_3^+$, cations). We are careful to divide the hydrogen ion concentration by a so-called "standard state" concentration—the agreed-upon value is 1 M—in order to ensure that when taking the log we have a unitless quantity. This step is often skipped in textbooks.

The integer 7 is often etched in our memory from school as the pH of water, but there is nothing special about the integral value of 7. Water has a neutral pH of about 7, with the exact value varying with temperature, ionic strength, and pressure. What is the pH inside the cell? Just like with other parameters describing the "state" of molecules and cells, the answer depends on physiological conditions and which compartment within the cell we are considering (that is, which organelle). Despite these provisos, crude generalizations about the pH can be a useful guide to our thinking.

An *E. coli* cell has a cytoplasmic pH of ≈7.2–7.8 (BNID 106518), which is measured as discussed below. This pH value is a result of both active and passive mechanisms required for the ability of *E. coli* to colonize the human gastrointestinal tract, which contains niches of pH ranging from 4.5 to 9 (BNID 106518). A passive mechanism for maintaining this pH relies on what is called the buffering capacity of the cell. The buffering capacity is the amount of a strong acid needed to decrease the pH by one unit. A characteristic value for the buffering capacity of the cell interior is 10–100 mM per pH unit (BNID 110750, 110775, 107126–107130), or by our rule of thumb, about 10–100 million protons need to be added per μm^3. This equivalently means that there are about 10–100 million ionizable groups in a cubic micron of cell material that will release a proton when the pH is decreased by one unit. This capacity is provided by the cell's metabolite pool—that is, a change in pH will result in a release or absorption of hydronium ions that counteracts the externally induced change in pH. As shown in the vignette entitled "What are the concentrations of free metabolites in cells?" (pg. 94), the main ingredients of the metabolite pool are glutamate, glutathione, and free phosphates. These metabolites have concentrations in the mM range—that is, millions of copies per bacterial volume. The pK_a values, which indicate at what pH value a molecule will tend to change its protonation state and thus release or absorb a proton (equivalent to a hydronium ion), are not far from the neutral range of pH 7 for phosphates and glutathione. Having the pK_a in that range ensures that these metabolites will tend to counteract changes in pH. Active mechanisms for controlling the hydrogen ion concentration include the use of transporters such as ATPases, which are driven by ATP hydrolysis to pump protons against their concentration gradient. These transporters are regulated such that the cell can actively involve them in order to sculpt the intracellular pH.

As a second depiction of an organism's characteristic pH range, budding yeast is reported to have a cytoplasmic pH of ≈7 in exponential growth on glucose, which decreases to ≈5.5 in stationary phase (BNID 110927, 107762, 109863). As shown in **Figure 2-11A**, these measurements were carried out using more fluorescent protein tricks, this time with a pH-sensitive fluorescent protein. By examining the ratio of the light intensity emitted by this protein at two distinct wavelengths, it is possible to calibrate the pH, as shown in **Figure 2-11B**. Yeast flourish when the external pH is mildly acidic. The process of transporting molecules into yeast cells is often based on co-transport with an incoming proton and is thus more favorable if the external pH is lower than the internal pH (BNID 109863). Pumping excess protons into the vacuole is a way of maintaining a cytoplasmic pH near 7. In the process the vacuole is acidified to a pH of ≈5.5–6.5. The same fluorescence measurements reveal that the yeast mitochondria in these conditions have a pH of 7.5. **Figure 2-11C** shows a case where the internal

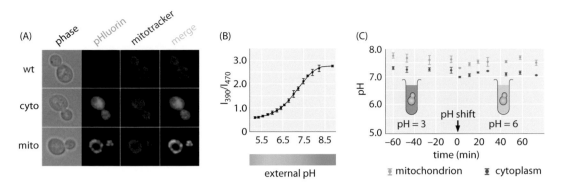

Figure 2-11 Measuring the pH of cells *in vivo* using pH-sensitive fluorescent protein. (A) Microscopy images using both phase imaging and fluorescence. (B) Ratio of intensity at two different wavelengths, 390 nm and 470 nm, can be used to calibrate the pH. (C) Time course for the pH as measured using the fluorescent probe. (Adapted from Orij R, Postmus J, Ter Beek A et al. [2009] *Microbiology* 155:268–278.)

pH of a yeast strain is kept almost constant under very different pH conditions in the surrounding medium. In another experiment, the internal pH shows a different dynamic behavior by closely following the external pH (BNID 110912). The reasons underlying the variation in these responses are still under study. Using such pH-sensitive probes in mammalian HeLa cells revealed that the cytoplasm and nucleus had a pH of ≈7.3, mitochondria had a pH of ≈8.0, ER had a pH of ≈7.5, and Golgi had a pH of ≈6.6 (BNID 105939, 105940, 105942, 105943). Tissues in animals ranging from the brain to muscles to the heart have a pH in the range of 6.5–7.5 (BNID 110768, 110769, 110770, 110771).

Even though hydrogen ions appear to be ubiquitous in the exercise sections of textbooks, their actual abundance inside cells is extremely small. To see this, consider how many ions are in a bacterium or mitochondrion of volume 1 μm³ at pH 7 (BNID 107271, 107272). Using the rule of thumb that 1 nM corresponds to ≈1 molecule per bacterial cell volume, and recognizing that pH 7 corresponds to a concentration of 10^{-7} M (or 100 nM), this means that there are about 100 hydrogen ions per bacterial cell at the typical pH values found in such cells, as worked out in more detail in the calculation shown in **Estimate 2-5**. This should be contrasted with the fact that there are in excess of a million proteins in that same cellular volume, each one containing several ionizable groups, each of which has a pK_a value close to 7 (and thus the tendency to gain or release a hydrogen ion).

How can so many reactions involving hydronium ions work with so few ions in the cell? To answer that question, we need to think about how long it takes an active site to find a charge required for a reaction. It is important to note two key facts—namely, cells have a strong buffering capacity (as a

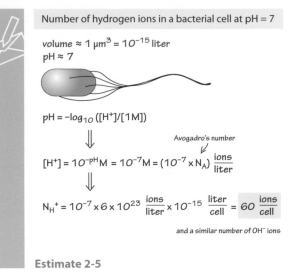

Number of hydrogen ions in a bacterial cell at pH = 7

$\text{volume} \approx 1 \; \mu m^3 = 10^{-15} \; \text{liter}$
$pH \approx 7$

$pH = -\log_{10}([H^+]/[1M])$

$\Downarrow$

Avogadro's number

$[H^+] = 10^{-pH} M = 10^{-7} M = (10^{-7} \times N_A) \dfrac{\text{ions}}{\text{liter}}$

$\Downarrow$

$N_{H^+} = 10^{-7} \times 6 \times 10^{23} \dfrac{\text{ions}}{\text{liter}} \times 10^{-15} \dfrac{\text{liter}}{\text{cell}} = 60 \dfrac{\text{ions}}{\text{cell}}$

and a similar number of OH^- ions

Estimate 2-5

result of metabolites and amino acid side chains) and the hopping time of charges between different water molecules is very short in comparison with the reaction times of interest.

If the 100 hydrogen ions we have estimated are present in each cell were all used up to alter the charge state of macromolecules, the pH still would not change significantly because there are literally millions of groups on proteins and metabolites (such as ATP) that would compensate by releasing ions as soon as the pH began to change. Hence, these 100 ions would be quickly replenished whenever they were consumed in reactions. This implies that there are orders of magnitude more ion-utilizing reactions that can take place. This capability is quantified by the cell's buffering capacity. But how does a reaction "find" the hydronium ion to react with if they are so scarce? The lifetime of a hydronium ion is extremely brief, about 1 picosecond (10^{-12} s, BNID 106548). Lifetime in this context refers to the "hopping" time scale when the charge will move to another adjacent water molecule (also called the Grotthuss mechanism). The overall effective diffusion rate is very high: $\approx 7000 \; \mu m^2/s$ (BNID 106702), a value that should be contrasted with the much lower diffusion rates for most biological molecules. The lifetime and diffusion values can be interpreted to mean that for every ion present in the cell, 10^{12} water molecules (on average) become charged very briefly every second. In an *E. coli* cell, there are about 10^{11} water molecules. Thus, every single ion "visits" every molecule 10 times per second, and for 100 ions per cell, every molecule will be converted to an ion 10^3 times per second, even if very briefly in each such case. As a result, an enzyme or reaction that requires such an ion will find plenty of them in the surrounding water, assuming the kinetics is fast enough to allow utilization before the ions neutralize to be formed somewhere else in the cell.

The scale of the challenge of keeping a relatively constant pH can be appreciated by thinking of the dynamic pools of metabolites inside a cell. Think of glucose being catabolized in glycolysis. In this process, internal electron rearrangements known as substrate-level phosphoryl-ation convert non-charged groups into a carboxyl acid group (COOH⁻). This conversion releases a hydrogen ion in the metabolites of the pro-cess, each having a concentration in the mM range. A 1 mM increase in concentration of such an intermediate releases about 10^6 protons per cell. This number of protons would cause the pH of the cell to drop to 3 (!) if not for the buffering capacity discussed above and the concur-rent changes in metabolite concentrations. This is but one example that illustrates the powerful and tightly regulated chemistry of hydrogen ions inside the cell.

What are the concentrations of different ions in cells?

Beginning biology students are introduced to the macromolecules of the cell (proteins, nucleic acids, lipids, and carbohydrates) as being the key players in cellular function. What is disturbingly deceptive about this picture is that it makes no reference to the many ionic species, without which cells could not function at all. Ions have a huge variety of roles in cells. Several of our favorites include the role of ions in electrical com-munication (Na^+, K^+, and Ca^{2+}), as cofactors in dictating protein function with entire classes of metalloproteins (constituting by some estimates at least one-fourth of all proteins), in processes ranging from photo-synthesis to human respiration (Mn^{2+}, Mg^{2+}, and Fe^{2+}), as a stimulus for signaling and muscle action (Ca^{2+}), and as the basis for setting up trans-membrane potentials that are then used to power key processes, such as ATP synthesis (H^+ and Na^+).

A census of the ionic charges in a mammalian tissue cell, as well as in the surrounding intercellular aqueous medium in the tissue, is shown in **Figure 2-12** (left and middle panels). The figure also shows the com position of another bodily fluid, the blood plasma, which is separated from tissues through the capillary walls. The figure makes it clear that in each region the sum of the negative ion charges equals the sum of the positive charges to a very high accuracy. This is known as the law

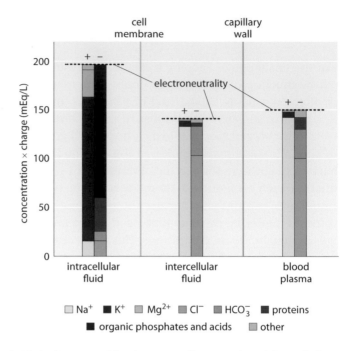

Figure 2-12 Ionic composition in mammalian organisms. Three distinct regions are characterized: the cellular interior ("intracellular fluid"), the medium between cells ("intercellular fluid"), and the blood plasma that is outside the tissue, beyond the capillary wall. The y-axis is in units of ionic concentration called Eq for "equivalents," which are equal to the ion concentration multiplied by its absolute charge. These units make it easy to see that the total amount of positive and negative charge is equal in each compartment, in line with the principle of electroneutrality. Even though it is not evident from the figure, the total free solute concentrations (sum of concentrations of both positive and negative components, not taking into account their charge) are the same in the intracellular and intercellular fluid. This reflects the fact that the two compartments are in osmotic balance. (Adapted from Andersen O [2013] Cellular electrolyte metabolism. In Encyclopedia of Metalloproteins [RH Kretsinger, VN Uversky & EA Permyakov eds], pp. 580–587, Springer. From BNID 110754.)

of electroneutrality. The relatively tiny deviations we might expect are quantified in the vignette entitled "What is the electric potential difference across biological membranes?" (pg. 196). Figure 2-12 also shows that blood ionic composition is very similar to that of the interstitial fluid. Yet, the composition of the cell interior is markedly different from the milieu outside the cell. For example, the dominant positive ion within the cell is potassium, with a concentration that is more than 10-fold higher than that of sodium. Outside the cell, the situation reverses, with sodium as the dominant positive ion. These and the other differences are carefully controlled by both channels and pumps, and we discuss some of their functional importance below.

ion conc. (mM)	sea water	E. coli	S. cerevisiae	mammalian cell (heart or RBC)	blood plasma	BNID
K$^+$	≈10	30–300	300	100	4	104049
Na$^+$	≈500	10	30	10	100–200	104050
Mg^{2+}	≈50	30–100 (bound); 0.01–1 (free)	50	10 (bound) 0.5 (free)	1	104983, 100770, 101953
Ca^{2+}	≈10	3 (bound);100 nM (free)	2 (bound)	10–100 nM (free)	2	100130, 110746, 111366
Cl$^-$	≈500	10–200 media dependent		5–100	100	105409, 110744
BNID	106594	105926, 107033, 107114, 111425	107752	103966, 107187	105409	

Table 2-3 Ionic concentrations in sea water, a bacterial cell, a yeast cell, a mammalian cell, and in the blood. Concentrations are all in units of mM. Values are rounded to one significant digit. Unless otherwise noted, concentration reported here include the total of both free and bound ions. Note that concentrations can change by more than an order of magnitude depending on cell type and physiological and environmental conditions, such as the medium osmolarity or external pH. Na$^+$ concentrations are especially hard to measure due to trapping and sticking of ions to cells. Most Mg^{2+} ions are bound to ATP and other cellular components. More BNIDs used to construct this table: 104083, 107487, 110745, and 110754.

Ion channels serve as passive barriers that can be opened or closed in response to environmental cues, such as voltage across the membrane, the concentration of ligands, or membrane tension. Pumps, by way of contrast, use energy in the form of protons or ATP in order to pump charged species against their concentration gradient. The differences in concentration mediated by these membrane machines can often be several orders of magnitude, and in the extreme case of calcium ions, correspond to a 10,000-fold greater concentration of ions outside of the cell than inside, as shown in **Table 2-3**. The dominant players in terms of abundance inside the cell are potassium (K$^+$), chloride (Cl$^-$), and magnesium (Mg^{2+}), although Mg^{2+} is mostly bound to ATP, ribosomes, and other macromolecules and metabolites such that its free concentration is orders of magnitude lower. Table 2-3 shows some typical ionic concentrations in bacteria, yeast, and mammalian cells. Some ion concentrations are regulated tightly, particularly toxic metal ions that are also essential for certain processes, but K$^+$ is also regulated by osmolarity, which is essential for growth. Other ions, such as Na$^+$, are less tightly regulated. One of the provocative observations that emerges from this table is that positive ions are much more abundant than negative ions. What is the origin of such an electric imbalance in the simple ions? Many of the metabolites and macromolecules of the cell are negatively charged. This negative charge is conferred by phosphate in small metabolites and DNA and by carboxylic groups on the acidic amino acids, such as the most abundant free metabolite, glutamate. Much more on these cellular players can be found in the vignette entitled "What are the concentrations of free metabolites in cells?" (pg. 94).

Potassium is usually close to equilibrium in animal and plant cells. Given that its concentration inside the cell is about 10- to 30-fold higher than outside the cell, how can it be in equilibrium? Assume we start with this

concentration difference across the membrane, and with no electric potential difference (there are counter ions on each side of the membrane to
balance the initial charges and they cannot move). As the potassium ions
diffuse down their concentration gradient, from the inside to the outside,
they quickly create an electric potential difference due to their positive net
charge (the net charge movement is miniscule compared to the ion concentrations on the two sides of the membrane, as discussed in the vignette
entitled "What is the electric potential difference across membranes?"; pg.
196). The potential difference will increase until its effect will exactly balance the diffusive flux, and this is when equilibrium will be reached. This
type of equilibrium is known as electrochemical equilibrium. Indeed, from
the equilibrium distribution we can infer that the cell has a negative electric
potential inside and by how much. The direction of the voltage difference
across the cell membrane is from positive (outside) to negative (inside), as
can be naïvely expected from the pumping of protons out of the cell, and as
discussed in quantitative terms in the vignette entitled "What is the electric
potential difference across membranes?" (pg. 196).

The concentrations described above are in no way static. They vary with the
organism and with the environmental and physiological conditions. To flesh
out the significance of these numbers, we examine a case study from neuroscience. For example, how different is the charge density in a neuron before
and during the passage of an action potential? As noted above, the opening
of ion channels is tantamount to a transient change in the permeability of
the membrane to charged species. In the presence of this transiently altered
permeability, ions rush across the membrane, as described in detail in the
vignette entitled "How many ions pass through an ion channel per second?"
(pg. 226). But how big a dent does this rush of charge actually make in the
overall concentrations? Muscle cells in which such depolarization leads to
muscle contraction often have a diameter of about 50 μm, and a simple estimate (BNID 111449) reveals that the change in the internal charge within the
cell as a result of membrane depolarization is only about a thousandth of a
percent (10^{-5}) of the charge within the cell. This exemplifies how minor relative changes can still have major functional implications.

What are the concentrations of free metabolites in cells?

The cell's canonical components of proteins, nucleic acids, lipids, and sugars are complemented by a host of small metabolites that serve a number of key roles. These metabolites are broadly defined as members of

the many families of molecules within cells having a molecular weight of less than 1000 Da. Recent measurements have made it possible to take a census of these metabolites in bacteria, as shown in **Table 2-4**. Perhaps the most familiar role for these metabolites is as the building blocks for the polymerization reactions leading to the assembly of the key macromolecules of the cell. However, their biochemical reach is much greater than this restricted set of reactions. These metabolites also serve as energy sources, key activity regulators, signal transducers, electron donors, and buffers of both pH and osmotic pressure.

An inventory of which metabolites are present and at what concentrations is of great interest since it provides a picture of the stocks available to the cell as reserves for building its macromolecules. In addition, this inventory tells us which compounds are most ubiquitous and how we should think about the various chemical reactions (both specific and nonspecific) of which they are part. The concentrations of some metabolites are easy to measure, whereas others are notoriously difficult. Thanks to advances in mass spectrometry, the comprehensive survey of cell metabolite concentrations detailed in Table 2-4 and **Figure 2-13** became possible. The table depicts the most abundant metabolites in *E. coli* during growth on M9 medium supplemented with glucose. These surveys do not include simple ions such as potassium and chloride, which we discuss separately. Another key property beyond the concentration is the turnover time of the metabolite pool, which we discuss separately in the vignette entitled "What is the turnover time of metabolites?" (pg. 228).

The molecular census of metabolites in *E. coli* reveals some overwhelmingly dominant molecular players. The amino acid glutamate wins out in Table 2-4 at about 100mM, which is higher than all other amino acids combined, as depicted in Figure 2-13. Our intuition and memory is much better with absolute numbers than with concentrations, so recall our rule of thumb that a concentration of 1 nM corresponds to roughly one copy of the molecule of interest per *E. coli* cell. Hence, 100 mM means that there are roughly 10^8 copies of glutamate in each bacterium. How many protein equivalents is this? If we think of a protein as being built up of 300 amino acids (aa), then these 10^8 glutamates are equivalent to roughly 3×10^5 proteins—that is, roughly 10% of the $\approx 3 \times 10^6$ proteins making up the entire protein census of the cell (BNID 100088). This small calculation also shows that the "standard conditions" of 1 M concentration often employed in biochemical thermodynamic calculations are unrealistic for the cell, because such a concentration will take up all the cell mass (or more for larger compounds). Glutamate is negatively charged, as are most of the other abundant metabolites in the cell. This stockpile of negative charges is balanced mostly by a corresponding positively charged stockpile of free potassium ions (K^+), which have a typical concentration of roughly 200 mM.

metabolite	mM	metabolite	mM
glutamate	96	S-adenosyl-L-methionine	0.18
glutathione	17	phosphoenolpyruvate	0.18
fructose 1,6-bisphosphate	15	threonine	0.18
ATP	9.6	FAD	0.17
UDP-N-acetylglucosamine	9.2	methionine	0.14
hexose-P	8.8	2,3-dihydroxybenzoicacid	0.14
UTP	8.3	NADPH	0.12
GTP	4.9	fumarate	0.11
dTTP	4.6	phenylpyruvate	0.090
aspartate	4.2	NADH	0.083
valine	4.0	N-acetylglucosamine-1P	0.082
glutamine	3.8	serine	0.068
6-phospho-D-gluconate	3.8	histidine	0.068
CTP	2.7	flavinmononucleotide	0.054
NAD	2.6	4-hydroxybenzoate	0.052
alanine	2.5	dGMP	0.051
UDP-glucose	2.5	glycerolphosphate	0.049
glutathionedisulfide	2.4	N-acetylornithine	0.043
uridine	2.1	gluconate	0.042
citrate	2.0	malonyl-CoA	0.035
UDP	1.8	cyclic AMP	0.035
malate	1.7	dCTP	0.034
3-phosphoglycerate	1.5	tyrosine	0.029
glycerate	1.4	inosine diphosphate	0.024
coenzyme A	1.4	GMP	0.024
citrulline	1.4	acetoacetyl-CoA	0.022
pentose-P	1.3	riboflavin	0.019
glucosamine-6-phosphate	1.2	phenylalanine	0.018
acetylphosphate	1.1	aconitate	0.016
gluconolactone	1.0	dATP	0.016
GDP	0.68	cytosine	0.014
acetyl-CoA	0.61	shikimate	0.014
carbamyl aspartate	0.59	histidinol	0.013
succinate	0.57	tryptophan	0.012
arginine	0.57	dihydroorotate	0.012
UDP-glucaronate	0.57	quinolinate	0.012
ADP	0.55	ornithine	0.010
asparagine	0.51	dAMP	0.0088
2-ketoglutarate	0.44	adenosine phosphosulfate	0.0066
lysine	0.40	myo-inositol	0.0057
proline	0.38	propionyl-CoA	0.0053
dTDP	0.38	ADP-glucose	0.0043
dihydroxyacetone phosphate	0.37	anthranilate	0.0035
homocysteine	0.37	deoxyadenosine	0.0028
CMP	0.36	cytidine	0.0026
isoleucine + leuicine	0.30	NADP+	0.0021
deoxyribose-5-P	0.30	guanosine	0.0016
AMP	0.28	adenine	0.0015
inosine monophosphate	0.27	deoxyguanosine	0.00052
PRPP	0.26	adenosine	0.00013
succinyl-CoA	0.23		
inosine triphosphate	0.20	Sum	231
guanine	0.19		

Table 2-4 Intracellular concentrations of the most abundant metabolites in glucose-fed, exponentially growing *E. coli* measured via mass spectrometry. (Adapted from Bennett BD, Kimball EH & Gao M [2009] *Nat Chem Biol* 5:593–599.)

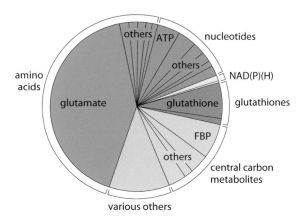

sum of concentrations ≈ 200 mM

Figure 2-13 The composition of free metabolites for an *E. coli* cell growing on glucose. Metabolites are colored based on their functional group. In each category, "other" refers to other metabolites in that category whose names are not shown due to small size; "various others." refers to other metabolites that are not part of any of the other categories, such as UDP-*N*-Ac-glucosamine; and FBP stands for fructose 1,6-bisphosphate. (Adapted from Bennett BD, Kimball EH & Gao M [2009] *Nature Chem Biol* 5:593–599.)

The second most abundant metabolite, glutathione, is the key regulator of the cell redox potential, strongly affecting protein structure by making and breaking sulfur bonds among cysteine residues. This key player is further discussed in the vignette entitled "What is the redox potential of a cell?" (pg. 189). The third most abundant metabolite, fructose 1,6-bisphosphate, is a central component of the carbon highway of the cell (that is, glycolysis), and the fourth most abundant metabolite is ATP, the main energy currency. Moving from the specific roles to the larger picture, one glaring feature revealed by the table is the broad range of concentrations found for these metabolites, which range from roughly 10^{-1} to 10^{-7} M. Given our rule of thumb that a concentration of 1 nM implies about one such molecule in the volume of an *E. coli* cell, this implies that the range of metabolite numbers inherent in these measurements is from as many as 10^8 copies of glutamate (as noted above) down to as few as 100 copies of the nucleoside adenosine, a millionfold range of concentrations.

As seen in Table 2-4, the total concentration of free metabolites is on the order of 250 mM. How can we put this number in perspective? One prominent example of where we have a feel for metabolite production is in the production of our favorite alcoholic beverages, where ethanol is produced by yeast. Yeast produces beer and wine through the fermentation of sugars to the alcohol, ethanol. Fermentation produces 2–3 ATP molecules per glucose molecule consumed, much less than the ≈30 that can be produced when using the tricarboxylic acid (TCA)

Alcohol concentration in fermenting brewer's yeast

yeast
(*S. cerevisiae*)

ethanol (CH_3CH_2OH, MW = 46 Da)
final concentration of ethanol ≈ 5%

$$\approx 50 \frac{g}{L} \Rightarrow c = \frac{50\ g/L}{46\ g/mol} \approx \boxed{1\ M}$$

Estimate 2-6

cycle. Yet, brewer's yeast still prefers to ferment, even when oxygen is available for the TCA cycle. One explanation for this odd behavior from the perspective of energy utilization is that fermentation with its associated excreted by-products creates an environment that is unsuitable for other organisms that inhabit the same niche. As such, we might speculate that by producing this alcohol, yeast effectively prevent bacteria from growing nearby. In **Estimate 2-6**, we provide a schematic of the numbers associated with fermentation, which serve as the basis for this speculative mechanism, although it awaits rigorous experimental examination. The alcohol content is typically ≈4% in beer and ≈12% in wine. Ethanol (CH_3CH_2OH), therefore, has a concentration of 40 g/L in beer and 120 g/L in wine. With a molecular weight of 46, we show in Estimate 2-6 that 5% is equivalent to ≈1 M concentration. Many bacteria are unable to grow in the presence of such high concentrations of alcohol,[‡] which affects what is called the "fluidity" of the cell membrane as well as the contents of the cell interior (ethanol is a small molecule to which the membrane is partially permeable). Brewer's yeast is adapted to these high ethanol concentrations and to low pH values, both of which inhibit the growth of most bacteria. Indeed, after the completion of the fermentative phase, yeast move to a phase of respiration that makes use of the extra energy capacity of ethanol with relatively little competition. It has been suggested that this is the reason that yeast choose this growth strategy.[§] We note, though, that this speculation on yeast growth strategy has been criticized[¶] as not being an evolutionary stable strategy against "cheater" mutants that do not produce ethanol, but would still enjoy the lower competition and thus could penetrate and overtake the ethanol-producing population.

[‡] Tamura K, Shimizu T & Kourai H (1992) Effects of ethanol on the growth and elongation of *Escherichia coli* under high pressures up to 40 MPa. *FEMS Microbiol Lett* 78:321–324.
[§] Piskur J, Rozpedowska E, Polakova S et al. (2006) How did *Saccharomyces* evolve to become a good brewer? *Trends Genet* 22:183–186.
[¶] Molenaar D, van Berlo R, de Ridder D & Teusink B (2009) Shifts in growth strategies reflect tradeoffs in cellular economics. *Mol Syst Biol* 5:323 (doi: 10.1038/msb.2009.82).

In summary, we return to the abundance of metabolites in Table 2-4. Why are some metabolites present in the cell at such high abundance while others are present at such minute concentrations? The starting point of an analysis of this question should focus on the costs and benefits of maintaining each of these metabolites at a given concentration. It is of interest to learn whether these concentrations are dictated by some adaptation that has been sculpted by evolution. Rationalizing the most abundant metabolites, such as glutamate, FBP, and glutathione, seems like a natural place to begin addressing this challenge. Surprisingly, we have been unable to find rigorously articulated and experimentally corroborated explanations for the relative concentrations of different metabolites, a befitting future challenge for cell and systems biologists.

What lipids are most abundant in membranes?

Cells are separated from the external world by complex membranes that are a rich combination of lipids and proteins. The same membrane sequestration strategy that separates the interior of cells from the rest of the world is also used for separating the cellular interior into a collection of membrane-bound organelles, such as the nucleus, the endoplasmic reticulum, the Golgi apparatus, and mitochondria. All of these membrane systems are host to a diverse collection of lipids that come in different shapes, sizes, and concentrations. There are literally hundreds of distinct types of lipid molecules found in these membranes and, interestingly, their composition varies from one organelle to the next, even though these distinct membrane systems are in communication through intracellular trafficking by vesicles. Even at a given moment in time, the plasma membrane is remarkably asymmetric, with different classes of lipids occupying the outer and cytosolic leaflets of the membrane, for example. The molecules making up membranes are often known for their dual relationship with the surrounding water molecules, since the hydrophilic head groups have a favorable interaction with the surrounding water, while the long-chain carbon tails incur a substantial free energy cost when in contact with water. This ambivalence provides a thermodynamic driving force for the formation of bilayers in which the hydrophobic tails are sequestered in the membrane interior, leaving the hydrophilic head groups exposed to the surrounding solution.

To see the functional implications of this great lipid diversity, we begin by examining some common ingredients from the kitchen. Both the fats and oils (olive, soy, etc.) we use to make delicious meals are made up of lipids. In general, the lipids in the fats we eat do not contain double bonds (they are saturated, meaning that their carbon tails have as many hydrogens bound to them as possible). This results in chain molecules that are long and straight, implying that they interact strongly with each other, making them solid at room temperature. By way of contrast, lipids in oils contain double bonds (they are known as unsaturated—that is, each carbon could have partnered up with more hydrogens) that create kinks in the molecules. They are thus hindered in their ability to form ordered structures and, as a result, are liquid at room temperature. An analogous situation occurs in biological membranes. Muscle cells with a high concentration of lipid chains that are unsaturated (oil-like) tend to be more fluid. The quantitative physiological and molecular implications of this fact are still under study.

Experimentally, the study of lipid diversity is a thorny problem. "Sequencing" a set of single or double bonds along a carbon backbone requires very different analytic tools than sequencing nucleotides in DNA or amino acids in proteins. Still, the "-omics" revolution has hit the study of lipids, too. The use of careful purification methods, coupled with mass spectrometry, has made inroads into the lipid composition of viral membranes, synaptic vesicles, and organellar and plasma membranes from a number of different cell types. To appreciate what is being learned in "lipidomic" studies, we first need to have an impression of the classification of the different lipid types. Learned committees of experts have attempted to tame the overwhelming chemical diversity of lipids by organizing them into eight categories—namely, fatty acyls, glycerolipids, glycerophospholipids, sphingolipids, sterol lipids, prenol lipids, saccharolipids, and polyketides. The classification criteria are based on the distinct chemistry of both the hydrophobic and hydrophilic pieces of these molecules. **Figure 2-14** shows the chemical structure for a representative from each of these categories as found in cell membranes. Simple rules of thumb about the geometry of these molecules that we can use to instruct our intuition are that the cross-sectional area of each such lipid range between roughly 0.25 and 0.5 nm^2 (BNID 106993), leading to a few million lipids per squared micron of membrane area. Their characteristic lengths are roughly 2 nm (BNID 105298), in line with the bilipid membrane being about 4–5 nm wide, as discussed in the vignette entitled "What is the thickness of the cell membrane?" (pg. 54). The mass of each lipid is usually in the range of 500–1000 Da (BNID 101838), somewhat larger than amino acids or nucleotides.

Because of the advances in lipidomic technologies, we are now at the point where it is becoming possible to routinely measure the concentrations of the array of different lipid types found in the various membranes of the cell in organisms ranging from single-celled prokaryotes all the way to the cells of our immune system. In broad brushstrokes, what has been learned

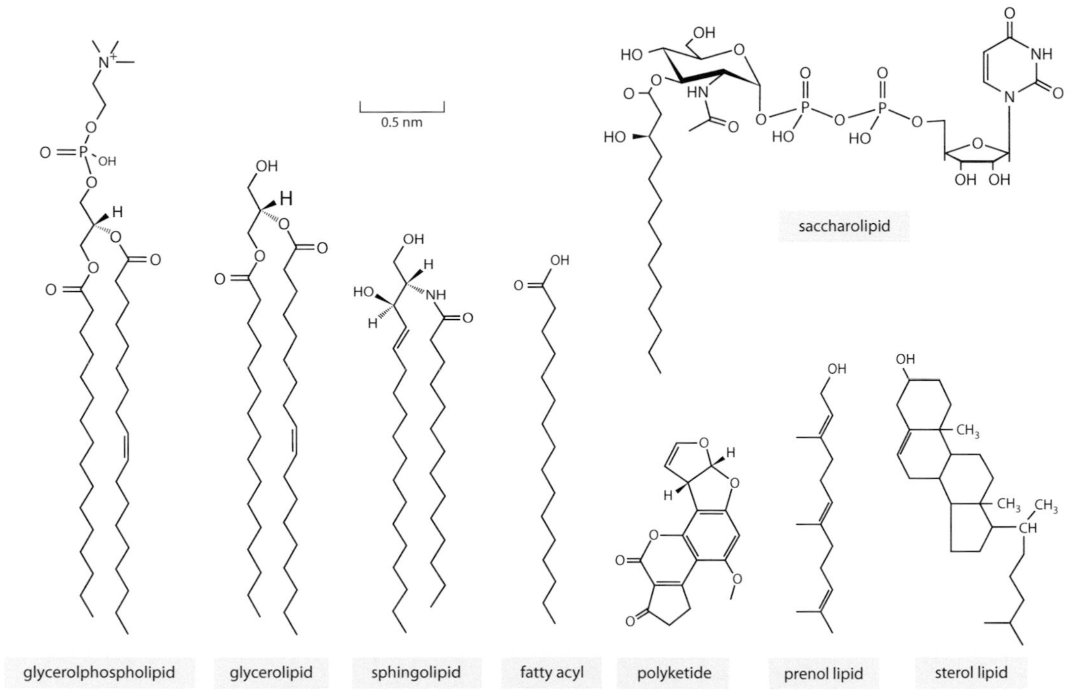

saccharolipid

| glycerolphospholipid | glycerolipid | sphingolipid | fatty acyl | polyketide | prenol lipid | sterol lipid |

0.5 nm

Figure 2-14 Diversity of membrane lipids. Structures of representatives from the key lipid types found in lipidomic surveys of biological membranes. (Adapted from Fahy E, Subramaniam S, Brown HA et al. [2005] *J Lipid Res* 46:839–861.)

is that in most mammalian cells, phospholipids account for approximately 60% of total lipids by number and sphingolipids make up another ≈10%. Nonpolar sterol lipids range from 0.1% to 40%, depending on cell type and which subcellular compartment is under consideration. The primary tool for such measurements is the mass spectrometer. In the mass spectrometer, each molecule is charged and then broken down, such that the masses of its components can be found and from that its overall structure reassembled. Such experiments make it possible to infer both the identities and the number of the different lipid molecules. Absolute quantification is based upon spiking the cellular sample with known amounts of different kinds of lipid standards. One difficulty following these kinds of experiments is the challenge of finding a way to present the data such that it is actually revealing. In particular, in each class of lipids there is wide variety of tail lengths and bond saturations. **Figure 2-15** shows the result of a recent detailed study of the phospholipids found in budding yeast. In Figure 2-15A, we see the coarse-grained distribution of lipids over the entire class of lipids found, while Figure 2-15B gives a more detailed picture of the diversity even within one class of lipids. **Figure 2-16** goes further and gives an organelle-by-organelle accounting of the lipid distributions found in a mammalian cell.

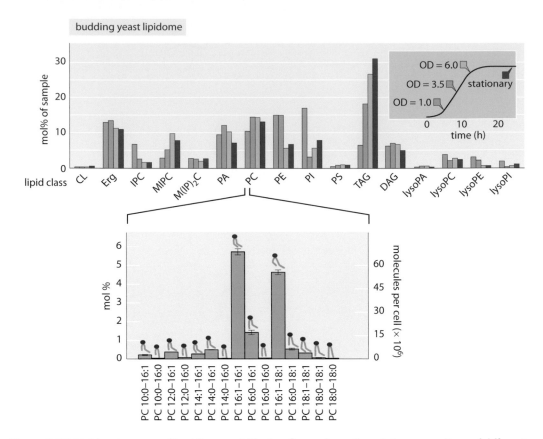

Figure 2-15 Lipidomic survey of budding yeast. The top figure shows the relative proportions of different lipid types as a function of the physiological state of the cells as determined by where they are along the growth curve (inset). The lower panel illustrates that for each lipid type shown in the top panel, there is an incredible diversity of chemically related lipids that differ in tail length and degree of saturation. CL: cardiolipin; Erg: Ergosterol; IPC: inositolphosphorylceramide; MIPC: mannosyl-inositol phosphorylceramide; M(IP)2C: mannosyl-di-(inositolphosphoryl) ceramide; PA: phosphatidic acid; PC: phosphatidylcholine; PE: phosphatidyl-ethanolamine; PI: phosphatidylinositol; PS: phosphatidylserine; TAG: Triacylglycerols; DAG: diacylglycerol; LPC: Lysophosphatidylcholine. (Top panel adapted from Klose C, Surma MA, Gerl MJ et al. [2012] *PLoS One* 7:e35063. Lower panel adapted from Ejsing CS, Sampaio JL, Surendranath V et al. [2009] *Proc Nat Acad Sci USA* 106:2136–2141.)

With the increasing sophistication of experimental methods in lipidomics, it is now even possible to trace out the life history over time of the lipid distribution in a particular cell type. How should we think about the significance of all of this lipid diversity for the underlying biological function of the cells and organelles that harbor such diversity? One of the reasons this lipid distribution is interesting is that these different membrane systems are constantly exchanging material as a result of the active trafficking processes that take place within cells. For example, communication between the endoplasmic reticulum (ER) and the Golgi apparatus through vesicle transport means that there is a flux of lipids from one organelle to the other. Yet differences in composition are somehow maintained. Together,

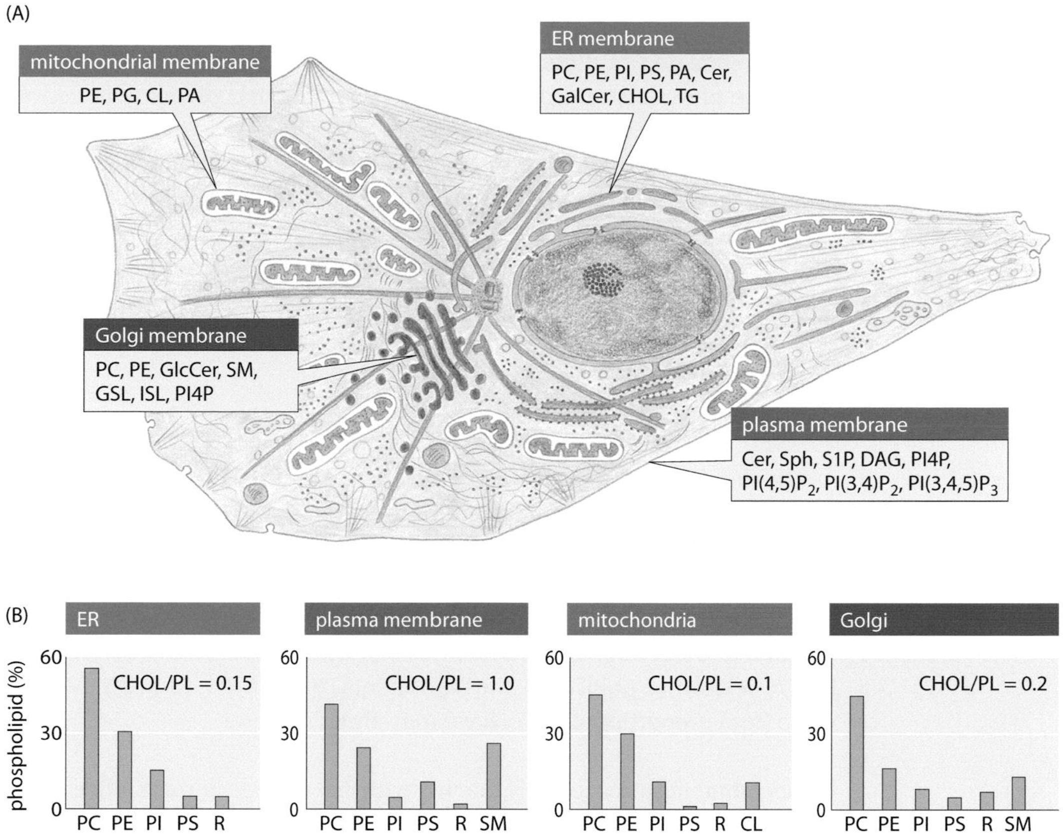

(A)

mitochondrial membrane
PE, PG, CL, PA

ER membrane
PC, PE, PI, PS, PA, Cer,
GalCer, CHOL, TG

Golgi membrane
PC, PE, GlcCer, SM,
GSL, ISL, PI4P

plasma membrane
Cer, Sph, S1P, DAG, PI4P,
$PI(4,5)P_2$, $PI(3,4)P_2$, $PI(3,4,5)P_3$

(B)

ER   CHOL/PL = 0.15
plasma membrane   CHOL/PL = 1.0
mitochondria   CHOL/PL = 0.1
Golgi   CHOL/PL = 0.2

phospholipid (%)

PC PE PI PS R   PC PE PI PS R SM   PC PE PI PS R CL   PC PE PI PS R SM

Figure 2-16 Lipid synthesis and steady-state composition of cell membranes. Lipid production is spread across several organelles. (A) The sites of synthesis for the major lipids. The main organelle for lipid biosynthesis is the endoplasmic reticulum (ER), which produces the bulk of the structural phospholipids and cholesterol. The lipid composition of different membranes also varies throughout the cell. (B) The composition out of the total phospholipid for each membrane type in a mammalian cell. As a measure of sterol content, the molar ratio of cholesterol to phospholipid is indicated. SM: sphingomyelin; R: remaining lipids. For more detailed notation, see previous figure caption. (Adapted from van Meer G, Voelker DR & Feigenson GW [2008] *Nat Mol Cell Biol* 9:112–124.)

the composition differences between different organelles and the changes in composition as cells make the transition between different characters (such as the change in polarity of epithelial cells during tissue formation) illustrate the exquisite control that is exercised over lipid concentrations, belying the idea of lipids as passive bystanders in the lives of cells. A second insight that emerges from such studies is revealed in Figure 2-15A, where we see that as a culture of yeast cells reach saturation, the distribution of lipids changes. One of the most interesting outcomes of that study on the flexibility of the yeast lipidome is the insight that triacylglycerols (TAG) increase in abundance. These lipids are important, both to sustain viability during starvation and to provide raw materials for the synthesis of new fatty acids when cells resume growth.

In light of these various quantitative and factual observations into the lipid composition of different cell types, the field is now faced with the challenge of understanding how all of this molecular diversity is tied to physiological functionality. In this book, we aim to give a sense of how the numbers in biology often make functional sense, and in the case of lipidomics, we await future research (and knowledgeable readers) to go beyond the descriptions given here.

How many proteins are in a cell?

As the dominant players in the cell in terms of both biomass and functionality, proteins get a large share of the attention in molecular and cell biology research. Yet, a small shift in emphasis to challenges of a more quantitative nature about these proteins raises all sorts of unanswered questions. For example, how many proteins are in a cell? That is, what is the total number of protein molecules, not the number of different types? Before reviewing published measurements, we can try to estimate this value from properties of the cell we may already know.

Protein content scales roughly linearly with cell volume or mass. Given that cell volume can change several fold, based on growth conditions or which specific strain was used, we will first analyze the number of proteins per unit cell volume (that is, the protein number density) and later multiply by cell volume to find the actual number of proteins per cell for our cell of interest.

Our first method for estimation is shown as a back-of-the-envelope calculation in **Estimate 2-7**, using rounded "generic" parameter values. The estimation relies on knowledge of the protein's mass per unit volume (denoted by c_p). The units of c_p are (g protein)/(mL cell volume) and this parameter has been reported for different cell types. We denote by l_{aa} the average length of the amino acids in a protein, and the average mass of an amino acid is given by m_{aa}. In light of these definitions, the number of proteins per unit volume is given by

$$N/V = c_p/(l_{aa} \times m_{aa}).\qquad(2.2)$$

In *E. coli* and other bacteria, we use an average protein length, l_{aa}, of 300 aa/protein, and in budding yeast, fission yeast, and human cells, we use the larger value of 400 aa/protein. Values are rounded to one significant figure (within about 10–20% accuracy), in line with variations in estimated values in the literature. The average lengths used were calculated

How many proteins are in a cell?

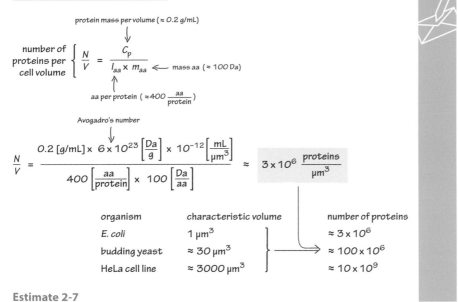

$$\text{number of proteins per cell volume} \left\{ \begin{array}{c} \frac{N}{V} = \frac{c_p \leftarrow \text{protein mass per volume } (\approx 0.2 \text{ g/mL})}{l_{aa} \times m_{aa} \leftarrow \text{mass aa } (\approx 100 \text{ Da})} \\ \uparrow \\ \text{aa per protein } (\approx 400 \frac{aa}{\text{protein}}) \end{array} \right.$$

$$\frac{N}{V} = \frac{0.2 \,[\text{g/mL}] \times 6 \times 10^{23} \left[\frac{\text{Da}}{\text{g}}\right] \times 10^{-12} \left[\frac{\text{mL}}{\mu\text{m}^3}\right]}{400 \left[\frac{aa}{\text{protein}}\right] \times 100 \left[\frac{\text{Da}}{aa}\right]} \approx 3 \times 10^6 \, \frac{\text{proteins}}{\mu\text{m}^3}$$

(Avogadro's number)

organism	characteristic volume	number of proteins
E. coli	1 μm³	≈ 3 × 10⁶
budding yeast	≈ 30 μm³	≈ 100 × 10⁶
HeLa cell line	≈ 3000 μm³	≈ 10 × 10⁹

Estimate 2-7

by weighting the protein lengths by their abundance in the cell. This takes into account issues such as high abundance proteins, which tend to be smaller than low abundance proteins.

Moving on to the protein concentration in the cells, reports are surprisingly scarce, with old measured values for c_p being 0.24 g/mL for *E. coli* and 0.28 g/mL for budding yeast (BNID 105938, 108879, 108263, 108874). Values are expected to be similar when the concentration values refer to either the total cell volume and protein complement, including membrane-associated proteins, or solely to cytoplasmic volume and proteins. Assuming an average amino acid mass of 100 Da and using some unit conversions, we arrive at

$$\left(\frac{N}{V}\right)_{E.coli} = \frac{0.24 \frac{\text{g}}{\text{mL}} \times 6 \times 10^{23} \frac{\text{Da}}{\text{g}} \times 10^{-12} \frac{\text{mL}}{\mu\text{m}^3}}{300 \frac{aa}{\text{protein}} \times 100 \frac{\text{Da}}{aa}} \approx 5 \times 10^6 \frac{\text{proteins}}{\mu\text{m}^3} \quad (2.3)$$

and

$$\left(\frac{N}{V}\right)_{yeast} = \frac{0.28 \frac{\text{g}}{\text{mL}} \times 6 \times 10^{23} \frac{\text{Da}}{\text{g}} \times 10^{-12} \frac{\text{mL}}{\mu\text{m}^3}}{400 \frac{aa}{\text{protein}} \times 100 \frac{\text{Da}}{aa}} \approx 4 \times 10^6 \frac{\text{proteins}}{\mu\text{m}^3}. \quad (2.4)$$

(This calculation is also shown schematically with generic parameter values in Estimate 2-7.) Though this is what we aimed for, the reader might be wondering about the value of c_p that was used. We can derive it based on other better-known properties—namely, cell density, water content, and protein fraction of dry mass. The total cell density, d, is about 1.1 g/mL (BNID 103875, 102239, 106439). The water content, which we denote by w, is $\approx 70\%$ in *E. coli* and $\approx 60\%$ in budding yeast by mass (BNID 105482, 103689). The protein fraction of the dry mass, p, is $\approx 55\%$ in *E. coli* and $\approx 40\%$ in yeast. The relationship between these quantities is

$$c_p = d \times (1 - w) \times p. \tag{2.5}$$

Plugging in the numbers, we find that

$$c_{p,E.\ coli} = 1.1\ \text{g/mL} \times (1 - 0.7) \times 0.55 = 0.19\ \text{g/mL} \tag{2.6}$$

and

$$c_{p,\text{yeast}} = 1.1\ \text{g/mL} \times (1 - 0.6) \times 0.4 = 0.18\ \text{g/mL}. \tag{2.7}$$

The resulting values are smaller than those quoted above by 20–40% and lead to estimates of $\approx 3 \times 10^6$ protein/μm^3 and $\approx 2 \times 10^6$ protein/μm^3 in *E. coli* and budding yeast, respectively.

We can now use characteristic volumes to obtain the number of proteins per cell rather than per unit cell volume. For an *E. coli* cell of 1 μm^3 volume, there is not much that has to be done, because 1 μm^3 is our unit of cell volume, and the two estimates give a range of 2–4 million proteins per cell. For a budding yeast cell of 40 μm^3 (haploid; BNID 100430, 100427), the two estimates give a range of 90–140 million proteins per cell. Extrapolating these protein densities to mammalian cells, a value of about 10^{10} proteins per cell is predicted for characteristic cell lines that have average volumes of 2000–4000 μm^3.

How do these values compare to previous reports in the literature? **Table 2-5** shows a compilation of values based on published proteome-wide studies. Notably, in many cases a total sum over all proteins was not reported and was inferred for our purposes by summing all measured abundances. Some of the total sums are in line with the general estimates above, mostly those for bacteria. In contrast, many of the values for eukaryotic cells, including yeast and mammalian cells, are as much as 10-fold lower than predicted. Whether this seeming discrepancy is due to calibration issues in the mass spectrometry studies that measured them or inaccuracies in the parameter values used in the

reported proteins per cell	cell volume (μm³)	proteins per volume (10⁶/μm³)	mismatch from calculation
M. pneumonia			
0.05×10^6	0.015	3	< 2-fold
L. interrogans			
$1.0–1.2 \times 10^6$ *	0.22	5	< 2-fold
E. coli			
2.36×10^6	0.86	2.7	< 2-fold
B. subtilis			
2.3×10^6 *	1.13	2.0	< 2-fold
1.3×10^6 *	0.62	2.1	< 2-fold
1.8×10^6 *	0.85	2.1	< 2-fold
S. aureus			
0.35×10^6 *	0.33	1.1	≈ 3-fold
0.27×10^6 *	0.23	1.2	≈ 3-fold
0.26×10^6 *	0.23	1.1	≈ 3-fold
budding yeast (haploid)			
50×10^6	≈ 30–40	1–2	≈ 2-fold
47×10^6 *	≈ 30–40	1–2	≈ 2-fold
53×10^6	≈ 30–40	1–2	≈ 2-fold
fission yeast			
60.3×10^6	≈ 100	0.6	≈ 5-fold
M. musculus (NIH3T3 cells)			
3×10^9 *	≈ 2000	1.5	< 2-fold
H. Sapiens (U2OS)			
$0.95–1.7 \times 10^9$ *	≈ 4000	0.2–0.4	≈ 10-fold
H. sapiens (HeLa)			
2.0×10^9 *	≈ 2000	1	≈ 3-fold
2.3×10^9 *	≈ 2000	1	≈3-fold

*Value for total proteins per cell was not explicitly reported and is based on summing the abundance values as reported in the supplementary material across the proteome.

Table 2-5 A collection of estimates on the number of proteins per cell based on various sources. In some cases, the number is inferred from supplementary information and was not reported as such. When cell volume was not reported in the study, literature values under similar conditions were used. Detailed references and measurement methods are detailed in Table 1 of Milo R [2013] *Bioessays* 35:1050–1055.

estimate remains to be learned.** We take this as an indication that there is a standing challenge for careful analysis in order to achieve definitive

** Milo R (2013) What is the total number of protein molecules per cell volume? A call to rethink some published values. *Bioessays* 35:1050–1055 (doi: 10.1002/bies.201300066).

answers for those interested in quantitatively mapping the cell's contents.

What are the most abundant proteins in a cell?

Even after reading several textbooks on proteins, one may still be left wondering which of these critical molecular players in the life of a cell is the most quantitatively abundant. Many of the biochemical and regulatory pathways that make up the life of a cell have been or are now being mapped with exquisite detail, and many of the nodes have essential roles. But a wiring diagram does not a cell make. To really understand the relative rates of the various components of these pathways, we need to know about the abundances of the various proteins and their substrates. Further, if one is interested in assessing the biosynthetic burden of these various molecular players, the actual abundance is critical. Similarly, the many binding reactions that are the basis for much of the busy biochemical activity of cells, whether specific binding of intentional partners or spurious nonspecific binding between unnatural partners, is ultimately dictated by molecular counts.

We begin with a consideration of the molecular census of the carbon-fixing enzyme Rubisco, the molecular gatekeeper between the inorganic and the organic worlds. This key molecular player is required at extremely high concentrations. Let's see how much and why. As schematically depicted in **Estimate 2-8**, the photon flux under full sun illumination that can be used to excite photosynthesis has been measured to be about 2000 microeinstein/$(m^2 \times s)$. An einstein is a unit referring to one mole of photons. About 30% of this flux is maximally utilized, and beyond that there is saturation of the photosynthetic apparatus. About 10 photons are required to supply the energy and reducing power to fix one carbon atom. A Rubisco monomer has a mass of 60 kDa (BNID 105007) and works at a relatively sluggish maximal rate of $\approx 1-3$ reactions per second per catalytic site. Combining these facts, as done in Estimate 2-8, we find that the cell needs $\approx 1-3$ g/m^2. Let's estimate the total protein content in a leaf. A characteristic leaf has a height of about 300 microns. The dry mass occupies $\approx 10\%$ (BNID 107837, 110839), because there are big water-filled vacuoles that take up most of the leaf volume, while giving it a large area for light interception. So we arrive at about 30 g/m^2 of dry weight. If the soluble proteins are about one-third of the total dry mass, then this leads to about 10 g/m^2 (BNID 107837, 107403). Given the value above of 3 g/m^2 of Rubisco, we conclude that about one-third of the soluble protein mass needs to be Rubisco. Indeed, the experimental determinations in C3 plants, such as wheat, potato, and

Fraction of leaf-soluble protein occupied by Rubisco

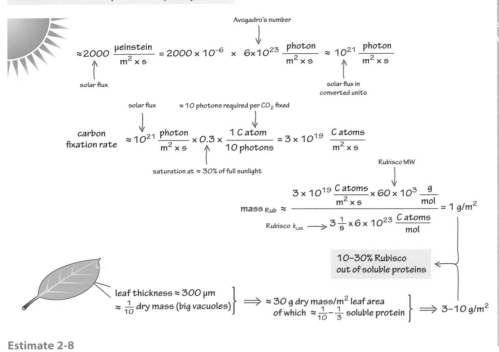

Estimate 2-8

tobacco, find that Rubisco constitutes in the range of 25–60% of all soluble proteins in leaf cells (BNID 101762).

What about other organisms? In the late 1970s, a unique catalog of the quantities of 140 proteins under different growth rates in *E. coli* was created using 2D gel electrophoresis and ^{14}C labeling (BNID 106195). Newer methods have recently enabled extensive protein-wide surveys of protein content using mass spectrometry, TAP labeling (BNID 101845), and fluorescent light microscopy (BNID 106257). A database (http://pax-db.org/) has been exploited to collect such data on protein abundances across organisms. Visualization of such data can be performed using Voronoi treemaps, as shown in **Figure 2-17** (for visualization of more datasets, see www.proteomaps.net). The picture emerging from these kinds of experiments shows several prominent players. Not surprisingly, ribosomal proteins and their ancillary components are highly abundant. The elongation factor EF-TU, responsible for mediating the entrance of tRNA to the free site of the ribosome, was characterized as the most abundant protein in the original 1978 catalog with a copy number of ≈60,000 proteins per bacterial genome. Values were given on a per genome basis, rather than per cell. Because the number of genome copies scales roughly as the cell volume, reporting values on a per genome basis corrects for the increase in

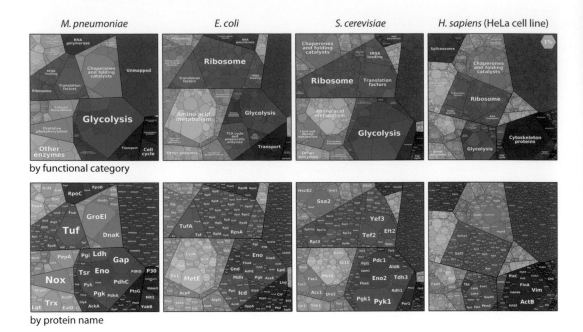

Figure 2-17 Proteomaps, a hierarchical presentation of the composition of a proteome using Voronoi treemaps. Each protein is associated with a polygon whose size is proportional to the abundance of that protein, thereby emphasizing highly expressed proteins. Functionally related proteins are placed in common subregions to show the functional makeup of a proteome at a glance. Four model cells are shown; the HeLa cell line was chosen for *H. Sapiens*. Upper row: depiction by functional category; lower row: depiction by protein name. The proteomes were measured under relatively rapid exponential growth. (Adapted from Liebermeister W, Noor E, Flamholz A et al. [2014] *Proc Natl Acad Sci USA* 111:8488–8493.)

cell size with growth rate. This absolute molecular count can be repackaged in concentration units using the rule of thumb that one molecule per bacterial cell volume is about 1 nM in concentration. Such a conversion leads to roughly a concentration of 100 μM for EF-Tu (BNID 104733). Recall that under different growth conditions, the cell size and thus total protein content can change several fold (see, for example, the vignette on yeast size, pg. 12), and this growth rate dependence of the protein census is especially important for ribosomal proteins.

Another contender for the title of most abundant protein is ACP, the acyl carrier protein, which plays an important role in fatty acid biosynthesis. This protein carries fatty acid chains as they are elongated. It is claimed to be the most abundant protein in *E. coli*, with about 60,000 molecules per cell (BNID 106194). However, in a recent high-throughput mass spectrometry measurement on minimal medium (BNID 104246), a value of ≈80,000 was reported, making it the third most abundant protein reported. The most abundant protein found in this particular survey of *E. coli* was RplL, a ribosomal protein (estimated at ≈110,000 copies per cell, which exists in four copies per ribosome, in contrast to other ribosomal proteins, which have only one copy per ribosome), and TufB (the elongation factor also known as EF-TU, estimated at ≈90,000 copies per cell). The next most abundant reported proteins are a component of the chaperone system, Gro-EL-Gro-ES, which is necessary for proper folding of many proteins, and GapA, a key enzyme in glycolysis.

Indeed, looking at a comparative functional view of protein abundance across several cell types, the proteins of glycolysis are the dominant fraction in the budding yeast (about one-quarter of the proteome in rich medium). Glycolysis serves as the backbone of energy and carbon metabolism, and the mass flux it carries is the largest in the cell.

Structural proteins can also be highly abundant. FimA is the major subunit of the 100–300 fimbria (pili) of *E. coli* (BNID 101473), which is used by sessile bacteria during their transition to stationary phase. Every pilus has about 1000 copies (BNID 100107), so a simple estimate leads us to expect hundreds of thousands of this repeating monomer on the outside of the cell. In vertebrate cells, actin, sometimes accounting for 5–10% of protein content, is often at the top of the list.

As noted above, protein content varies based on growth conditions and gene induction. For example, *LacZ*, the gene responsible for breaking lactose into glucose and galactose, is usually repressed and the protein has only a small number of copies (10–20, BNID 106200), but under full induction it was characterized to have a concentration of 50 μM (BNID 100735)—that is, about 50,000 copies per cell.

If we looked at the sum total over all organisms, what would be the most abundant protein on Earth? This title is usually ascribed to Rubisco. Indeed, it carries out the task of fixing carbon, which is done on a massive scale across the planet and supports all actions of the biosphere. Since starting this book, we have had second thoughts. In a paper we wrote,[††] we tried to give a sense of the ubiquity of Rubisco by normalizing it on a per-person basis. This gave about 5 kg of Rubisco protein per-person (though clearly Rubisco, though supporting our diets, is not a human protein). In several reports, collagen, a connective tissue protein that is localized extracellularly, was found to account for about 30% of the protein mass in humans (BNID 109730, 109731). In a 70-kg human, with two-thirds water and half of the rest protein, this gives about 10 kg total protein, suggesting as much as 3 kg of collagen. That might be a somewhat inflated value, but collagen is not only in humans. What is the

[††] Phillips R & Milo R (2009) A feeling for the numbers in biology. *Proc Natl Acad Sci USA* 106:21465–21471 (doi: 10.1073/pnas.0907732106).

largest biomass of animals on Earth? It is actually our livestock in the form of cows, pigs, poultry, and so on, at a total mass of about 100 kg per person (BNID 111482; more than 20 times the mass of all wild land mammals!). Livestock have a similar collagen concentration to humans (BNID 109821), so these numbers suggest that collagen should displace Rubisco as the titleholder for the most abundant protein on Earth. Even for the title in the category of catalytic proteins, rather than "boring" structural proteins, the race is still open. Given the immense mass of bacteria on Earth and the accumulating proof from proteomics and metagenomics for the ubiquity of glycolytic proteins, they are also prime contenders for the title of the most abundant protein.

How much cell-to-cell variability exists in protein expression?

It is tempting to discuss the absolute numbers or concentrations of expressed proteins within cells by assigning a single value, as opposed to speaking about distributions. Many methods for the measurement of protein quantity, such as measuring fluorescence using a spectrophotometer, supply only a single number that is an average over an entire population of cells. With the advent of quantitative microscopy and flow cytometry, both of which relied on the discovery of green fluorescent protein (GFP), the role of variability has also moved to center stage. Functional roles for variability have already been shown in processes, such as environmental responses, where differences from one cell to the next effectively implement bet-hedging, permitting some subset of a population to best adapt to some environmental insult. Yet the full implications and importance for the lifestyles of various organisms is still a hot area of research.

If we perform an experiment in which single-cell microscopy is used to query the fluorescence in thousands of different cells, as exemplified in **Figure 2-18**, a first stage in representing the data is to plot the distribution. **Figure 2-19** gives an example of such a distribution for the case of mRNAs. Many biological quantities display the log-normal distribution, where the characteristic bell-shaped distribution is achieved when plotting the histogram in log scale. Different underlying mechanisms can result in such a distribution.[‡‡] For example, a first-order kinetic parameter that is normally distributed and appears in the exponent of an autocatalytic growth process

---

[‡‡] Koch AL (1966) The logarithm in biology. 1. Mechanisms generating the log-normal distribution exactly. *J Theor Biol* 12:276–290.

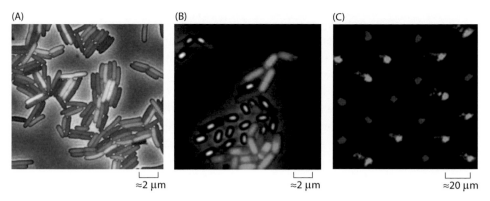

Figure 2-18 Examples of cell-to-cell variability in gene expression. (A) *E. coli* cells with identical promoters, resulting in the production of fluorescent proteins with different colors. Noise results in a different relative proportion of red and green protein in each cell. (B) *B. subtilis* cells that are genetically identical adopt different fates, despite the fact that they are subjected to identical conditions. The green cells are growing vegetatively, the white cells have sporulated, and the red cells are in the "competent" state. (C) Drosophila retina revealing different pigments, as revealed by staining photoreceptors with antibodies to different photopigments. The green-sensitive photopigment Rh6 is in green and the blue-sensitive photopigment Rh5 is in blue. (A, and B, adapted from Eldar A & Elowitz MB [2010] *Nature* 467:167–173. C, adapted from Losick R & Desplan C [2008] *Science* 320:65–68.)

will lead to a log-normal distribution. Alternatively, any characteristic that is the result of the multiplication of many other random processes is expected to be log-normally distributed due to the central limit theorem. A take-home lesson is that one has to be very careful in making claims about the mechanism that gives rise to a given distribution. The reason is

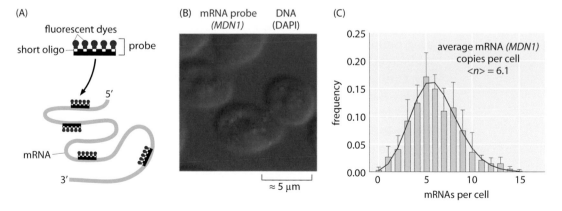

Figure 2-19 Measuring single-cell variability of mRNA levels in budding yeast. (A) Cartoon showing how probes are designed to target different regions of an mRNA copies molecule of interest. (B) Fluorescence microscopy image of yeast cells revealing the number of mRNA copies per cell. (C) Histogram showing the number of mRNAs per cell for a particular gene (MDN1) of interest in yeast. (Adapted from Zenklusen D, Larson DR & Singer RH [2008] *Nat Struct Mol Biol* 15:1263–1271.)

that often many different mechanisms can lead to the same generic distribution. Usually the next stage in characterization and data reduction is to calculate the statistics of a distribution—usually the mean and standard deviation. The level of variability in the population is usually given in terms of the coefficient of variation, the CV, which equals the ratio of the standard deviation to the mean. Alternatively, the Fano factor is the ratio of the variance (that is, the standard deviation squared) to the mean. This is of interest since it is known that for processes of a general form known as a Poisson process, the variance is predicted to be equal to the mean (a Fano factor equal to 1), thus serving as a baseline expectation on the kind of noise that might be found for some promoters.

What is known about the actual levels of cell–cell variation in protein expression? Measurements using fluorescent proteins have been the main tool for answering this question. Figure 2-18 shows how two-color experiments visually reveal the disparities in expression in bacteria. In this case, the *lacI* promoter was used to drive the expression of YFP and CFP genes integrated at opposing locations along the circular *E. coli* genome. In quantifying this variability, we first have to note the approximately twofold change in size and content through the cell cycle. This is often corrected for by calculating a value normalized to the cell size. The amount of variability was quantified as having a characteristic CV for bacteria of ≈0.4 (BNID 107859), which could be further broken down into differences among cells and differences within a cell among identical promoters.

In human cells, similar measurements were undertaken with the CV values for a set of 20 proteins measured during the cell cycle. It was found that the CV was quite stable throughout the cell cycle, while among proteins the values ranged from 0.1 to 0.3 (BNID 107860). As a rule of thumb, a log-normal distribution with a CV of ≈0.3 will have a ratio of ≈2 between the cells at the 90th percentile and the 10th percentile of expression intensity. One can go beyond the static "snapshot" level of variation to ask how quickly there is mixing within the population in which a cell that was a relatively low expresser becomes one of the high expressers, as shown in **Figure 2-20**. Measuring such dynamics is based on time-lapse microscopy, and the mixing time or memory time scale is quantified by the autocorrelation function that measures the average correlation between the levels at times t and $t + \tau$, where τ denotes the time difference between the measurements. For protein levels in human cells, the memory time— the interval at which half of the correlation was lost—was between one and three generation times (BNID 108977, 107864), with some proteins mixing faster and others more slowly. Proteins with long mixing times can cause epigenetic behavior, where cells with identical genetic makeup respond differently, for example, to chemotherapy treatment.

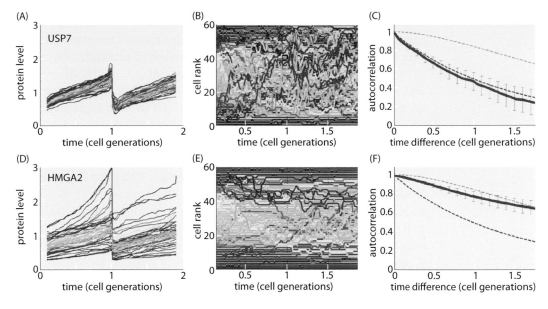

Figure 2-20 Variability and memory of protein levels in human cells. Different proteins have different levels of variability as well as differing rates of mixing within the population range. (A), (D) Time courses of fluorescent reporter levels indicating the levels of a protein over two cell cycles and showing the degree of variability among cells from the same cell line. The protein in the upper panel (USP7) is much less variable than the protein in the lower panel (HMGA2). (B), (E) Cells are ranked by level of expression of the tagged protein, and their dynamics over time is made clear using a color code based on their level at the beginning of the first cell cycle. (C), (F) The rate of mixing of the protein levels within the cell population quantified by the autocorrelation function of protein levels as a function of time difference. Mixing times range from about one cell cycle to over two cell cycles. (Adapted from Sigal A, Milo R, Cohen A et al. [2006] *Nature* 444:643–646.)

What are the concentrations of cytoskeletal molecules?

Just as there is a battery of macromolecules that participate in the flow of information between proteins and DNA, there is also a wide collection of different molecules that dictate when and where the molecules of the cytoskeleton will be assembled into the filamentous networks that criss-cross cells. When thinking about the question of cell motility, leading this cast of molecular players is the protein actin, a soluble protein with a run-of-the-mill ≈40-kDa mass, but which forms rigid filamentous assemblies with long persistence lengths of about 10 μm (BNID 106830) that are crucial for propelling cells forward.

As shown in the vignette on How big are the cell's filaments? (pp. 57), the leading edge of a motile cell (such as a keratocyte) is characterized by a dense and branched network of actin filaments, which create protrusions (such as filopodia and lamellipodia). These protrusions are peppered with sites of adhesion between the cell and external solid substrate. These sites of adhesion have a characteristic diameter of 100–300 nm (BNID 102267) and an average lifetime of 20 s (BNID 102266). They serve as anchors for the mesmerizing cellular dynamics revealed in time-lapse images of motile cells crawling on surfaces.

How much actin does it take to set up such a network? Similarly, how many attendant proteins are there to make sure that such filaments are "constructed" at the right time and place? One way to begin to answer such questions is through simple estimates based upon inspecting electron microscopy images of typical filaments at the leading edge of motile cells. Since the size of a typical monomer is roughly 5 nm and the filaments themselves are characterized by micron-scale lengths, each filament is made up of hundreds of actin monomers. Though electron microscopy images provide a compelling structural vision of the leading edge of a motile cell (see figure 1-48), they leave us wondering about the host of other molecular partners that control the spatiotemporal patterns of filament formation. Other methods (and cell types) have been used to take the molecular roll call of the many proteins implicated in cytoskeletal network formation.

A powerful model system for investigating questions about the dynamics of the actin cytoskeleton is provided by the fission yeast, *Schizosaccharomyces pombe*. One of the reasons that these eukaryotic cells are so helpful is that their uses of actin are centered on the formation of three specialized classes of structures, as shown in **Figure 2-21**. The first class of actin structure is that associated with intracellular transport, a signature feature of eukaryotes, and in fission yeast, it is the cargo-carrying molecular motors that move along this network of actin filaments that mediate this process. A second of the primary functions of the actin cytoskeleton is to mediate the fission process, whereby one mother cell divides into two daughters through the formation of a contractile ring at the cell middle. Finally, actin is a key player in the endocytosis process, where the formation of dense actin patches provides part of the force-generating machinery that makes membrane invaginations possible. These fission yeast cells were used to take a careful census of the actin cytoskeleton that provides the absolute numbers and concentrations of both actin and its accessory proteins.

To get a sense of the number of molecular copies of the cytoskeletal proteins and their various accessory proteins, the systematic fusion of fluorescent proteins to each and every actin-related protein and the calibration of the fluorescence signal using antibody techniques permitted a direct measurement of protein copy numbers, as shown in **Figure 2-22**. Specifically, by

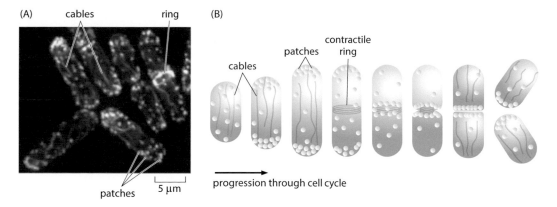

Figure 2-21 The actin cytoskeleton in fission yeast. (A) Fluorescence microscopy image of the various actin structures found in the fission yeast. (B) Schematic of the time variation of the distribution of actin over the cell cycle. During the cell division process, actin normally invested in patches and cables is retasked to form the contractile ring. (Adapted from Kovar DR, Sirotkin V & Lord M [2011] *Trends Cell Biol* 21:177–187.)

measuring overall fluorescence levels and then exploiting calibration factors to convert intensities into molecular counts, it was possible to determine the molecular census for an entire suite of actin-related proteins. As reported in **Table 2-6**, the numbers per cell range from just over 1 million copies of actin monomers per cell (about 1% of the proteome, making it one

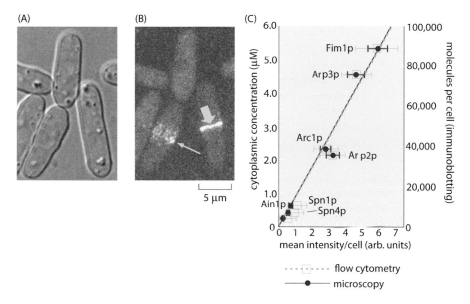

Figure 2-22 Molecular census of the actin cytoskeleton in fission yeast. (A) Phase contrast images of fission yeast cells. (B) Fluorescence images of myosin inside the yeast cells. (C) Calibration of the census. The number of molecules per cell, as determined from immunoblotting, shows a linear relation with the average fluorescence per cell. (Adapted from Wu JQ & Pollard TD [2005] *Science* 310:310–314.)

protein	cytoplasmic concentration (μM)	copies per cell (volume of 92 μm³)	percent concentrated at actin patch/spindle pole/cell division site
actin Act1p	63±11	1,400,000	>13
actin patch proteins			
Arp2	2.9	47,000	10
Arp3	4.1	67,000	7
ARPC1	2.5	40,000	15
ARPC3	2.4	39,000	12
ARPC5	1.9	31,000	12
capping protein Acp2p	1.2	19,000	17
fimbrin Fim1p	5.3	87,000	15
spindle pole body proteins			
SPB protein Sad1p	0.2	3300	31
polo kinase Plo1p	0.3	6600	1
SIN kinase Cdc7p	0.2	4000	5
cytokinesis proteins			
anillin-like Mid1p	0.09	2100	40
myosin-II Myo2p *kan*	0.5	7300	27
myosin-II ELC Cdc4p	4.8	77,000	22
myosin-II RLC Rlc1p	0.6	9600	18
IQGAP Rng2p *kan*	0.2	2700	35
mYFP-Cdc15p *kan*	2.1	36,000	21
formin Cdc12pII	0.04	600	11
UCS protein Rng3pII	0.1	1900	3
Rng3p in *myo2-E1*II	0.3	6800	30
alpha-actinin Ain1p	0.2	3600	8
myosin-II Myp2p	0.4	6100	21
septin Spn1p	0.6	10,000	35
septin Spn4p	0.5	8100	34
anillin-like Mid2p	0.1	1800	NA
protein kinase C Pck2p	0.3	4300	13
Rho GEF Rgf1p	0.3	4300	5
Rho GEF Rgf3p	0.2	3200	4
chitin sythase Ch2p	0.1	2100	3

Table 2-6 Concentrations of actin and actin-related proteins in *S. pombe*. Values are rounded to two significant digits. (Adapted from Wu JQ & Pollard TD [2005] *Science* 310:310–314.)

of the most abundant in the cell; see also the vignette entitled "What are the most abundant proteins in a cell?"; pg. 108) to somewhat less than 1000 copies of the actin filament capping protein formin.

What are the host of different actin-related proteins all for? One of the hallmark features of "living matter" is the exquisite control that is exercised over cellular processes. That is, most biological processes only happen when and where they are supposed to. In the case of actin polymerization, there is a battery of control proteins for coordinating the actin polymerization process. These include proteins that cap monomers (thus forbidding them from participating in filament formation), proteins that communicate with membrane lipids that tell actin to form filaments near these membranes in order to form the protrusions at the leading edge, proteins that bind to preexisting filaments and serve as branching sites to send off new filaments in a different direction, and so forth. As seen in the table, there are more than 50 such proteins, and they occur with different concentrations covering a range from tens of nM to several µM.

How can we rationalize the numbers detailed in Table 2-6? One of the immediate impressions that comes from inspection of the data is that there are in some cases orders of magnitude differences in the quantities of different proteins. For example, while there are in excess of a million actin monomers, there are only roughly 50,000 copies of the protein complex that regulates the actin cytoskeleton, Arp2/3, and only 600 copies of the regulatory protein formin. These numbers make intuitive sense, because a given filament might only be decorated by one Arp2/3 complex or formin dimer. These abundances might be further reasoned out by imagining several different categories of molecules. First, it is not surprising that actin is in a category by itself since it is the fundamental building block for constructing the long filaments involving tens to thousands of monomers each. The second category of molecules includes those that are required at a stoichiometry of one or a few per filament or patch, such as the capping and branching proteins. These would be expected to be found with tens of thousands of molecules per cell, as we discuss below. Finally, we might expect that regulatory proteins could be found in quantities of less than one copy per filament. Referring to Table 2-6, we see that most factors, such as motor components (myosin) or branching molecules (Arp), come with copy numbers in the many thousands, while regulatory proteins (kinases) are in the few thousands. To estimate the number of filaments and monomers, it is useful to think of the interphase stage of the cell cycle, when much of the actin is tied up in the formation of several hundred actin patches distributed across the cell, with each such patch containing more than 100 small filaments built up from 10–100 monomers. To construct all of these patches requires more than 500,000 actin monomers, corresponding to nearly half of the pool of utilized monomers. During mitosis, this balance is shifted, since at this stage in the cell cycle nearly half of the actin is now invested in constructing the contractile ring at the center of the cell. This ring is responsible for pinching the two daughter cells apart. The actin invested in the construction of this ring can be reasoned out by noting that there are roughly 2000 filaments making up these rings, with each such ring roughly 0.5 µm in length, implying that hundreds of thousands of monomers are implicated in the formation of these rings.

These are only several examples of the rich and complex cytoskeletal architectures found in living cells. As can be seen during cell division for eukaryotic cells, there is also an equally fascinating network of microtubule filaments that are key to separating the newly formed chromosome copies into the daughter cells. Microtubules also form molecular highways on which traffic is shuttled around by cargo-carrying molecular motors. Similar rationale might be provided for the microtubule-related census, though current experimental attempts to characterize the microtubule cytoskeleton lag behind efforts on the actin-based system.

All told, the cytoskeleton is one of the most critical features of cellular life, and just as we need to know about the concentration of transcription factors to understand how they regulate genetic decision making, the concentrations of cytoskeletal proteins and their accessory factors is critical to developing a sense of the highly orchestrated dynamics found in cells.

How many mRNAs are in a cell?

Given the central place of gene regulation in all domains of biology, there is great interest in determining the census of mRNA in cells of various types. We are interested both in specific genes and in the entire transcriptome as a function of environmental conditions and developmental stage. Such measurements provide a direct readout of the instantaneous regulatory state of the cell at a given time and, as such, give us a powerful tool to analyze how cellular decision making is implemented. We begin with an exercise of the imagination to see if we can use a few key cellular facts to estimate the number of mRNAs. Given our knowledge of the kinetics of the processes of the central dogma during one cell cycle, the number of mRNA molecules per cell can be worked out as shown in **Estimate 2-9**. The essence of the estimate is to recognize that over the course of the cell cycle, the number of proteins must be doubled through protein synthesis. This protein synthesis is based, in turn, on the distribution of mRNA molecules that are present in the cell. As shown in this back-of-the-envelope calculation, we can derive an estimate for rapidly dividing cells of 10^3–10^4 mRNA molecules per bacterial cell and 10^5–10^6 mRNA molecules per the 3000 μm^3 characteristic size of a mammalian cell.

Modern techniques have now largely superseded those leading to the classic census numbers. One approach is oriented towards "genome-wide" measurements, in which an attempt is made to size up the number

How many mRNA molecules are in a cell?

We start with the characteristic protein concentration in cells:

$$\frac{N_{protein}}{V} \approx 3 \times 10^6 \text{ proteins/}\mu m^3$$

rate of protein production per cell

$$R = \frac{N_{protein}}{\tau}$$

doubling time

bacteria
$V \approx 1 \mu m^3$, $\tau \approx 1\,h$

$$\approx \frac{3 \times 10^6 \text{ proteins}}{3000\,s} \approx 10^3 \text{ proteins/s}$$

mammalian cell
$V \approx 3000\,\mu m^3$, $\tau \approx 24\,h$

$$\approx \frac{10^{10} \text{ proteins}}{10^5\,s} \approx 10^5 \text{ proteins/s}$$

rate of translation
$v \approx 10\text{–}50 \text{ bases/s}$

rate of protein production per mRNA

$$r = \frac{v}{d} \approx 0.1\text{–}1 \text{ protein/mRNA/s}$$

distance between ribosomes $\to d \approx 30\text{–}300 \text{ bases}$

ribosomes touching

few ribosomes per mRNA (supported by observations)

$$N_{mRNA} = \frac{R}{r}$$

bacteria
$$\approx \frac{10^3 \text{ proteins/s}}{0.1\text{–}1 \text{ protein/mRNA/s}} \approx 10^3\text{–}10^4 \text{ mRNA/bacterial cell}$$

mammalian cell
$$\approx \frac{10^5 \text{ proteins/s}}{0.1\text{–}1 \text{ protein/mRNA/s}} \approx 10^5\text{–}10^6 \text{ mRNA/mammalian cell}$$

Estimate 2-9

of mRNAs across the entire transcriptome. These results are based upon the method of RNA-Seq, where individual mRNAs are sequenced from the cell lysate. Since the count is based on the frequency of sequence reads, it requires calibration. To that end, the sample is spiked with mRNA standards whose quantities are known prior to sequencing. An example of such a result is shown in **Figure 2-23**, which reports on the distribution of mRNAs in *E. coli* grown under both rich- and minimal-media conditions. The result of this study is that the number of transcripts per cell (≈ 8000 mRNA copies/cell) for cells grown in LB media is roughly threefold larger than the number of transcripts per cell (≈ 3000 mRNA copies/cell) for cells grown in minimal media. Given that the number of genes is in excess of 4000, this implies that the mean copy number in LB is on the order of one per cell. For most genes the number of copies is even lower, meaning that in most cells there is not even a single copy, while a few cells have one or two copies. We find this striking fact to be a beautiful example of how biological numeracy informs our intuition.

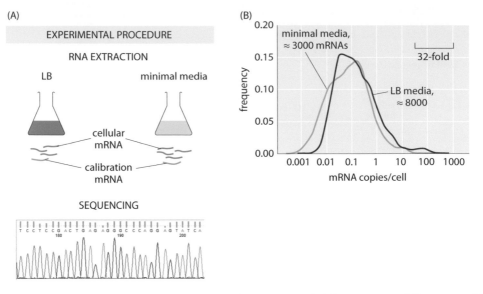

Figure 2-23 Using sequencing to find the number of mRNAs per cell. (A) mRNA is carefully extracted from cells and mixed with the synthesized mRNA that serves for calibration. Deep sequencing enables counting the number of copies of each mRNA type and from this the total number of mRNAs can be inferred. (B) mRNA counts for *E. coli* grown in rich media and minimal media. (Data courtesy of Zoya Ignatova.)

Reproducibility is one of the most important issues at the center of the biological numeracy called for in our book, especially when different methods are brought to bear on the same problem. A very useful alternative for taking the mRNA census is built around direct counting by looking at the individual mRNA molecules under a microscope. Specifically, a gene-by-gene decomposition using techniques, such as single-molecule fluorescence in-situ hybridization (FISH), complements the RNA-Seq perspective described above. FISH provides a window onto the mRNA spatial and cell-to-cell distribution for a particular species of mRNA molecule, as shown here in **Figure 2-24** and as shown in Figure 2-23 of the vignette entitled "How much cell-to-cell variability exists in protein expression?" (pg. 112). The idea in this case is to design probes that bind to the mRNA of interest through complementary base pairing. Each such probe harbors a fluorophore and, hence, when the fixed cells are examined in the microscope, the intensity of the fluorescence of these probes is used as a readout of the number of mRNAs. As seen in Figure 2-24B, the number of mRNAs per cell is generally between 0.1 and 1, with several outliers having both smaller and larger mRNA counts. As has also been emphasized throughout the book, different conditions result in different numbers, and the FISH results recreated here correspond to slow growth.

What about the mRNA census in other cell types besides bacteria? Again, both sequencing methods and microscopy have been brought to bear

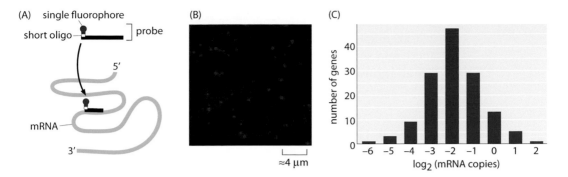

Figure 2-24 Fluorescence microscopy approach to taking the mRNA census. Using fluorescence in-situ hybridization (FISH), it is possible to use fluorescence intensity of specific probes that hybridize to an mRNA of interest to count these mRNA. (A) Schematic of the single-molecule probes used to label mRNA. (B) Fluorescence image of a field of *E. coli* cells with the mRNA for a specific gene imaged. (C) Histogram of the mean number of mRNA in *E. coli* for a number of genes. (Adapted from Taniguchi Y, Choi PJ, Li GW et al. [2010] *Science* 329:533–538.)

on these questions in yeast and other eukaryotes. **Figure 2-25** presents results for the mRNA census in both budding and fission yeast. The total number of mRNA per cell is in the range of 20,000–60,000 in exponentially growing budding and fission yeast (BNID 104312, 102988, 103023, 106226, 106763). As with our earlier results for bacteria, here too we find that each gene generally only has a few mRNA molecules present in the cell at any one time. The vignette entitled "What is the protein-to-mRNA ratio in cells?" (pg. 124) provides a window into the amplification factor that attends a given mRNA as it is turned into the proteins of the cell in the

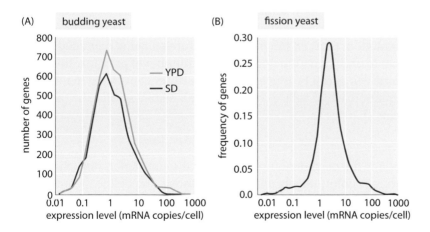

Figure 2-25 mRNA distributions in yeast. (A) mRNA distribution in budding yeast grown in rich media (YPD) and minimal media (SD) as measured using PCR. (B) mRNA distribution in fission yeast measured using RNA-Seq. Total number is estimated to be ~40,000. (A, adapted from Miura F, Kawaguchi N, Yoshida M et al. [2008] *BMC Genomics* 9:574. B, adapted from Marguerat S, Schmidt A, Codlin S et al. [2012] *Cell* 151:671–683.)

process of translation. For "typical" mammalian cells, a quoted value of 200,000 mRNA per cell (BNID 109916) is in line with our simple estimate above and shows that scaling the number of mRNA proportionally with size and growth rate seems to be a useful first guess.

What is the protein-to-mRNA ratio?

The central dogma hinges on the existence and the properties of an army of mRNA molecules that are transiently brought into existence in the process of transcription and often, shortly thereafter, degraded away. During the short time that they are found in a cell, these mRNAs serve as a template for the creation of a new generation of proteins. The question posed in this vignette is this: On average, what is the ratio of the translated message to the message itself?

Though many factors control the protein-to-mRNA ratio, the simplest model points to an estimate in terms of just a few key rates. To see that, we need to write a simple "rate equation" that tells us how the protein content will change in a very small increment of time. More precisely, we seek the functional dependence between the number of protein copies of a gene (p) and the number of mRNA molecules (m) that engender it. The rate of formation of p is equal to the rate of translation times the number of messages, m, since each mRNA molecule can itself be thought of as a protein source. However, at the same time new proteins are being synthesized, protein degradation is steadily taking proteins out of circulation. Further, the number of proteins being degraded is equal to the rate of degradation times the total number of proteins. These cumbersome words can be much more elegantly encapsulated in the following equation, which tells us how in a small instant of time the number of proteins changes as

$$p(t + \Delta t) = p(t) - \alpha p(t)\Delta t + \beta m \Delta t \tag{2.8}$$

where α is the degradation rate and β is the translation rate (though the literature is unfortunately torn between those who define the notation in this manner and those who use the letters with exactly the opposite meaning).

We are interested in the steady-state solution—that is, in what happens after a sufficiently long time has passed and the system is no longer changing. In that case,

$$dp/dt = 0 = \beta m - \alpha p. \tag{2.9}$$

This tells us, in turn, that the protein-to-mRNA ratio is given by

$$p/m = \beta/\alpha. \qquad (2.10)$$

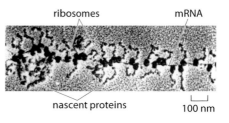

Figure 2-26 Ribosomes on mRNA as beads on a string. (Courtesy of Donald Fawcett/Visuals Unlimited, Inc.)

This is not the same as the number of proteins produced from each mRNA, because this value requires us to also know the mRNA turnover rate, which we take up at the end of the vignette. What is the value of β? A rapidly translated mRNA will have ribosomes decorating it like beads on a string, as captured in the classic electron micrograph shown in **Figure 2-26**. Their distance from one another along the mRNA is at least the size of the physical footprint of a ribosome (≈ 20 nm; BNID 102320, 105000), which is the length of about 60 base pairs (length of nucleotide ≈ 0.3 nm; BNID 103777), equivalent to ≈ 20 aa. The rate of translation is about 20 aa/s. It thus takes at least one second for a ribosome to proceed along the mRNA a distance equal to its own physical footprint on the mRNA, implying a maximal overall translation rate of $\beta = 1$ s^{-1} per transcript.

The effective degradation rate arises not only from the degradation of proteins, but also from a dilution effect as the cell grows. Indeed, of the two effects, often the cell division dilution effect is dominant, and hence the overall effective degradation time, which takes into account the dilution, is about the time interval of a cell cycle, τ. We thus have

$$\alpha = 1/\tau. \qquad (2.11)$$

In light of these numbers, the ratio p/m is therefore 1 s^{-1}/(1/τ) = τ. For *E. coli*, τ is roughly 1000 s, so $p/m \sim 1000$. If the mRNAs are not transcribed at the maximal rate, the ratio will be smaller. Let's perform a sanity check on this result. Under exponential growth at medium growth rate, *E. coli* is known to contain about 3 million proteins and 3000 mRNA (BNID 100088, 100064). These constants imply that the protein-to-mRNA ratio is ≈ 1000, precisely in line with the estimate given above. We can perform a second sanity check based on information from previous vignettes. In the vignette entitled "Which is bigger, mRNA or the protein it codes for?" (pg. 43), we derived a mass ratio of about 10:1 for mRNA to the proteins they code for. In the vignette entitled "What is the macromolecular composition of the cell?" (pg. 128), we mentioned that protein is about 50% of the dry mass in *E. coli* cells, while mRNA is only about 5% of the total RNA in the cell, which is itself roughly 20% of the dry mass. This implies that mRNA is about 1% of the overall dry mass. So the ratio of mRNA to protein should be about 50 times 10, or 500 to 1. From our point of view, all of these sanity checks hold together very nicely.

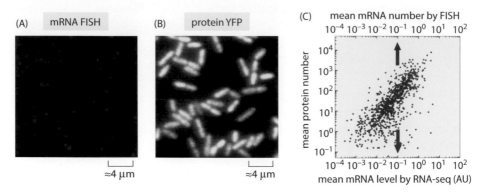

Figure 2-27 Simultaneous measurement of mRNA and protein in *E. coli*. (A) Microscopy images of mRNA level in *E. coli* cells. (B) Microscopy images of protein in *E. coli* cells. (C) Protein copy number vs. mRNA levels as obtained using both microscopy methods like those shown in part (A) and using sequencing-based methods. (Adapted from Taniguchi Y, Choi PJ, Li GW et al. [2010] *Science* 329:533–538.)

Experimentally, how are these numbers on protein-to-mRNA ratios determined? One elegant method is to use fluorescence microscopy to simultaneously observe mRNAs using FISH and their protein products, which have been fused to a fluorescent protein. **Figure 2-27** shows microscopy images of both the mRNA and the corresponding translated fusion protein for one particular gene in *E. coli*. Figure 2-27C shows results using these methods for multiple genes and confirms a 100- to 1000-fold excess of protein copy numbers over their corresponding mRNAs. As seen in that figure, not only is direct visualization by microscopy useful, but sequence-based methods have been invoked as well.

For slower-growing organisms, such as yeast or mammalian cells, we expect a larger ratio, with the caveat that our assumptions about maximal translation rate are becoming ever more tenuous, and with that, our confidence in the estimate decreases. For yeast under medium to fast growth rates, the number of mRNA was reported to be in the range of 10,000–60,000 per cell (BNID 104312, 102988, 103023, 106226, 106763). Because yeast cells are ≈ 50 times larger in volume than *E. coli*, the number of proteins can be estimated as larger by that proportion, or 200 million. The ratio p/m is then $\approx 2 \times 10^8 / 2 \times 10^4 \approx 10^4$, in line with the experimental value of about 5000 (BNID 104185, 104745). For yeast dividing every 100 minutes, this is on the order of the number of seconds in its generation time, in agreement with our crude estimate above.

As with many of the quantities described throughout the book, the high-throughput, genome-wide craze has hit the subject of this vignette as well. Specifically, using a combination of RNA-Seq to determine the mRNA copy numbers, and mass spectrometry methods and ribosomal profiling

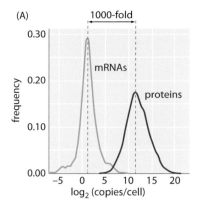
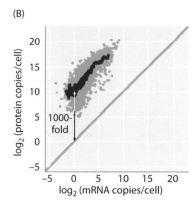

Figure 2-28 Protein-to-mRNA ratio in fission yeast. (A) Histogram illustrating the number of mRNA and protein copies as determined using sequencing methods and mass spectrometry, respectively. (B) Plot of protein abundance and mRNA abundance on a gene-by-gene basis. Recent analysis suggests that the protein levels have been underestimated and a correction factor of about a fivefold increase should be applied, thus making the ratio of protein to mRNA closer to 10^4. (Adapted from Marguerat S, Schmidt A, Codlin S et al. [2012] *Cell* 151:671–683. Recent analysis from Milo R [2013] *Bioessays* 35:1050–1055.)

to infer the protein content of cells, it is possible to go beyond the specific gene-by-gene estimates and measurements described above. As shown in **Figure 2-28** for fission yeast, the genome-wide distribution of mRNA and protein confirms the estimates provided above, showing more than a thousandfold excess of protein to mRNA in most cases. Similarly, in mammalian cell lines, a protein-to-mRNA ratio of about 10^4 is inferred (BNID 110236).

So far, we have focused on the total number of protein copies per mRNA and not on the number of proteins produced per production burst occurring from a given mRNA. This so-called burst size measurement is depicted in **Figure 2-29**, showing for the protein β-galactosidase in *E. coli* that the distribution of observed burst sizes quickly decreases from the common handful to much fewer cases of more than 10.

Finally, we note that there is a third meaning to the question that entitles this vignette—namely, how many proteins are made from an individual mRNA before it is degraded? For example, in fast-growing *E. coli*, mRNAs are degraded roughly every three minutes, as discussed in the vignette entitled "How fast do RNAs and proteins degrade?" (pg. 244). This time scale is some 10–100 times shorter than the cell cycle time. As a result, to move from the statement that the protein-to-mRNA ratio is typically 1000 to the number of proteins produced from an mRNA before it is degraded, we need to divide the number of mRNA lifetimes per cell cycle. We find that in this rapidly dividing *E. coli* scenario, each mRNA gives rise to about 10–100 proteins before being degraded.

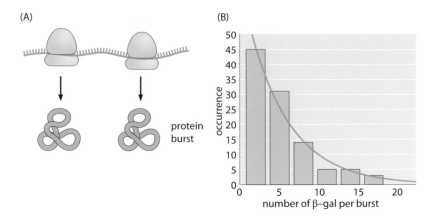

Figure 2-29 Dynamics of protein production. (A) Bursts in protein production resulting from multiple rounds of translation on the same mRNA molecule before it decays. (B) Distribution of burst sizes for the protein beta-galactosidase in *E. coli*. (Adapted from Cai L, Friedman N & Xie XS [2006] *Nature* 440:358–362.)

A recent study[§§] suggests revisiting the basic question of this vignette. Careful analysis of many studies on mRNA and protein levels in budding yeast, the most common model organism for such studies, suggests a non-linear relationship, where genes with high mRNA levels will have a higher protein-to-mRNA ratio than mRNAs expressed with low copy numbers. This suggests the correlation between mRNA and protein does *not* have a slope of 1 in log–log scale, but rather a slope of about 1.6, which also explains why the dynamic range of proteins is significantly bigger than that of mRNA.

What is the macromolecular composition of the cell?

Molecular biology aims to explain cellular processes in terms of the individual molecular players, resulting in starring roles for certain specific proteins, RNAs, and lipids. By way of contrast, a more holistic view of the whole cell or organism was historically the purview of physiology. Recently, the latter integrative view has been adopted by systems biology, which completes the circle by returning with the hard-won mechanistic knowledge from molecular biology to a holistic view of the molecular

[§§] Csárdi G, Franks A, Choi DS et al. (2015) Accounting for experimental noise reveals that mRNA levels, amplified by post-transcriptional processes, largely determine steady-state protein levels in yeast. *PLoS Genet* 11:e1005206 (doi: 10.1371/journal.pgen.1005206).

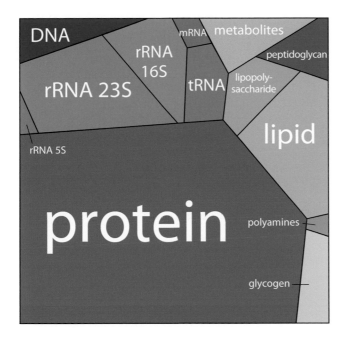

Figure 2-30 A Voronoi tree diagram of the composition of an *E. coli* cell growing with a doubling time of 40 min. Each polygon area represents the relative fraction of the corresponding constituent in the cell dry mass. Colors are associated with each polygon such that components with related functional roles have similar tints. The Voronoi tree diagram visualization method was developed in order to represent whole genome measurements from microarrays or proteome quantitation.

interlinkages that give rise to whole-cell behavior. A critical starting point for thinking globally about the cell is to understand the relative abundance of its different constituents.

Such a bird's-eye view of the composition of the cell is given in **Figure 2-30** for the case of *E. coli* during exponential growth with a doubling time of 40 minutes. Part of the figure is dominated by the usual suspects, with proteins making up just over half of the cellular content. More surprisingly, despite their critical role as gatekeepers of gene expression, mRNAs constitute only a small fraction when analyzed in terms of absolute mass, comprising only about 1% of the dry mass. The figure is based on a compilation of information determined for the cell composition of an *E. coli* recreated in **Table 2-7** (BNID 104954). This compilation first appeared in the classic textbook, *Physiology of the Bacterial Cell*, a prime example of a biological text that shows the constructive obsession with numeracy that characterized the early days of bacterial physiology. Protein is evaluated at ≈55% of the cell dry weight, followed by RNA at ≈20%, lipid at ≈10%, and DNA at ≈3% (the rest being polysaccharides, metabolites, ions, etc.). Similar efforts in budding yeast revealed that proteins constitute in the range

macromolecule	percentage of total dry weight	weight per cell (fg)	characteristic molecular weight (Da)	number of molecules per cell
protein	55	165	3×10^4	3,000,000
RNA	20	60		
23S rRNA		32	1×10^6	20,000
16S rRNA		16	5×10^5	20,000
5 S rRNA		1	4×10^4	20,000
transfer		9	2×10^4	200,000
messenger		2	1×10^6	1400
DNA	3	9	3×10^9	2
lipid	9	27	800	20,000,000
lipopolysaccharide	3	9	8000	1,000,000
peptidoglycan	3	9	$(1000)_n$	1
glycogen	3	9	1×10^6	4000
metabolites and cofactors pool	3	9		
inorganic ions	1	3		
total dry weight	100	300		
water (70% of cell)		700		
total cell weight		1000		

composition rules of thumb
- carbon atoms $\sim 10^{10}$
- 1 molecule per cell gives $\sim$1 nM conc.
- ATP required to build and maintain a cell over a cell cycle $\sim 10^{10}$
- glucose molecules needed per cell cycle $\sim 3 \times 10^9$ (2/3 of carbons used for biomass and 1/3 used for ATP)

Table 2-7 Overall macromolecular composition of an average *E. coli* cell in aerobic balanced growth at 37°C in glucose minimal medium, with doubling time of 40 minutes and 1 pg cell wet weight ($\approx$0.9 µm³ cell volume). (From BNID 104954). Modifications included increasing cell dry weight from 284 fg to 300 fg and total cell mass from 950 fg to 1000 fg, as well as rounding other values to decrease the number of significant digits such that values reflect expected uncertainties ranges. Under different growth rates, the volume and mass per cell can change several fold. The relative composition changes with growth rate but not as significantly. For a given cell volume and growth rate, the uncertainty in most properties is expected to be on the order of 10–30% standard deviation. Original values refer to B/r strain, but to within the uncertainty expected, the values reported here are considered characteristic of most common *E. coli* strains. Data sources can be found at BNID 111490. An independent source for slower growth rates can be found at BNID 111460. (Adapted with modifications from Neidhardt FC et al. [1990] Physiology of the Bacterial Cell. Sinauer Associates.)

of 40–50% of the cell dry mass, RNA $\approx$ 10%, and lipid $\approx$ 10% (BNID 111209, 108196, 108198, 108199, 108200, 102327, 102328). In mammalian cells, the fraction taken by RNA decreases to about 4%, while the fraction of lipids increases (BNID 111209).

What is the logic behind these values? Ribosomal RNA, for example, even though quite monotonous in terms of its diversity, comprises two-thirds of the ribosome mass, and given the requirements for constant protein synthesis, must be abundant. Ribosomal RNA is actually more than an order of magnitude more abundant than all mRNA combined. At the same time, mRNA is rapidly degraded with a characteristic half-life of about 4 minutes (BNID 104324) versus the very stable rRNA, which shows degradation (*in vitro*) after only several days (BNID 108023, 108024). Because of the fast degradation of mRNA, the overall synthesis of mRNA required by the cell is not so small and amounts to about one-half of the rRNA synthesis (at 40-min doubling time; BNID 100060). As another example for rationalizing the cell composition, the protein content, which is the dominant constituent, is suggested to be limited by crowding effects. Crowding more proteins per cytoplasm unit volume would hamper processes such as diffusion, which is already about 10-fold slower inside the cell than in pure water. We discuss such effects in the vignette entitled "What are the time scales for diffusion in cells?" (pg. 211). The average protein concentration in the cytoplasm is already such that the average protein has a water hydration shell of only ≈ 10 water molecules separating it from the adjacent protein hydration shell.

The amount of lipid in a "typical cell" can be deduced directly from the surface area of the membrane, though for eukaryotes, the many internal membranes associated with organelles need to be included in the estimate. Let's see how such an estimate works for the spherocylindrical, cigar-shaped *E. coli*. At a diameter of ≈1 μm, and for a characteristic growth rate where the overall length is ≈2 μm (1 μm cylinder and two half spherical caps of 1 μm diameter each), the surface area is an elegant $A = 2\pi$, or ≈6 μm². The volume is also a neat geometrical exercise that results in $V = 5\pi/12$, or ≈1.3 μm³ (though we often will choose to discuss it as having a 1 μm³ volume for simplicity, where order of magnitude estimations are concerned). As discussed in the vignette entitled "What is the thickness of the cell membrane?" (pg. 54), the lipid bilayer is about 4 nm thick (while larger values often mentioned might stem from elements sticking out of the membrane). The volume of the membrane is thus about 6 μm² × (4×10^{-3} μm) = 0.024 μm³. At ≈70% water and ≈30% dry mass of density ≈1.3 (BNID 104272), the overall density is ≈1.1 (BNID 103875), and the dry mass has a volume of about 1.3 μm³ × 1.1 g/cm³ × 0.3/1.3 g/cm³ ≈0.33 μm³. So the lipid bilayer occupies a fraction of about 7% of the dry mass. There are two lipid bilayers—the outer membrane and the cell membrane—so we should double this value to ≈14%. Noting that proteins decorating the membrane occupy between one-quarter and one-half of its area (BNID 105818), we are reasonably close to the empirically measured value of ≈9%.

How does the composition change for different growth conditions and in various organisms? Given that the classic composition for *E. coli* was attained already in the 1960s and 1970s, and that today we regularly read about quantitation of thousands of proteins and mRNA, we might have expected the experimental response to this question to be a standard exercise. The methods for quantifying protein are mostly variants of that developed by Lowry in 1951. The paper announcing these methods, which after the first submission had been returned for drastic cuts by the journal, apparently became the most highly cited paper in the history of science, with more than 200,000 citations. For all their virtues and citations, the methods in that work tended to be limited in their accuracy when applied to the full complement of cells, often turning into finicky biochemical ordeals. For example, other cell constituents such as glutathione, the main redox balancer of the cell, may influence the reading. As a result, comprehensive characterization of the cellular census for different conditions is mostly lacking. This situation limits our ability to get a true physiological or systems view of the dynamic cell and awaits revisiting by biologists merging good experimental technique with a quantitative approach.

MACHINES AND SIGNALS

What are the copy numbers of transcription factors?

Transcription factors are the protein sentinels of the cell, on the lookout to decide which of the many genes hidden within the DNA should be turned into an mRNA message at a given time. On the order of 200–300 distinct kinds of transcription factors (that is, coded by different genes) exist in model bacteria such as *E. coli* (BNID 105088, 105089), with ≈1000 distinct kinds in animal cells (BNID 105072, 109202). Those enamored with simple model biological systems will delight to learn of parasites, such as *mycoplasma pneumoniae* or *buchnera aphidicola*, which seem to have only four distinct transcription factors (BNID 105075). Transcription factors are key players in regulating the protein composition of the cell, which they often do by binding DNA and actively interacting with the basal transcription apparatus, either activating or repressing transcription. Because they are prime regulators, they have been heavily studied, but in stark contrast to their ubiquity in published papers, their actual concentrations inside cells are usually quite low. Their concentration depends strongly on the specific protein, cell type, and environmental conditions, but as a rule of thumb, the concentrations of such transcription factors are in the nM range, corresponding to only 1–1000 copies per cell in bacteria or 10^3–10^6

in mammalian cells. This is in stark contrast to the most abundant proteins, such as glycolytic proteins or elongation factors, which will tend to occur with many thousands of copies in bacteria and many millions in mammalian cells. Not surprisingly, the cellular concentrations of transcription factors are often comparable to the Kd values of these proteins for DNA binding. Often, those transcription factors that occur at lower concentrations are specific and engaged in regulating only a few genes (for example, LacI, which regulates the lactose utilization operon), whereas those at higher concentrations have many genes as their targets and are sometimes known as global regulators (for example, the protein CRP, which modulates carbon source utilization in bacteria).

Given the central role the Lac repressor (LacI) plays in undergraduate molecular biology courses as the paradigm of gene regulation, it might come as a surprise that it usually appears with only about 10 tetrameric copies per cell (equivalent to a concentration of ≈ 10 nM; BNID 100734). Interestingly, nonspecific affinity to the DNA causes $\approx 90\%$ of LacI copies to be bound to the DNA at locations that are not the cognate promoter site and only at most several copies to be freely diffusing in the cytoplasm (both forms are probably important for finding the cognate target, as has been shown in elegant theoretical studies). Further, these small copy numbers have inspired important questions about how living cells manage (or exploit) inevitable stochastic fluctuations that are associated with such small numbers. For example, if the partitioning of these proteins upon cell division is strictly random, with such small numbers there is a chance that some daughter cells will be without a copy of some transcription factor at all.

Though LacI is the model transcription factor, most transcription factors show higher concentrations of tens to hundreds of nM, as can be seen in **Figure 2-31** (BNID 102632, 104515). The results shown in the figure were obtained using a beautiful recent method, which is one of several that has turned DNA sequencing into a legitimate biophysical tool for performing molecular censuses. In this case, the idea is that fragments of mRNA that have been protected by translating ribosomes are sequenced. The density of these ribosomal footprints tells us something about the rate of protein synthesis, which through careful calibrations makes it possible to quantify the number of proteins per cell. There are many interesting nuances associated with this data. For example, as shown in the figure, the distributions of copy numbers of activators and repressors are different, with activators on average having lower copy numbers than repressors. A second intriguing observation that emerges from these proteome-wide results is that transcription factors that are subject to allosteric control by ligand binding have on average much higher copy numbers than those that are ligand-independent.

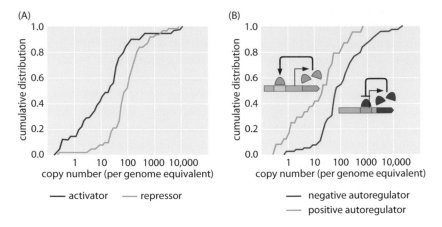

Figure 2-31 Measured copy numbers of transcription factors in *E. coli*. (A) Cumulative distributions for both activators and repressors, showing that activators typically occur between 1 and 100 copies per cell, whereas repressors generally occur between 10 and 1000 copies per cell. (B) Cumulative distributions for autoregulators. (Adapted from Li GW, Burkhardt D, Gross C & Weissman JS [2014] *Cell* 157:624–635.)

Even more effort than in the bacterial case has been invested in what is arguably the most studied protein of all time, p53 (with another key contender being hemoglobin), a transcription factor that is claimed to be involved in over 50% of cases of cancer (BNID 105092). Its name, like many other proteins, arises from its original characterization in gels, where it migrated as a protein of mass 53 kDa. Today we know it actually has a mass of 44 kDa, and its slow migration is due to many bulky proline residues, but the name persists. This critical transcription factor helps mediate the decision of a cell to perform programmed cell death versus continued proliferation, critically affecting tumor growth. It has a characteristic concentration of ≈100 nM (corresponding to ≈100,000 molecules in a mammalian MCF7 breast cancer cell line; BNID 100420). Transcription factors modulate transcription by changing their binding properties to DNA through interaction with signals coming from receptors, for example. Mutations in the DNA of cancer cells change p53 binding properties to the downstream genes it regulates, often stopping cell death from occurring, thus leading to uncontrolled growth.

Table 2-8 lists examples of the census of a variety of other transcription factors and an order-of-magnitude characterization of their absolute copy numbers. Given that transcription factors are such a big part of the daily life of so many researchers, this table aims to make it easier to develop intuitive rules of thumb for quantitative analysis. What can the absolute numbers or concentrations teach us? They are essential when we want to analyze the tendency for sequestering of transcription factors in complexes or inhibition by regulators, or to consider the effect of nonspecific binding to DNA, or to reckon the response time for triggering a transcriptional program, since in

organism	transcription factor	copies per cell order of magnitude	BNID
E. coli	LacI (carbon utilization)	10^1–10^2	100734
E. coli	AraC (carbon utilization)	10^2	105139
E. coli	ArcA (general aerobic respiration control)	10^4	102632
S. cerevisiae	Gal4 (carbon utilization)	10^2	109208
S. cerevisiae	Tfb3 (general transcription initiation factor)	10^3	109208
S. cerevisiae	Pho2 (phosphate metabolism)	10^4	109208
D. melanogaster, anterior blastoderm nuclei	bicoid (development)	10^4	106843
D.melanogaster, S2 cells	GAGA zinc finger	10^6	106846
mouse/rat macrophage	glucocorticoid, thyroid and androgen receptors associated zinc fingers	10^4	106899
mouse/rat macrophage	NF-kappaB p65	10^5	106901
H. sapiens cell lines	P53 (growth and apoptosis)	10^4–10^5	100420
H. sapiens cell lines	glucocorticoid, estrogen, steroid receptors associated zinc fingers	10^4–10^5	106904, 106906, 106911
H. sapiens cell lines	STAT6	10^4–10^5	106914
H. sapiens cell lines	NF-kappaB p65	10^5	106909
H. sapiens cell lines	Myc (global chromatin structure regulation)	10^5	106907

Table 2-8 Absolute copy numbers from a number of different organisms. Values are rounded to closest order or magnitude. For more values, see Biggin MD [2011] *Dev Cell* 21:611–626; BNID 106842).

each of these cases the formation of molecular partnerships depends upon the concentrations of the relevant molecular actors. We advocate keeping characteristic orders of magnitude such as those shown in the table at one's disposal, but we also remember that the number of such factors often varies both in space and time. This is especially clear in the case of developmental patterning, where often it is the spatial variation in transcription factor concentrations that lays down the patterns that ultimately become the body plan of the animal. For example, the gradient along the anterior–posterior axis of the fly embryo of the transcription factor bicoid (shown in the table) is a critical ingredient in the patterning of the fly, with similar proteins shaping humans.

What are the absolute numbers of signaling proteins?

Bacteria move in a directed fashion to regions with more nutrients. Neutrophils, as the assassins of the immune system, chase down bacterial invaders by sniffing out chemical signals coming from their prey.

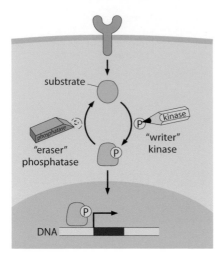

Figure 2-32 Schematic of a generic signaling network. A membrane receptor at the cell surface (orange) releases a substrate. The substrate is modified by the addition of a phosphate group by a kinase. The addition of the phosphate group localizes the protein to the nucleus (brown), where it then acts as a transcription factor. Removal of the phosphate group is mediated by a phosphatase.

Photoreceptors respond to the arrival of photons by inducing signaling cascades that we interpret as the act of seeing. The cells in developing embryos take on different fates, depending upon where they are within the organism. To accomplish these tasks, cells are guided by a host of molecular sentinels whose job is to receive signals about the external world and to make decisions based upon those inputs. The conceptual architecture of the signaling modules that carry out these kinds of responses is indicated schematically in **Figure 2-32**. As the diagram indicates, multiple molecular players implement the response to signals, and the answers to questions about signal amplification, specificity, and feedback can all depend upon the number of copies of each of the molecular partners.

One of the conceptual threads that will run through our entire discussion of signaling is that proteins are modified by the addition ("write") and removal ("erase") of chemical groups, such as phosphate groups or methyl groups. Though we will use this notation in several of the figures in this vignette, the reader should not think that the addition of the group necessarily corresponds to the active form of the modified protein. In many instances, the signaling event corresponds to the removal of a phosphate group, and the unphosphorylated conformation is the active form. For example, in the case of the chemotaxis signaling molecule CheY, in some organisms the phosphorylated form triggers the motor to change direction, whereas in other organisms the unphosphorylated form directs this response. To the best of our knowledge, whether there is an evolutionary advantage to one or the other tactic still awaits clarification.

One of the defining characteristics of signaling proteins is that depending upon environmental conditions, the concentration of the relevant signaling molecule, or of the active form, can vary dramatically. As a result, the very feature of these proteins that makes them most interesting stands in the way of giving a precise and definitive answer to the question of the "generic" number of such signaling proteins within cells. Hence, we adopt the strategy of providing a collection of examples that serve to paint a picture of the relevant ranges of signaling protein concentrations, mindful of the dependence of the resulting census on the conditions to which the cell has been subjected.

To provide a quantitative picture of the molecular census of signaling molecules, we resort to some of the most celebrated signaling systems, as

(A)   bacterial two-component signaling

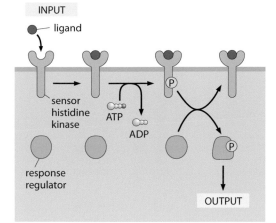

INPUT

ligand

sensor
histidine
kinase

ATP

ADP

response
regulator

OUTPUT

(B)   eukaryotic MAP kinase cascade

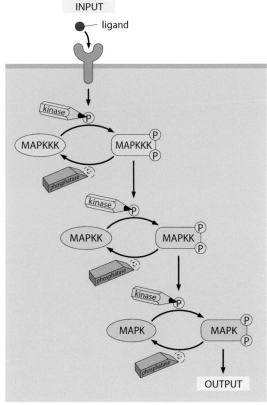

INPUT

ligand

kinase

MAPKKK

MAPKKK

phosphatase

kinase

MAPKK

MAPKK

phosphatase

kinase

MAPK

MAPK

phosphatase

OUTPUT

Figure 2-33 Model signaling pathways. (A) Two-component signaling systems in bacteria. The membrane receptor is a kinase that phosphorylates a soluble messenger molecule, which is activated by phosphorylation. (B) MAP-kinase pathway. The MAPKKK phosphorylates the MAPKK, which phosphorylates the MAPK molecule, which then induces some output.

indicated schematically in **Figure 2-33**. Perhaps the simplest of cell signaling pathways is found in bacteria and goes under the name of two-component signal transduction systems (see Figure 2-33A). These pathways are characterized by two key parts—namely, a membrane-bound receptor that receives signals from the external environment, but that also harbors a domain (a histidine kinase) on the cellular interior, and a response regulator that is chemically modified by the membrane-bound receptor. Often, these response regulators are transcription factors that require phosphorylation in order to mediate changes in gene expression. In *E. coli*, there are over 30 such two-component systems (BNID 107848). Figure 2-33B shows a similarly central signal transduction system in eukaryotes known as the MAP-kinase pathway. Like their bacterial counterparts, these pathways make it possible for some external stimulus, such as a pheromone or high osmolarity, to induce changes in the regulatory state of the cell.

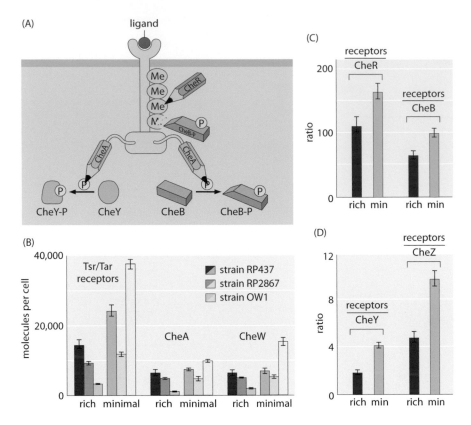

Figure 2-34 Census of the molecules of the bacterial chemotaxis signaling pathway. (A) Schematic of the molecular participants involved in bacterial chemotaxis. (B) Number of chemotaxis receptor molecules and number of CheA and CheW (which connects the Tsr/Tar receptors to CheA) molecules. Results are shown for different strains and for different growth media. (C) Ratio of number of receptors to CheR and CheB for both rich and minimal media. (D) Ratio of number of receptors to CheY and CheZ (the phosphatase of CheY) for both rich and minimal media. (B, C, and D adapted from Li M & Hazelbauer GL [2004] *J Bacteriol* 186:3687–3694.)

Probably the most well studied of all bacterial two-component systems is that associated with bacterial chemotaxis. This signaling system detects chemoat-tractants in the external medium, resulting in changes to the tumbling fre-quency of the motile cells. As will be discussed in the vignette entitled "What are the physical limits for detection by cells?" (pg. 175), the chemoreceptors have exquisite sensitivity and very broad dynamic range. **Figure 2-34A** shows the wiring diagram that implements this beautiful pathway. One of the ways that the stoichiometric census of these signaling proteins is made is using bulk methods in which a population of cells is collected and broken open and their contents allowed to interact with antibodies against the protein of interest. By comparing the amount of protein fished out by these antibodies to those measured using purified proteins of known concentration, it is pos-sible to perform a calibrated measurement of the quantity of protein, such as that reported in **Figure 2-34B** for the two-component system relevant to bacterial chemotaxis. Despite as much as a 10-fold difference in the absolute numbers of molecules per cell, depending upon strain and growth condition, the relative concentrations of these different molecules are maintained at nearly constant stoichiometric ratios.

Recent years have seen the emergence of DNA sequencing, not only as a genomic tool, but also as a powerful and quantitative biophysical tool that provides a window onto many parts of the molecular census of a cell. Indeed, these methods have been a powerful addition to the arsenal of techniques being used to characterize the processes of the central dogma, such as the number of mRNA molecules per cell and the number of proteins. The way these methods work is to harvest cells for their mRNA, for example, and then to sequence those parts of the mRNA that are "protected" by ribosomes. The abundance of such protected fragments provides a measure of the rate of protein synthesis on the gene corresponding to that mRNA. In the context of two-component signaling systems, the molecular census of more than 20 of these systems has been taken using this method known as ribosome profiling. As shown in **Figure 2-35**, the histidine kinases usually come with tens to hundreds of copies per cell, while their corresponding response regulators come in much higher quantities of about an order of magnitude more molecules per cell.

But why should we care about these absolute numbers? Binding partnerships between different molecular species depend upon their concentrations.

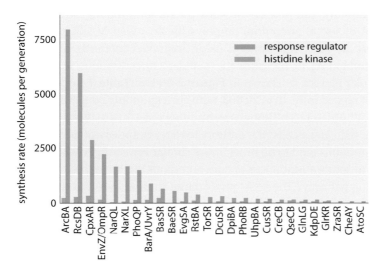

Figure 2-35 Molecular census for two-component signaling systems in *E. coli*. These two-component systems consist of a membrane-bound histidine kinase and a soluble response regulator. The figure shows the number of molecules of both the kinase and response regulator for many of the *E. coli* two-component systems. (Adapted from Li GW, Burkhardt D, Gross C & Weissman JS [2014] *Cell* 157:624–635.)

Biological action, in turn, often depends upon the binding events that induce conformational change, whether in the context of chemoattractants in the bacterial medium or of acetylcholine and the gating of the ion channels of the nervous system. This suggests that our sole effort should focus on a proper concentration census of the cell. We agree that concentrations should be the top priority; however, we find that absolute numbers are often a helpful basis for gaining intuition for the cellular milieu—a kind of "feeling for the organism," as phrased by Barbara McClintock, one of the heroines of 20th-century genetics. Let's compare our cognitive capabilities for dealing with concentrations versus absolute numbers. We have all learned early in life to differentiate between a thousand and a million. We have by now developed an intuition about such values that we do not have in dealing with, say, µM versus mM. With this familiarity and intuition regarding absolute values, we suggest there comes an almost automatic capability to make mental notes of such orders of magnitude. We thus rarely confuse a thousand with a million or a billion, whereas we have witnessed many cases where mM was confused with µM or nM. In this spirit we make a point in the next part of this vignette to drive home the rule of thumb that a characteristic number of copies for many signaling molecules per mammalian cell is about a million, even though 1 µM provides a more biochemically meaningful characterization.

To continue to build this kind of quantitative intuition, we consider another extremely well-characterized signaling system found in yeast (see Figure 2-33B). The process of yeast pheromone mating, the *S. cerevisiae* version of sexual attraction, employs the so-called MAPK pathway. This pathway in yeast was studied using improved methods of quantitative immunoblotting to measure the cellular concentrations of the relevant molecular players, as shown in **Figure 2-36** and **Table 2-9**. Copy numbers per cell ranged from 40 to 20,000, with corresponding concentrations in the range of 1 nM to 1 µM. Though the budding yeast is two orders of magnitude smaller than HeLa cells (the authors used a volume of ≈ 30 µm³), we see the concentrations tend to be much more similar across organisms. How much do the absolute abundances or concentrations matter for the function of the signaling pathway? The yeast pheromone study shows that the concentration of the scaffolding protein (Ste5, at about 500 copies per cell, ≈ 30 nM) dictates the cell's behavior by mediating a tradeoff between the dynamic range of the signaling system and the maximal output response.[¶¶]

MAPK pathways are also important in multicellular organisms, providing a model pathway of signal transduction intimately related to growth regulation

[¶¶] Thomson TM, Benjamin KR, Bush A et al. (2011) Scaffold number in yeast signaling system sets tradeoff between system output and dynamic range. *Proc Nat Acad Sci USA* 108:20265–20270 (doi: 10.1073/pnas.1004042108).

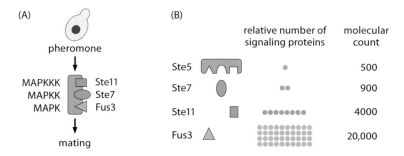

Figure 2-36 Census of proteins in a yeast signaling system. (A) Schematic of the MAPK pathway associated with the mating response in yeast. (B) Molecular count of the various molecules in the mating response pathway. (Adapted from Thomson TM, Benjamin KR, Bush A et al. [2011] *Proc Nat Acad Sci USA* 108:20265–20270.)

and many other processes. One of the upstream proteins associated with these pathways is the Ras protein. In HeLa cells and 3T3 fibroblasts, this protein was measured to have 10^4–10^7 copies under various conditions (BNID 101729). The close to three order-of-magnitude variation reveals a broad range of viable concentrations. Ras interacts with Raf, estimated at about 10^4 copies per cell, which interacts with Mek at roughly 10^5–10^7 copies, which

protein	name	molecules/cell	standard error	concentration (nM)
G protein-coupled receptor	Ste2	7000	400	400
G-α	Gpa1	2000	300	130
G-β	Ste4	2000	100	110
PAK kinase	Ste20	4000	500	200
scaffold	Ste5	500	60	30
MAPKKK binding partner	Ste50	1000	100	70
MAPKKK	Ste11	4000	90	200
MAPKK	Ste7	900	70	50
MAPK	Fus3	20,000	3000	1100
MAPK	Kss1	20,000	2000	1200
MAPK	Hog1	6000	400	300
scaffold/MAPKK	Pbs2	2000	200	140
MAPK phosphatase	Msg5	40	3	2
cell cycle inhibitor	Far1	200	20	14
transcriptional activator	Stw12	1400	40	80
transcriptional repressor	Dig1	5000	500	300
transcriptional repressor	Dig2	1000	80	70

Table 2-9 Abundances of signaling molecules associated with the MAPK cascade in budding yeast before pheromone addition. Abundances are based on quantitative immunoblotting. Concentration was calculated assuming a cell volume of 29 fL. The standard error indicates the uncertainty on the number of molecules per cell as estimated in this specific experiment. Values were rounded to one significant digit. (Adapted from Thomson TM, Benjamin KR, Bush A et al. [2011] *Proc Nat Acad Sci USA* 108:20265–20270; From BNID 107680).

interacts in turn with Erk, measured at 10^6–10^7 copies. For a HeLa cell with a characteristic median volume of $\approx$3000 fL, these copy numbers translate into concentrations from $\approx$10 nM to $\approx$10 μM, assuming a homogenous distribution over the cell volume. Other pathways, such as those of Wnt/beta-catenin (BNID 101958) or TGF-beta, show similar concentration ranges. An example of an outlier with respect to typical concentrations is Axin in the Wnt/beta-catenin pathway, whose concentration is estimated to be in the pM range (BNID 101951). Localization effects can have a dramatic effect by increasing effective concentrations. One example is the import of transcription factors from the cytoplasm to the nucleus, where the absolute number does not change, but the local concentration increases relative to its value in the cytoplasm by several fold, which leads the transcription factor in the nucleus to activate or repress genes without its overall cellular concentration changing. Another example is the effect of scaffolding proteins that hold target proteins in place next to each other, thus facilitating interaction as in the MAPK cascade mentioned above. The importance of high local concentration effects led Müller Hill to refer to it as one of the main ingredients of life.[***] These and more recent studies highlight that it is not only the average concentration or absolute numbers that matter, but rather how these signaling proteins are spatially organized within the cell (BNID 110548).

One of the important conclusions to emerge from these studies is an interesting juxtaposition of large variability in overall numbers of signaling molecules, depending upon both strain and growth conditions, coupled with a roughly constant ratio of the individual molecular players. Very often it is found that a some-fold change in the concentration is the key determinant of the underlying function and the property to which the circuits of signal transduction seem to be tuned. Though cell-to-cell variability will often show a twofold difference in absolute value, a temporal change of twofold in the ratio of components will be quickly detected and elicit a strong response. Numbers like those described here call for a theoretical interpretation, which will provide a framework to understanding, for example, the relative abundances of receptors and their downstream partners.

How many rhodopsin molecules are in a rod cell?

Responses in signaling pathways depend critically upon how many molecules there are to respond to the signal of interest. The concentrations of molecules, such as rhodopsin in photoreceptor cells, determine the light

---

[***] Müller-Hill B (2006) What is life? The paradigm of DNA and protein cooperation at high local concentrations. *Mol Microbiol* 60:253–255.

intensity that can be detected in vertebrate eyes. Beyond this, the number of rhodopsins also helps us understand how frequently a given rod cell will spontaneously fire in the dark. Though our focus on rhodopsin might seem highly specialized, we find it an informative case study, because the signaling cascade associated with vision is one of the best characterized of human signaling cascades. Further, it exhibits many generic features found in signaling events of many other kinds. Some of the key molecular players found here include G-coupled receptors and ligand-gated ion channels, molecules in signaling cascades that are ubiquitous throughout the living world. **Figure 2-37** shows how the molecules in the outer segment of a photoreceptor respond to the arrival of a photon, which is absorbed by

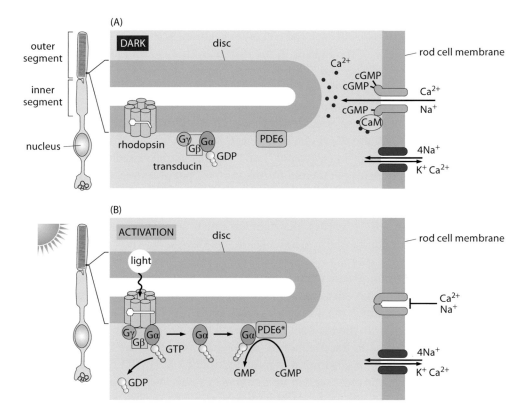

Figure 2-37 Signal transduction in the retina. (A) In the dark, the rhodopsin is in the inactive state and ions are free to cross the rod cell membrane. In the dark, the cGMP phosphodiesterase PDE6 is inactive, and cGMP is able to accumulate inside the rod cell. Cyclic GMP binds to a ligand-gated ion channel (dark green), which is permeable to both sodium ions and calcium ions. Calcium is transported back out again by an exchanger (shown in brown), which uses the energy from allowing sodium and potassium ions to run down their electrochemical gradients to force calcium ions to be transported against their gradient. (B) Activation of rhodopsin by light results in the hydrolysis of cGMP, causing cation channels to close. When a photon activates a rhodopsin protein, this triggers GTP-for-GDP exchange on transducin, and the activated α subunit of transducin then activates PDE6, which cleaves cGMP. The ligand-gated channels close, and the transmembrane potential becomes more negative. (Adapted from Stockman A, Smithson HE, Webster AR et al. [2008] *J Vis* 8:1–10.)

the retinal pigment, covalently but reversibly held by the opsin protein, together making up the rhodopsin molecule.

In this vignette, we use a collection of estimates to work out the number of rhodopsins in a photoreceptor cell. We begin by estimating the number of membrane discs in the outer segment of a rod cell. As seen in both the electron microscopy image and the associated schematic in the vignette entitled "How big is a photoreceptor?" (pg. 17), the rod outer segment is roughly 25 μm in length and is populated by membrane discs that are roughly 10 nm thick and 25 nm apart. This means there are roughly 1000 such discs per rod outer segment. Given that the rod cell itself has a radius of around 1 μm, this means that the surface area per disc is roughly 6 μm^2, resulting in an overall membrane disc area of 6000 μm^2. One crude way to estimate the number of rhodopsins in each rod cell outer segment is to make a guess for the areal density of rhodopsins in the disc membranes. Rhodopsins are known to be tightly packed in the disc membranes, and we can estimate their mean spacing as 5–10 nm (that is, about one to two diameters of a characteristic protein), corresponding to an areal density of $\sigma = 1/25 - 1/100$ nm^{-2}. In light of these areal densities, we estimate the number of rhodopsins per membrane disc to be between $(6 \times 10^6 \text{ nm}^2) \times (1/25 \text{ nm}^{-2}) \approx 2 \times 10^5$ and $(6 \times 10^6 \text{ nm}^2) \times (1/100 \text{ nm}^{-2}) = 6 \times 10^4$. The actual reported numbers are $\approx 10^5$ rhodopsins per membrane disc or $\approx 10^8$ per photoreceptor (BNID 108323), which is on the order of the total number of proteins expected for such a cell volume, as discussed in the vignette entitled "How many proteins are in a cell?" (pg. 104). This tight packing is what enables the eye to be able to function so well at extremely low light levels.

For a molecule that absorbs photons, such as retinal, it is convenient to define the effective cross section that quantifies its absorption capacity. Concretely, the absorption cross section is defined as that area perpendicular to the incident radiation such that the photon flux times that area is equal to the number of photons absorbed by the molecule. The cross section for absorption of retinal is about 1 Å^2 (BNID 111337)—that is, 10^{-2} nm^2. The cross section thus connects a physical property (absorption ability) with the geometrical notion of area (of the photoreceptor). Here is an example of why this is useful. With 10^8 rhodopsins per photoreceptor, we arrive at a total absorption cross section of (10^8 rhodopsin/photoreceptor) $\times$ (10^{-2} nm^2/rhodopsin) = 10^6 nm^2/photoreceptor = 1 μm^2. Each photoreceptor cell has a geometrical cross-sectional area of about 4 μm^2/photoreceptor, as discussed in the vignette entitled "How big is a photoreceptor?" (pg. 17). From the similarity between the cross section for absorption and the actual cross-sectional area of the photoreceptor, we can infer that the concentration of rhodopsins is of the correct order of magnitude to efficiently absorb all photons arriving (at least under low illumination levels when reactivation of rhodopsin following absorption does not become

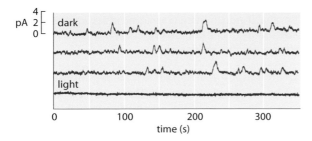

Figure 2-38 Spontaneous isomerization of retinal. The top three traces correspond to the current measured from a single photoreceptor as a function of time in the dark. The lower trace shows the current in the light and demonstrates that the channels are closed in the presence of light. (Adapted from Baylor DA, Matthews G & Yau KW [1980] *J Physiol* 309:591–621.)

limiting). This does not mean there are no lost photons. Indeed, to achieve superior night vision, many nocturnal animals have a special layer under the retina called the *tapetum lucidum*, which acts as a reflector to return the photons that were not absorbed by the rhodopsins back to the retina for another opportunity to be absorbed. This increases their ability to hunt prey at night and enables naturalists to find hyena and wolves (and also domestic cats) at night from a distance, by looking for eyeshine, which is the reflection of a flashlight.

The membrane census of rhodopsin also sheds light on the rate at which rod cells suffer spontaneous thermal isomerizations of the pigment retinal. As shown in **Figure 2-38**, measurements of the currents from individual rod photoreceptors exhibit spontaneous isomerizations. Reading off of the graph, we estimate roughly 30 spontaneous events over a period of 1000 s, corresponding to a rate of once per 30 s. We know that the total rate is given by

$$\text{rate} = (\text{number of rhodopsins}) \times (\text{rate/rhodopsin}). \qquad (2.12)$$

Armed with the number of rhodopsins estimated above, we deduce that the rate of spontaneous isomerization per rhodopsin is once about every 100 years.

The signaling cascade that follows the absorption of a photon only starts with the isomerization of a retinal molecule. Once the rhodopsin molecule has been thus activated, it sets off a signaling cascade within the rod cell that amplifies the original signal, as already shown in Figure 2-37 and elaborated on more quantitatively in **Figure 2-39**. In particular, once the rhodopsin has been activated, it encounters a membrane-bound G-protein-coupled receptor and activates its alpha subunit. Over a period of 100 ms, one activated rhodopsin will create $\approx 10^3$ of

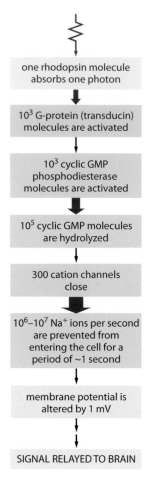

Figure 2-39 Signal amplification is achieved at several steps of the pathway, such that the energy of one photon eventually triggers a net charge change of about one million sodium ions.

these activated alpha subunits (Gα), as explained in detail in the wonderful book, *The First Steps in Seeing*, by R. W Rodieck. These molecules bind another molecule known as phosphodiesterase, which can convert cyclic guanosine monophosphate (cGMP) into guanosine monophosphate (GMP). The significance of this molecular reaction is that the cGMP gates the cGMP channels in the rod cell membrane that lead to the change in membrane potential upon excitation. Hence, the activation of the receptor via photons leads to a closing of the channels and a change in the membrane potential. The census of the various molecular players in this signaling cascade is shown in **Figure 2-40**. Though we can depict the molecular details and the associated copy numbers as an advanced Rube Goldberg machine, the fitness advantage of this specific design, beyond the obvious need to amplify a small

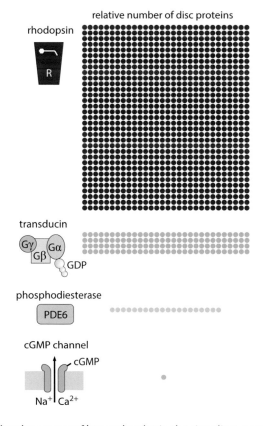

Figure 2-40 Molecular census of key molecules in the signaling cascade in the retina.

signal, is still quite a mystery even to researchers in the field. As a parting note, we consider another amazing number related to the function of the rod cell. Every pigment molecule, once it absorbs a photon, is photobleached, and it takes about 10 min and a sequence of biochemical steps after transport to a separate organelle to fully regenerate (BNID 111399, 111394). The inventory of pigments in the rod cell has to compensate for this long delay.

How many ribosomes are in a cell?

One of the familiar refrains in nearly all biology textbooks is that proteins are the workhorses of the cell. As a result, cells are deeply attentive to all the steps between the readout of the genetic information hidden within DNA and the expression of active proteins. One of the ways that the overall rhythm of protein production is controlled is through tuning the number of ribosomes. Ribosomes are one of the dominant constituents in cells especially in rapidly dividing cells. The RNA making up these ribosomes accounts for ≈85% of the cell's overall RNA pool (BNID 106421). Though DNA replication, transcription, and translation are the three pillars of the central dogma, within the proteome, the fraction dedicated to DNA polymerase (BNID 104123) or RNA polymerase (BNID 101440) is many times smaller than the tens of percent of the cell protein dedicated to ribosomes (BNID 107349, 102345). As such, there is special interest in the abundance of ribosomes and the dependence of this abundance on growth rate. The seminal work of Schaechter et al. established early on the far-from-trivial observation that the ribosomal fraction is a function of the growth rate and mostly independent of the substrate—that is, different media leading to similar growth rates tend to have similar ribosomal fractions.[†††] Members of the so-called "Copenhagen school" (including Schaechter, Maaloe, Marr, Neidhardt, Ingraham, and others) continued to make extensive quantitative characterizations of how the cell constituents vary with growth rate that serve as benchmarks decades after their publication and provide a compelling example of quantitative biology long before the advent of high-throughput techniques.

Table 2-10 shows the number of ribosomes in *E. coli* at different doubling times. It is also evident in the table how the cell mass and volume depend strongly on growth rate, with faster dividing cells being much larger. As calculated in

[†††] Schaechter M, Maaloe O & Kjeldgaard NO (1958) Dependency on medium and temperature of cell size and chemical composition during balanced grown of *Salmonella typhimurium. J Gen Microbiol* 19:592–606.

doubling time (min)	ribosomes per cell	dry mass per cell (fg)	ribosome dry mass fraction (%)	ribosome fraction x doubling time (min)
24	72,000	870	37	9.0
30	45,000	640	32	9.5
40	26,000	430	27	11
60	14,000	260	24	14
100	6800	150	21	20

Table 2-10 Number and fraction of ribosomes as a function of the doubling time. Values are rounded to one significant digit. Ribosomes per cell and dry mass per cell from Bremmer H & Dennis PP [1996] *Escherichia coli* and *Salmonella*: Cellular and Molecular Biology [FC Neidhardt ed] ASM Press. Ribosome dry mass fraction is calculated based on a ribosome mass of 2.7 MDa. (BNID 100118.)

the fourth column of the table, and schematically in **Estimate 2-10**, at a fast doubling time of 24 min, the 72,000 ribosomes per cell represent over one-third of the dry mass of the cell (BNID 101441, 103891). Accurate measurements of this fraction from the 1970s are shown in **Figure 2-41**.

Several models have been set forth to explain these observed trends for the number of ribosomes per cell. In order to divide, a cell has to replicate its protein content. If the translation rate is constant, there is an interesting deduction to be made. We thus make this assumption even though the translation rate varies from ≈ 20 aa/s in *E. coli* at fast growth rate to closer to ≈ 10 aa/s under slow growth (BNID 100059). Think of a given cell volume in the cytoplasm. Irrespective of the doubling time, the ribosomes in this volume have to produce the total mass of proteins in the volume within a cell cycle. If the cell cycle becomes, say, three times shorter, then the necessary ribosome concentration must be three times higher to complete the task. This tacitly assumes that the polymerization rate is constant, that active protein degradation is negligible, and that the overall protein content does not change with growth rate. This is the logic underpinning the prediction that the ribosomal fraction is proportional to the growth rate. Stated differently, as the doubling time becomes shorter, the required

Fraction of ribosomes in cell dry mass at close to maximal growth rate

at 24 min doubling time:

$\approx 70,000$ ribosomes (MW$_{ribo}$ = 2.7 MDa)
cell dry mass ≈ 1 pg

$$\text{ribosome dry mass fraction} = \frac{70,000 \times 2.7 \times 10^6 \frac{g}{mol}}{1 \times 10^{-12}g \times 6 \times 10^{23} \frac{1}{mol}} \approx \frac{1}{3}$$

Estimate 2-10

ribosomal fraction is predicted to increase such that the ribosomal fraction times the doubling time is a constant, reflecting the total proteome concentration. The analysis also suggests that the synthesis rate scales as the growth rate squared, because the time to reach the required ribosome concentration becomes shorter in proportion with the doubling time. How well does this toy model fit the experimental observations?

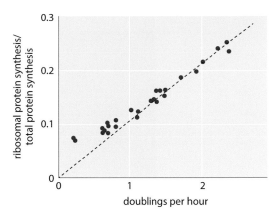

Figure 2-41 Fraction of ribosomal protein synthesis rate out of the total cell protein synthesis. Measurements were performed on cultures in balanced growth and thus the relative rate is similar to the relative abundance of the ribosomal proteins in the proteome. (Adapted from Ingraham JL, Neidhardt FC & Schaechter M [1990] Physiology of the Bacterial Cell. Sinauer Associates.)

As shown in the right column of Table 2-10 and in Figure 2-41, the ratio of ribosome fraction to growth rate is relatively constant for the faster growth rates in the range of 24–40 min, as predicted by the simple model above, and the ratio is not constant at slow growth rates. Indeed, at slower growth rates, the ribosome rate is suggested to be slower (BNID 100059). More advanced models[†††] consider different constituents of the cells (for example, a protein fraction that is independent of growth rate, a fraction related to the ribosomes, and a fraction related to the quality of the growth medium), which result in more nuanced predictions that fit the data over a larger range of conditions. Such models provide a large step towards answering the basic question of what governs the maximal growth rates of cells.

Traditionally, measuring the number of ribosomes per cell was based on separating the ribosomes from the rest of the cell constituents, measuring what fraction of the total mass comes from these ribosomes and then, with conversion factors based on estimations of cell size and mass, ribosomal molecular weight, and so on, inferring the abundance per cell. Recently, a more direct approach is available that is based on explicitly counting individual ribosomes. In cryo-electron microscopy, rapidly frozen cells are visualized from many angles to create what is known as a tomographic 3D map of the cell. The known structure of the ribosome is then used as a template that can be searched in the complete cell tomogram. This technique was applied to the small, spiral-shaped prokaryote *Spiroplasma melliferum*. As shown in **Figure 2-42**, in this tiny cell, 10–100 times smaller than *E. coli* by volume (BNID 108949, 108951) and slower in growth, researchers counted on average 1000 ribosomes per cell (BNID 108945). Similar direct counting

---

[†††] For example, Scott M, Gunderson CW, Mateescu EM et al. (2010) Interdependence of cell growth and gene expression: origins and consequences. *Science* 330:1099–1102 (doi: 10.1126/science.1192588).

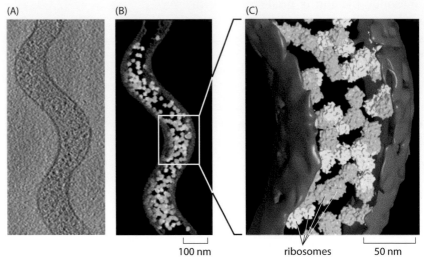

Figure 2-42 Cryo-electron tomography of the tiny bacterium, *Spiroplasma melliferum*. Using algorithms for pattern recognition and classification, components of the cell such as ribosomes were localized and counted. (A) Single cryo-electron microscopy image. (B) 3D reconstruction showing the ribosomes that were identified. Ribosomes labeled in green were identified with high fidelity, while those labeled in yellow were identified with intermediate fidelity. (C) Close-up view of part of the cell. (Adapted from Ortiz JO, Förster F, Kürner J et al. [2006] *J Struct Biol* 156:334–341.)

efforts have been made using the super-resolution techniques that have impacted fluorescence microscopy, as shown in **Figure 2-43**, where a count was made of the ribosomes in *E. coli*. A comparison of the results from these

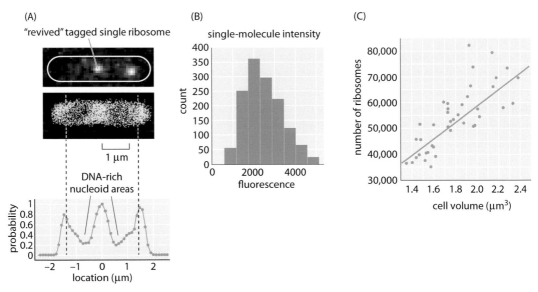

Figure 2-43 Counting and localizing ribosomes inside cells using single-molecule microscopy. (A) Two ribosomes identified from the full super-resolution image shown below. (B) Single-molecule intensity distribution. (C) Number of ribosomes as a function of cellular volume. (Adapted from Bakshi S, Siryaporn A, Goulian M et al. [2012] *Mol Microbiol* 85:21–38. With permission from John Wiley & Sons.)

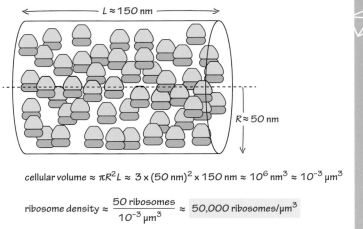

How many ribosomes in a cellular volume?

$L \approx 150$ nm

$R \approx 50$ nm

cellular volume $\approx \pi R^2 L \approx 3 \times (50 \text{ nm})^2 \times 150 \text{ nm} \approx 10^6 \text{ nm}^3 \approx 10^{-3} \text{ µm}^3$

ribosome density $\approx \dfrac{50 \text{ ribosomes}}{10^{-3} \text{ µm}^3} \approx 50{,}000 \text{ ribosomes/µm}^3$

Estimate 2-11

two methods is made in **Estimate 2-11**, where a simple estimate of the ribosomal density is made from the cryo-electron microscopy images, and this density is then scaled up to a full *E. coli* volume, demonstrating an encouraging consistency between the different methods.

ENERGY

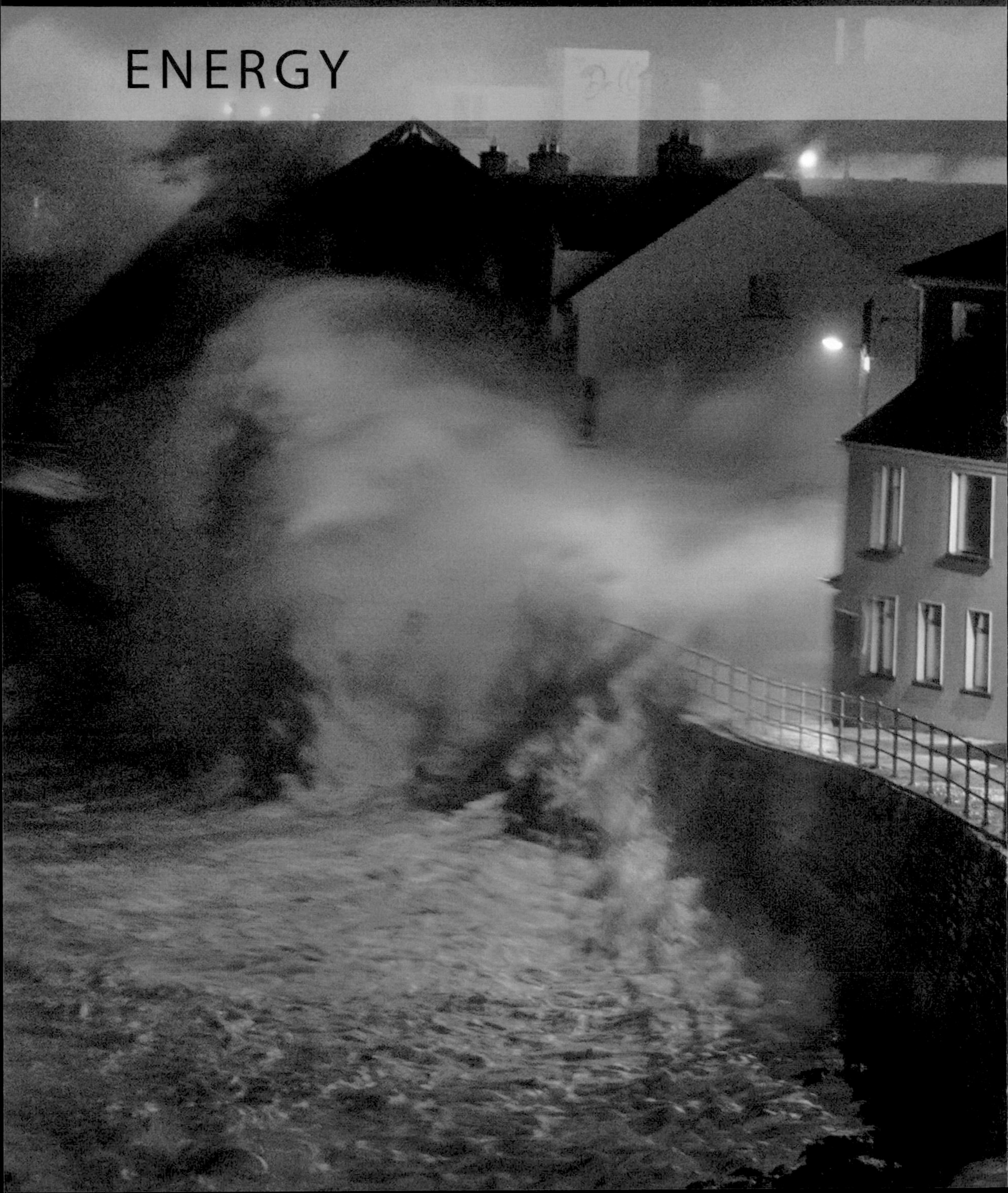

Chapter 3: Energies and Forces

Energy and force are two of the great unifying themes of physics and chemistry. But these two key concepts are crucial for the study of living organisms as well. In this chapter, we use a series of case studies to give a feeling for both the energy and force scales that are relevant in cell biology (**Figure 3-1**).

In the first part of the chapter, we consider some of the key energy currencies in living organisms, what sets their scale, and what such energy is used for. One overarching idea is that the fundamental unit of energy in physical biology is set by the energy of thermal motions—namely, $k_B T$, where k_B is the Boltzmann constant and T is the temperature in kelvin. Our discussion of thermal energy centers on the way in which many biological processes reflect a competition between the entropy and the energy, a reminder that free energy G is written as

$$G = H - TS, \tag{3.1}$$

where H is the enthalpy and S is the entropy. Whether we think of the spontaneous assembly of capsid proteins into viruses or the binding of chemoattractant to a chemoreceptor, the competing influences of entropy and energy determine the state of the system. Like everyone else, we then acknowledge the primacy of ATP as the energy currency of the cell. This discussion is followed by an examination of two of the other key energy currencies—namely, the storage of energy in transmembrane potentials and the origins of reducing power in compounds such as NADPH. We then

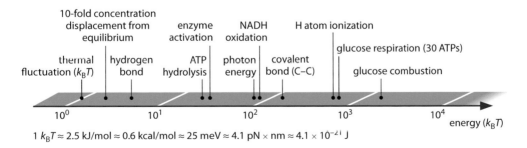

$1\ k_B T \approx 2.5\ \text{kJ/mol} \approx 0.6\ \text{kcal/mol} \approx 25\ \text{meV} \approx 4.1\ \text{pN} \times \text{nm} \approx 4.1 \times 10^{-21}\ \text{J}$

Figure 3-1 Range of characteristic energies central to biological processes. Energies range from thermal fluctuations to combustion of the potent glucose molecule. In glucose respiration we refer to the energy in the hydrolysis of the 30 ATP molecules that are formed during the respiration of glucose.

turn to the study of the redox potential and the amazing series of molecular partnerships that have been formed in the oxidation–reduction reactions in the cell.

In the second part of the chapter, we complement our studies of energy by exploring the way in which energy is converted into useful work through the application of forces. Our study of forces begins by considering how both molecular motors and cytoskeletal filaments exert forces in processes ranging from vesicle transport to chromosome segregation to cell division to the motion of cells across surfaces. This is followed by a discussion of the physical limits of force-generating structures such as cytoskeletal filaments. How much force can an actin filament or a microtubule support before it will rupture?

When writing this chapter, it became apparent to us that some of the energies, like those of a photon or combustion of a sugar, are easy to pinpoint accurately. Others are trickier because they depend upon the concentration of the various molecular players, such as the case of the hydrolysis of ATP. Finally, there are the cases where it is very hard to even define, let alone provide a concrete value. Examples of these subtle cases include the energy of a hydrogen bond, the free energies associated with the hydrophobic effect, and the entropic cost of forming a complex of two molecules. While it is easy to clearly define and separate the length of a biological object from its width, it is much harder to separate, say, the energy arising from a hydrogen bond from the other interactions, such as those with the surrounding water. Together, the case studies presented in this chapter acknowledge the importance of energy in biological systems and attempt to give a feeling for energy transformations that are necessary for cell growth and survival.

BIOLOGY MEETS PHYSICS

What is the thermal energy scale and how is it relevant to biology?

Molecules are engaged in incessant random motions as a result of their collisions with the molecules of the surrounding medium, as will be described in more detail in our discussion of Brownian motion in the vignette entitled "What are the time scales for diffusion in cells?" (pg. 211).

What was not clear when Robert Brown made the discovery of the motions that now bear his name is that his observations struck right to the heart of one of the most important organizing principles in all of biology—namely, the way in which the competition between deterministic and stochastic energies dictates phenomena in nearly all the molecular processes of life.

The physical consequences arising from thermal forces are familiar to us all. For example, think of the drop-off in the density of air as a function of altitude. The height dependence of the density of air reflects an interplay between the force of gravity, which implies an increasing potential energy investment as the molecules rise higher in the atmosphere, and the entropy benefit, which comes from allowing the molecules to explore a larger volume by increasing their altitude.

The simplest way to analyze the effects of the competition between energetic and entropic contributions to free energy in the setting of molecular and cell biology is to equate the deterministic energy of interest to $k_B T$, which reflects the thermal energy scale. To see this play out in the familiar everyday example of the density as a function of the altitude, this strategy corresponds to equating the potential energy of the molecule, given by mgh, to the thermal energy $k_B T$, as noted in **Table 3-1**. Following this idea and solving for h, we estimate a length scale of

$$h = k_B T/mg \approx 10 \text{ km,} \qquad (3.2)$$

length scale name	energetic term	entropic term	equation	characteristic value	BNID
atmospheric concentration decay length	gravitational	occupation of spatial states	$mgh = k_B T \Rightarrow h_{th} = \dfrac{k_B T}{mg}$ m: mass h: height g: acceleration due to gravity	8 km	111406
persistence length	bending	number of states of polymer chain	$E \times I/\xi_P = k_B T \Rightarrow \xi_P = \dfrac{EI}{k_B T}$ E: Young's modulus I: moment of inertia	DNA: 50 nm actin: 10 μm microtubule: 1–10 mm	103112, 105505, 105534
Bjerrum length	electrostatic interaction	occupation of spatial states	$kq^2/l_B = k_B T \Rightarrow l_B = \dfrac{kq^2}{k_B T}$ q: charge k: Coulomb's constant	0.7 nm	106405
Debye length	electrostatic interaction	occupation of spatial states	$2c_\infty \lambda_D^2 q^2/\varepsilon_0 D = k_B T$ $\Rightarrow \lambda_D = \sqrt{\dfrac{\varepsilon_0 D k_B T}{2c_\infty q^2}}$ c_∞: salt conc. D: dielectric constant	1 nm (at 100 mM monoionic conc.)	105902

Table 3-1 Length scales that emerge from the interplay of deterministic and thermal energies.

which is indeed a good estimate for the height at which the density of the atmosphere is reduced significantly from its density at the surface of the Earth (more precisely, by a factor of e, the natural logarithm—that is, ≈threefold). In statistical mechanics, the balance struck between energy and thermal fluctuations is codified through the so-called Boltzmann distribution, which tells us that the probability of a state with energy E is proportional to $\exp(-E/k_{B}T)$, illustrating explicitly how the thermal energy governs the accessibility of microscopic states of different energy.

In the cellular context, several important length scales emerge as a result of the interplay between thermal and deterministic energies (for examples, see Table 3-1). If we think of DNA (or a cytoskeletal filament) as an elastic rod, then when we equate the bending energy and the thermal energy, we find the scale at which spontaneous bending can be expected as a result of thermal fluctuations, also known as the persistence length. For DNA this length has been measured at roughly 50 nm (BNID 103112), and for the much stiffer actin filaments it is found to be 10 μm (BNID 105505). The interplay between Coulomb interactions and thermal effects for the case of charges in solution is governed by another such scale called the Bjerrum length. It emerges as the length scale for which the potential energy of electrical attraction is equated to $k_{B}T$ and represents the distance over which electrostatic effects are able to dominate over thermal motions. For two opposite charges in water, the Bjerrum length is roughly 0.7 nm (BNID 106405).

The examples given above prepare us to think about the ubiquitous phenomena of binding reactions in biology. When thinking about equilibrium between a bound state and an unbound state, as in the binding of oxygen to hemoglobin, a ligand to a receptor, or an acid H-A and its conjugated base A^{-}, there is an interplay between energies of binding (enthalpic terms) and the multiplicity of states associated with the unbound state (an entropic term). This balancing act is formally explored by thinking about the free energy ΔG. Thermodynamic potentials such as the Gibbs free energy take into account the conflicting influences of enthalpy and entropy. Though often the free energy is the most convenient calculational tool, conceptually, it is important to remember that the thermodynamics of the situation is best discussed with reference to the entropy of the system of interest and the surrounding "reservoir." Reactions occur when they tend to increase the overall entropy of the world. The enthalpic term, which measures how much energy is released upon binding, is a convenient shorthand for how much entropy will be created outside of the boundaries of the system as a result of the heat released from that reaction.

In light of these ideas about the free energy, we can consider how binding problems can be thought of as an interplay between enthalpy and entropy. The entropy that is gained by having an unbound particle is in the common limit of dilute solutions proportional to $-\ln(c/1\text{ M})$, where c is the concentration. We are careful not to enrage the laws of mathematics, which do not allow taking a logarithm of a value that has units, so we divide by a standard concentration. The lower the concentration c, the higher the gain in entropy from adding a particle (note the minus sign in front of the logarithm). But how do we relate the entropy gained and the enthalpy? This linkage is made once again through the quantity $k_B T$. Note that $k_B T \ln([c]/1\text{ M})$ has units of energy, and stands for the entropic contribution to the free energy gained upon liberating the molecule of interest from its bound configuration. This will be compared to the energy released in the binding process. Whichever term is bigger will govern the direction the process will proceed, towards binding or unbinding. When the two terms are equal, we reach a state of equilibrium with equal propensity for both bound and unbound states. By computing the condition for equality to hold, we can determine the critical concentration at which the entropic and enthalpic terms exactly balance.

Guided by this perspective, we now explore one of the classic case studies for every student of biochemistry. In particular, we examine how the pK_as of amino acids can be understood as a competition. As shown in **Estimate 3-1**, this competition can be understood as a balance between the entropic advantage of freeing up charges to let them wander around in solution and the energetic advantages dictated by interactions, such as Coulomb's law, which tends to keep opposite charges in close proximity. The pK_a is defined as the pH where an ionizable group (releasing H^+) is exactly half ionized and half neutral. So the place of c in the equation for the entropy is taken by 10^{-pK_a} M. Armed with the understanding of this connection between pK_a and the entropic term, we can better appreciate the significance of pK_a as a tuning parameter. Note from above that the entropy change upon liberating a molecule (or ion) at concentration c is

$$\Delta S = -k_B \ln(c/1\text{ M}). \tag{3.3}$$

For $c = 10^{-pK_a}$, this means the entropy change is given by

$$\Delta S = -pK_a \times k_B \ln(10). \tag{3.4}$$

In particular, if the pK_a is higher by one unit, the entropic term required to balance the energy of interaction (the enthalpic gain) is higher by a value of $k_B T \cdot \ln(10)$. If we express $k_B T \cdot \ln(10)$ in units of kJ/mol, we get 6 kJ/mol, the same value mentioned in the rule of thumb connecting concentration ratios and energies. If the interaction is a purely electrostatic one, we can interpret it using Coulomb's law. The binding energy of two opposite charges in water increases by about $k_B T \cdot \ln(10)$ when the distance between the charges changes from 0.3 nm to 0.15 nm (both being characteristic

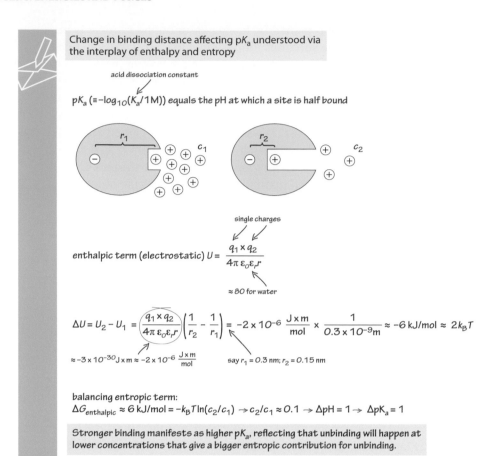

Estimate 3-1

interatomic distances). We thus note that if the charges are 0.15 nm closer (the difference of 0.3 and 0.15 nm), then the pK_a will be one unit higher. This shows how in equilibrium, the stronger attractive interaction coming from the closer distance between the charges will lead to half ionization at a lower concentration of the separated charges in the solvent. Lower concentration means there is a higher associated entropic gain per separated charge, and this higher gain is required to balance the attractive force. This is a manifestation, though quite abstract, we admit, of how forces and energies relate to concentrations.

It is interesting to observe how many physical processes have energies that are similar at the nanometer length scale, as depicted in **Figure 3-2**. This makes the life of biomolecules intriguing, because rather than having one major process dominating their interactions, such as, say, gravitation for astronomical length scales, they are governed by an intricate interplay between, for example, electrostatic repulsion and attraction forces, mechanical deformations, thermal energy, and chemical bonds energies.

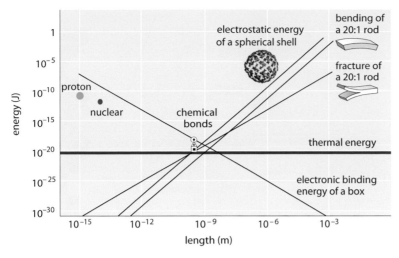

Figure 3-2 The convergence of the energies associated with many physical phenomena to a similar range at the nanometer length scale. (Adapted from Phillips R & Quake S [2006] *Phys Today* 59:38–43.)

What is the energy of a hydrogen bond?

Hydrogen bonds are ubiquitous and at the heart of many biological phenomena, such as the formation of the alpha helix and beta sheet secondary structures in proteins, as shown in **Figure 3-3**. Similarly, the binding of base

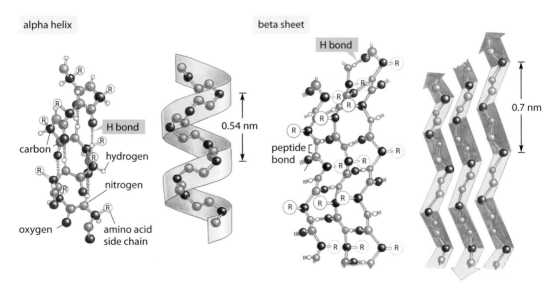

Figure 3-3 Hydrogen bonding and protein secondary structure. Alpha helix and beta sheet structures both depend upon hydrogen bonding, as labeled in both of the schematics. (Adapted from Alberts B, Johnson A, Lewis J et al. [2015] Molecular Biology of the Cell, 6th ed. Garland Science.)

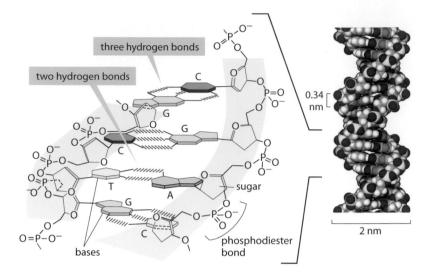

Figure 3-4 Base pairing in the DNA double helix. Illustration of how bases are assembled to form DNA, a double helix with two "backbones" made of the deoxyribose and phosphate groups. The four bases form stable hydrogen bonds with one partner such that A pairs only with T and G pairs with C. A space-filling atomic model approximating the structure of DNA is shown on the right. The spacing between neighboring base pairs is roughly 0.34 nm. (Adapted from Alberts B, Johnson A, Lewis J et al. [2015] Molecular Biology of the Cell, 6th ed. Garland Science.)

pairs in DNA that holds the double helix together is based on every adenine forming two hydrogen bonds with thymidine and every cytosine forming three hydrogen bonds with guanine, as depicted in **Figure 3-4**. The binding of transcription factors to DNA is often based on the formation of hydrogen bonds, thus reflecting a nucleic acid–protein form of hydrogen bonding. These bonds govern the off-rate for transcription factor unbinding and thus the dissociation constant (with the on-rate often being diffusion limited and thus nonspecific to the binding site). In addition, hydrogen bonds are often central to the function of catalytic active sites in enzymes.

Because of the high frequency of hydrogen bonding in the energy economy of cells, it is natural to wonder how much free energy is associated with the formation of these bonds. Indeed, the energy scale of these bonds, slightly larger than the scale of thermal energies, is central to permitting the transient associations so typical of macromolecular interactions and that would be completely forbidden if these bonds were based upon covalent interactions instead. Though the length of hydrogen bonds is quite constant at ≈0.3 nm (BNID 108091), their energies defy simple and definitive characterization. This provides a challenging and interesting twist on this most basic of biological interactions. One of the ways to come to terms with the nuance in the free energy of hydrogen bonding is to appreciate that the members of a hydrogen bond can interact with their environment in many different ways. If a hydrogen bond is broken, the two members will form alternative hydrogen bonds with the surrounding solvent (water). But this raises the following question: If the dissolution of a hydrogen bond results in the formation of other hydrogen bonds, what is the source of any associated free-energy change? In fact, such bonding rearrangements alter the level of order in the solvent, and thus the entropy can be the dominant free-energy contribution.

Given the strong context dependence of the strength of hydrogen bonding, this becomes one of those cases in our book where order-of-magnitude thinking is more useful and honest than the attempt to provide one definitive number. A rule-of-thumb range for the energies associated with hydrogen bonds is 6–30 kJ/mol ($\approx$2–12 $k_B T$) (BNID 105374, 103914, 103913).

To get a better sense of the magnitude of hydrogen bond energies, we consider biology's iconic molecule of DNA. From the moment of its inception, the structure of DNA implied stories about how the molecule works. That is, through stacks of base pairs, the double helix hides within itself a model of how it might be replicated. As noted above, A-T pairing is characterized by two hydrogen bonds, whereas C-G base pairing is characterized by three such bonds, as shown in Figure 3-4. One of the first things that any student learns when joining a molecular biology lab is how to use a PCR machine, and one of the first bits of training that goes with it is programming that machine to go through its rhythmic changes in temperature as the double helix is melted and annealed again and again. What sets the temperatures used? In a word, the A-T content of the sequence of interest, reflecting in turn the number of hydrogen bonds that have to be disrupted and the strength of base stacking, another important biophysical process governing DNA stability (BNID 111667).

An even more compelling example of the magnitude of the base pairing effect is to use it to think about the specificity of codon–anticodon recognition in translation. The triplet pairing rule for tRNAs to recognize their mRNA partners is based upon each of the three bases pairing with its appropriate partner. But let's see how much discriminatory power such bonding is worth. For example, what happens when a C-G base pair is replaced by an incorrect C-T "base pair"? Now, many things change, but at least one hydrogen bond that should be present is no longer there. The Boltzmann distribution tells us how to evaluate the relative probability of different events as

$$p(1)/p(2) = \exp(-\Delta E/k_B T), \qquad (3.5)$$

where ΔE is the energy difference between those two states. If $\Delta E \approx -6$ kJ/mol ($= -2.3\, k_B T$), a lower end value for hydrogen bond energies, this implies a 10-fold difference in the two probabilities resulting already from only this hydrogen bond difference. The actual fidelity in codon–anticodon recognition is much higher and requires the energy-driven mechanism of kinetic proofreading. The beauty of this simple estimate is that it shows how the machinery of the Boltzmann distribution can be used to connect changes in hydrogen bonding energies to different levels of molecular discrimination.

The importance of hydrogen bonds lies not only in their energy, which leads to favorable binding, but also in their strong dependence on conformation. A slight change in the angle or distance between the relevant atoms and the energy will change drastically. For example, as shown in **Figure 3-5**, a hydrogen bond is the key element for specificity in a transporter protein that has to differentiate between two chemical groups of very similar size and charge—namely, phosphate and arsenate. A change

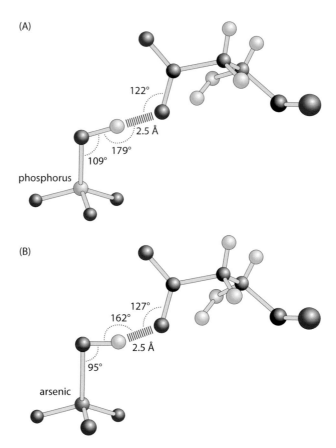

Figure 3-5 Hydrogen bonding angles for a phosphate-binding protein are close to optimal in the phosphate-bound structure, but are distorted with arsenate. (A) A close-up view of the short hydrogen bond between oxygen of a bound phosphate and the carboxylate of aspartate. The binding angles are close to the canonical optimal values. (B) The same bond in the arsenate-bound structure has a distorted suboptimal interaction angle. This difference can readily account for the ≈ 500-fold difference in favor of phosphate binding over arsenate. (Adapted from Elias M, Wellner A, Goldin-Azulay K et al. [2012] *Nature* 491:134–137.)

in angle and distance can result in orders-of-magnitude differences in the binding strength, making these bonds a key element conferring specificity. The spatial dependence is much weaker for other free-energy contributions, such as the hydrophobic effect discussed in the next vignette.

What is the energy scale associated with the hydrophobic effect?

Water is a polar material. This means that water molecules have a charge distribution that results in a net dipole moment. As a result of this important feature, when foreign molecules that do not themselves have such a dipole moment (such as hydrophobic amino acids, for example) are placed in water, the resulting perturbation to the surrounding water molecules incurs a free-energy cost. Interestingly, from a biological perspective, this free-energy cost is one of the most important driving forces for wide classes of molecular interactions, many of which lead to the formation of some of the most famed macromolecular assemblies, such as lipid bilayers and viruses.

This energetic effect, most commonly observed in water, is called the hydrophobic effect. Though the hydrophobic effect is extremely subtle and depends upon the size of the solute, for large enough molecules the hydrophobic effect can be approximated as being proportional to the area of the interface (the so-called interfacial energy). Specifically, for sufficiently large solutes in water, the energy penalty arising from adding a nonpolar area within water can be approximated as an interfacial energy of $\approx 4\ k_\mathrm{B}T/\mathrm{nm}^2$ or $\approx 10\ \mathrm{kJ/mol/nm}^2$ (BNID 101826). There is a long and rich theoretical tradition associated with trying to uncover the origins of this free-energy penalty, and for our discussion we adopt a particularly simple heuristic perspective, cognizant of the fact that a full theoretical treatment is fraught with difficulties. The argument goes that when a molecule with a "hydrophobic interface" is placed in water, the number of conformations (shown in **Figure 3-6**) of the surrounding water molecules is decreased. Since these water molecules have fewer accessible states, they have lower entropy, and hence the situation is less favorable in terms of free energy. Using this simple model suffices to estimate the free-energy scale associated with hydrophobic interactions that was presented above, though it breaks down for small solutes, where the hydrogen bonding network can readjust itself around the solute. In relation to the simplified conformations shown in Figure 3-6, it is suggested that next to the interface only, say, three of the six will be possible, because the others will not have a way to make hydrogen bonds. This illustrates how the conformation of the tetrahedral

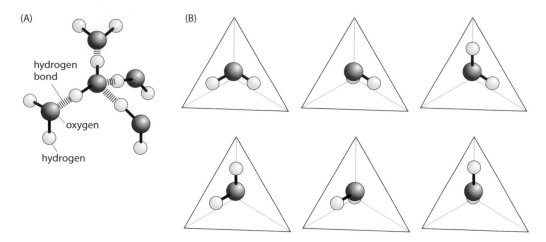

(A)

hydrogen
bond

oxygen

hydrogen

(B)

Figure 3-6 Simplified model of the hydrophobic effect. (A) A simplified model for possible orientations of water molecules in a tetrahedral network. (B) Each image shows a different arrangement of the water molecule that permits the formation of hydrogen bonds with neighboring water molecules. The hydrogen bonds are oriented in the directions of the vertices that are not occupied by hydrogens in the figure. Formation of a hydrophobic interface deprives the system of the ability to explore all of these different states, thus reducing the entropy. (Adapted from Dill K & Bromberg S [2011] Molecular Driving Forces, 2nd ed. Garland Science.)

network of hydrogen bonds is thought to be compromised when a nonpolar molecule is placed in solution. If there are 10 water molecules per nm^2 of interface (each ≈ 0.3 nm in size), and the number of conformations for each molecule decreased by a factor of 2, the free-energy cost is $10 \times k_B T \times \ln(2)/nm^2$, which is within a factor of two of the measured value.

To get a better feeling for these numbers, consider an O_2 molecule dissolved in water. We can estimate the area of contact with the surrounding water by thinking of a box ≈ 0.2 nm on each side. This gives an area of $\approx 6 \times 0.2^2 \approx 0.2$ nm^2, resulting in a free-energy penalty of about 1 $k_B T$. Every $k_B T$ translates to an equilibrium concentration in water that is lower by a factor of e (2.718...), based on the Boltzmann distribution, which states that a difference of energy E translates into a decreased occupancy of $e^{-E/k_B T}$. Nonpolar metabolites, which are an order of magnitude larger in area, will have a prohibitively large free-energy cost and are thus insoluble in water. A parallel challenge exists with polar compounds, such as peptides, RNA, and metabolites that have small occupancy in hydrophobic environments and are thus restrained from transferring across the nonpolar, hydrophobic lipid bilayer membranes of cells.

The hydrophobic effect can play a significant role in determining the binding affinity of a metabolite to an enzyme, as shown schematically in **Estimate 3-2**. A methyl group has a surface area of about 1 nm^2. A nonpolar

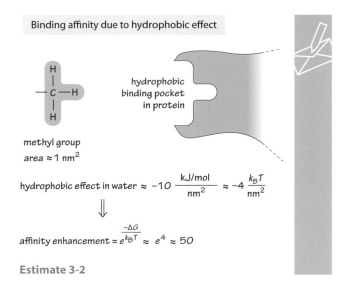

Binding affinity due to hydrophobic effect

methyl group
area ≈ 1 nm²

hydrophobic
binding pocket
in protein

$$\text{hydrophobic effect in water} \approx -10 \ \frac{\text{kJ/mol}}{\text{nm}^2} \approx -4 \ \frac{k_B T}{\text{nm}^2}$$

$$\text{affinity enhancement} = e^{\frac{-\Delta G}{k_B T}} \approx e^4 \approx 50$$

Estimate 3-2

surface initially exposed to water that gets buried within a hydrophobic binding pocket has a predicted stabilizing free-energy gain of ≈10 kJ/mol/ nm². For a methyl group, we thus find a free-energy difference of ≈10 kJ/ mol ≈4$k_B T$, which translates into an affinity enhancement of ≈e⁴≈ 50-fold, as derived in the calculation in Estimate 3-2.

These same types of arguments can be made for many of the most important macromolecules in the cell, ranging from the contributions to the driving force for protein folding to the basis of protein–protein contacts, such as those between the repeated subunits that make up a viral capsid, to the ways in which lipids congregate to form lipid bilayers. The lipid effect is best illustrated through the way in which the critical micelle concentration (that is, that concentration at which free lipids will no longer be tolerated in solution and they come together to make little spheres) depends upon both the lengths and number of tails in the lipid molecule. In this case, the hydrophobic cost for a given lipid scales linearly as n, the number of carbons in its tail. The corresponding critical micelle concentration depends exponentially on this value of n, as observed experimentally.

How much energy is carried by photons used in photosynthesis?

Nuclear reactions taking place 150 million kilometers away in the sun's interior are used to drive the bustling activity of life observed on planet

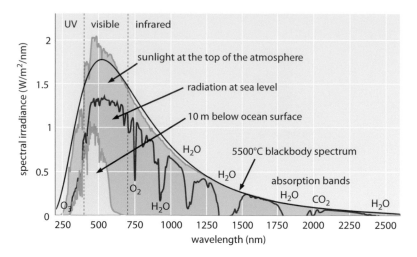

Figure 3-7 Spectrum of solar irradiation. The different curves show the radiation due to a blackbody at a temperature of ≈5500°C, the radiation density at the top of Earth's atmosphere, and the radiation density at sea level. The various absorption peaks due to the presence of the atmosphere are labeled with the relevant molecular species. (Adapted from the National Renewable Energy Laboratory.)

Earth. The energy that drives these biological reactions is heralded by the arrival of packets of light from space known as photons. In this vignette, we interest ourselves in how much energy is carried by these photons.

Even though the nuclear reactions deep within the sun are taking place at temperatures in excess of a million kelvin, during the journey of a photon from the sun's interior to its surface, it is absorbed and reemitted numerous times and is only emitted for the last time near the sun's surface, where the temperature is much lower. The sun's emission spectrum is thus that of a blackbody at ≈5500°C (**Figure 3-7**; BNID 110208, 110209). However, as a result of our own atmosphere, the photons reaching Earth's surface do not reflect a perfect blackbody spectrum since several wavelength bands get absorbed, as shown in Figure 3-7, resulting in a spectrum full of peaks and troughs.

The overall process taking place in photosynthesis, which serves as the energetic basis for our biosphere, is depicted in **Figure 3-8**. To drive photosynthesis, a photon must be energetic enough to excite photosynthetic pigments. In particular, this excitation refers to the fact that an electron in the pigment needs to get shifted from one molecular energy level to another. These pigments are coupled, in turn, to the photochemical machinery that converts electromagnetic energy into chemical energy by producing charge separation. This charge separation takes several forms. One contribution comes from an imbalance of protons across membranes,

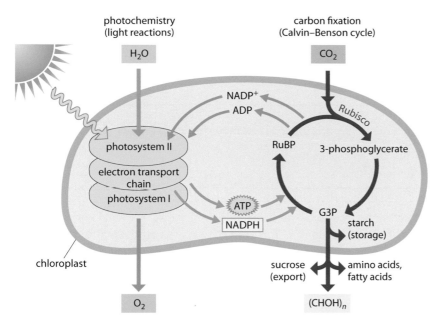

Figure 3-8 The flow of energy in the biosphere. Energy coming from the sun in the form of photons is stored through photochemical reactions in ATP and NADPH, while producing oxygen from water. These energy currencies are then used in order to fix inorganic carbon by taking carbon dioxide from the air and transforming it into sugars that are the basis for biomass accumulation and long-term energy storage in the biosphere.

which drive the ATP synthases—the molecular machines that synthesize ATP, resulting in an end product of ATP itself. The second form of charge separation manifests itself in the form of reducing power, the term for the transient storage of electrons in carriers such as NADPH, which are used later for stable energy storage in the form of sugars produced in the Calvin–Benson cycle. The redox reactions that drive the production of this reducing power are themselves driven by the light-induced excitation of pigments.

The vast majority (>99.9%[*]) of these photochemical transformations are performed by the most familiar and important pigment of them all—namely, chlorophyll. In the chlorophyll molecule, an electron moves to an excited energy level as a result of the absorption of a photon with wavelength ≤ 700 nm. To convert wavelength into energy, we exploit the Planck relation,

$$E = h\nu, \tag{3.6}$$

[*] Raven JA (2009) Functional evolution of photochemical energy transformations in oxygen-producing organisms. *Funct Plant Biol* 36:505–515.

which relates the photon energy E to its frequency ν via Planck's constant h. We then use the fact that the frequency times the wavelength λ is equal to the speed of light

$$\nu\lambda = c. \tag{3.7}$$

If we work using nm units to characterize wavelengths, this relation can be rewritten as

$$E = hc/\lambda \approx 1240/\lambda, \tag{3.8}$$

where energy is expressed in eV (electron volts). We thus get an energy scale of 1.8 eV at 700 nm. This unit of energy is equivalent to the energy that an electron will gain when moving across a potential difference of 1.8 volts. To transform into more familiar territory, it is equivalent to $\approx$180 kJ/ (mol photons) or $\approx$70 $k_B T$/photon. This is equivalent, in turn, to several times the energy associated with the hydrolysis of ATP, or the transfer of protons across the cell's membrane, and is thus quite substantial at the molecular scale.

Using Figure 3-7, we can estimate the overall energy flux associated with the incident photons and how many such photons there are. A crude but simple approximation is to replace the actual spectrum by a rectangle of width $\approx$1000 nm (between 300 nm and 1300 nm) and height $\approx$1 W/m^2/nm. Based on these values, the area under the curve is roughly 1000 W/m^2, which is quite close to the measured value for the mean incident power per unit area measured at the Earth's surface. As an aside, 1 kW is the average electrical power consumption per person in the Western world.

How many photons make up this steady stream of incident radiation? If we make yet another simplifying assumption—namely, that all photons have the same energy as the 180 kJ/(mol photons) we calculated above for a 700-nm photon, then we estimate that there are $\approx$1000 W/m^2/180 kJ/mol $\approx$ 5 mmol photons/s·m^2. The unit corresponding to 1 mole of photons per square meter bears the name of Albert Einstein, and our estimates show us that the number of photons incident on a 1 m^2 area each second is $\approx$5000 μ einsteins, or about 3×10^{21} photons. More than half of these photons are actually invisible to us because they are located in the infrared wavelength range (above 700 nm). Photons are absorbed by pigments such as chlorophylls that have effective cross sections for absorption of about 10^{-21} m^2 (BNID 100339). Given the photon flux on the order of 10^{21} photons/m^2, we infer about one excitation per chlorophyll per second.

The process of photosynthesis is the basis for humanity's usage of land and freshwater resources through the practice of agriculture. This conversion of light energy into mostly grains is performed at an efficiency that in well-cultivated conditions reaches about 1% (BNID 100761). Though it sounds low, we should appreciate the number of hurdles faced along the way. About one-half the incident energy occurs at infrared wavelengths and does not excite the chlorophyll molecules that convert the photons to excited electrons. Short wavelengths excite the chlorophyll, but the energy beyond the minimal excitation energy is quickly dissipated, causing on the average loss of another factor of two. The light and dark reactions usually get saturated at about one-tenth of the maximal sun intensity, and a process of photoinhibition diverts that energy into heat. Of the energy harvested and stored as sugar, about one-half is used by the plant to support itself through respiration. Finally, the harvest index, which is the fraction of biomass that can be consumed, is rarely more than one-half. A crude rule of thumb is that 1 m^2 will produce about 1 kg per year of edible dry mass.

Researchers working on photosynthesis are often asked: "Can you make a human that relies on photosynthesis and does not have to eat?" The short answer is no, and that is where the discussion usually ends. But let's entertain the possibility of covering the skin with photosynthetic tissue. The human skin is about 1 m^2 in area (BNID 100578). At a characteristic efficiency of 1%, this will yield under the peak 1000 W/m^2 noon sun about 10 W, which is still an order of magnitude lower than human requirements, as discussed in the vignette entitled "What is the power consumption of a cell?" (pg. 99).

What is the entropy cost when two molecules form a complex?

Biology is driven by molecular interactions. Our understanding of the constant flux back and forth between molecules with different identities is largely a story about free-energy differences between reactants and products, as all science students learn in their first chemistry course. However, the cursory introduction to these matters experienced by most students casts aside a world of beautiful subtleties that center on the many ways in which the free energy of a molecular system is changed as a result of molecular partnerships. Here we focus on the contribution to the free energy resulting from the entropy changes when molecules bind.

In this vignette, we address a simple conceptual question—namely, when two molecules A and B interact to form the complex AB, how large is the entropy change as a result of this interaction? The free energy G has the generic form

$$G = H - TS, \tag{3.9}$$

where H is the enthalpy and S is the entropy.

We see that in a simple case in which there is no enthalpy change, the entire free-energy balance is dictated by entropy. If a reaction increases the entropy, there is a corresponding negative free-energy change, signaling the direction in which reactions will spontaneously proceed. A deep though elusive insight into these abstract terms comes from one of the most important equations in all of science—namely,

$$S = k_B \ln W \tag{3.10}$$

which tells us how the entropy of a system S depends upon the number of microstates available to it, as captured by the quantity W. An increase in entropy thus reflects an increase in the number of microstates of the system. Assuming the system has the same chance to be in any microstate, spontaneous jiggling in the space of possible states will indeed lead the system to move to the condition with the most states—that is, with the highest entropy. At the risk of being clear to only those who had especially clear teachers (a substitute is Dill and Bromberg's excellent book, *Molecular Driving Forces*), we note that even the term representing the enthalpy change in the free energy is actually also an entropy term in disguise. Concretely, this term reflects the heat released outside of the system where it will create entropy. This effect is included in the calculation of the free energy, because it is a compact way of computing the entropy change of the "whole world" while focusing only on the system of interest.

A ubiquitous invocation of these far-reaching ideas is in understanding binding interactions. In these cases, there is a competition between the entropy available to the system when ligands are jiggling around in solution and the enthalpy released from the bonds created upon their binding to a receptor, for example. When a ligand has a dissociation constant of, say, 1 μM, it means that half the receptors will be bound with ligands at that concentration. When the concentration of a ligand is lower, the ligand in solution will have a larger effective volume to occupy with more configurations and thus will favor the unbound state. As a result, the receptor or enzyme will be in a state of lower fractional occupancy. At the other extreme, when the ligand concentration is higher than the dissociation constant, the ligand when unbound has a more limited space of configurations to explore in solution and the binding term will prevail, resulting in higher occupancy of the bound state. This is the statistical mechanical way of thinking about the free energy of binding as a strict competition between entropic and enthalpic terms.

What fundamentally governs the magnitude of the entropic term in these binding reactions? This is a subject notorious for its complexities, and we only touch on it briefly here. The entropy change upon binding is usually calculated with reference to the standard-state concentration of $c_0 = 1$ M (which can be thought of as a rough estimate for the effective concentration when bound) and is given by

$$\Delta S = -k_B \ln (c/c_0), \tag{3.11}$$

where c is the prevailing concentration of the ligand. Specifically, this formula compares the number of configurations available at the concentration of interest to that when one particle binds to the receptor at that same concentration. We now aim to find the actual magnitude of the entropy change term estimated by using the same expression

$$\Delta S = -k_B \ln (c/c_0). \tag{3.12}$$

If ligand–receptor binding occurs at concentration $c = 10^{-n}$ M, the entropy change is given by

$$\Delta S = nk_B \ln 10 \approx 10\text{--}20 \, k_B T \tag{3.13}$$

for $n \approx 4\text{--}8$; that is, 10 nM–100 μM. Using more sophisticated theoretical tools, this entropy change has been estimated for ligands binding to proteins to have values ranging from $\approx 6\text{--}20 \, k_B T \approx 15\text{--}50$ kJ/mol (BNID 109148, 111402, 111419), a range generally in line with the simple estimate sketched above. For protein–protein binding, a value under standard conditions of $40 \, k_B T \approx 100$ kJ/mol was estimated (BNID 109145, 109147). These calculations were partially derived from analyzing gases, because fully accounting for solvation effects is a big unresolved challenge. Inferring the value from experiments is challenging, but several efforts result in values of $\approx 6\text{--}10 \, k_B T \approx 15\text{--}25$ kJ/mol (BNID 109146, 111402) for cases ranging from the polymerization of actin, to the binding of oxygen to hemoglobin, as well as the interaction of biotin and avidin.

As discussed above, binding is associated with an entropic cost that is offset by enthalpic gain. An important consequence of this interplay is the ability to build extremely strong interactions from several interactions to the same substrate, which are each quite weak. In the first interaction, the entropic term offsets the binding energy, creating only a modest dissociation constant. But if a second binding interaction of the very same substrate occurs concurrently with the first one, the entropic term was already "paid," and the associated free-energy change will be much more substantial. Consider the case of binding of the actin monomer to the actin filament built of two protofilaments, and thus resulting in two concurrent binding interactions. The binding to each protofilament is independently quite weak, with a dissociation constant of 0.1 M, but the joint dissociation constant is 1 μM, because the ≈10 $k_B T$ entropic term is not offsetting the binding energy twice but only once. This effect, also referred to in the term avidity, is at the heart of antibodies binding specifically and tightly to antigens, as well as many other cases, including transcription factors binding to DNA, viral capsid formation, and so on.

How much force is applied by cytoskeletal filaments?

Force generation by cytoskeletal filaments is responsible for a diverse and important set of biological processes ranging from cell motility to chromosome segregation. These distinct mechanical functions are implemented by rich and complex structures (for example, branched, crosslinked, etc.) in which the filaments are linked together in various arrangements. Whether we think of the forces exerted by actin filaments at the leading edge of motile cells or the complicated arrangement of forces applied by microtubules during the process of chromosome segregation, understanding the basis and limits of force generation is a central pillar of modern cell biology.

Recent years have seen a steady stream of clever ideas for measuring how force is generated by the filaments of the cytoskeleton. Specifically, a series of beautiful measurements have made it possible to query the forces applied by bundles of cytoskeletal filaments and, more amazingly, of individual filaments engaged in the process of polymerization. Like with many force measurements, conceptually the idea is to use the deflection of a calibrated spring to read out the forces. Measurements on cytoskeletal filaments have exploited such generalized springs in several different ways. First, in optical traps, laser light can be used to trap a micron-sized bead, which can then be used as a "spring" to read out the forces of growing cytoskeletal filaments as they push against it. For small displacements of the bead away from the laser focus, there is a linear restoring force tending to push the bead back to the focus. As a result, bead displacement can serve as a surrogate for force itself. Using a setup like that shown in **Figure 3-9A**, the force generation due to individual filaments has been

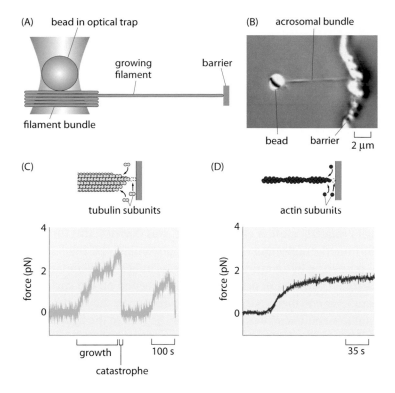

Figure 3-9 Optical-trap measurement of the force of polymerization. (A) Schematic of the use of an optical trap to measure forces as the filament grows into an obstacle. (B) Microscopy image of a bead (2 μm diameter) with attached acrosomal bundle. (C) Time evolution of the force during growth and catastrophe of microtubules. (D) Force buildup over time for actin polymerization. (A, and C, adapted from Kerssemakers JW, Munteanu EL, Laan L et al. [2006] *Nature* 442:709–712. B, and D, adapted from Footer MJ, Kerssemakers JW, Theriot JA & Dogterom M [2007] *Proc Natl Acad Sci USA* 104:2181–2186.)

measured directly by permitting the cytoskeletal filament to crash into a barrier during the process of polymerization. As the elongation proceeds, the restoring force exerted by the bead increases until it reaches the maximal force that can be overcome by the polymerization of the filament— the so-called stall force. At this point, the dynamics resets, as shown in **Figure 3-9B** in what is known as a shrinkage catastrophe. Such measurements result in a characteristic force scale of order 5 pN, comparable to the forces exerted by the more familiar translational motors such as myosin and kinesin. This value can be compared to the energy driving filament construction usually based on ATP or GTP hydrolysis. Such hydrolysis reactions provide roughly 20 $k_B T$ of free energy per nucleotide hydrolyzed and should be compared to the work done by a force of 5 pN acting over a monomer extension length, which is about 4 nm—that is, 20 pN × nm. In our tricks of the trade introduction, we refer to the rule of thumb that $k_B T$ is roughly equal to 4 pN nm, and thus the filament force acting over the 4-nm distance corresponds to a free energy of about 5 $k_B T$. Given that the energy conversion is imperfect, this seems like a very reasonable correspondence between the free energy available from nucleotide hydrolysis and the work done by the polymerizing filament.

Often, the behavior of cytoskeletal filaments is dictated by their collective action rather than by the properties of individual filaments. A veritable army of different proteins can alter the arrangements of cytoskeletal filaments by capping them, crosslinking them, nucleating branches, and a host of other alterations, thus shaping their force-generating properties. To measure the collective effects that emerge when more than one cytoskeletal filament is acting in concert, another clever "spring" was devised. This time the spring results from the deflection of a small (approximately 20-micron) cantilever when pushed on by an array of filaments, as indicated schematically in **Figure 3-10**. A collection of actin filaments is seeded on the surface beneath the cantilever and then, as the filaments polymerize, they make contact with the cantilever and bend it upwards. As can be seen in the figure, when many such filaments work together, the resulting force scale is tens of nN rather than several pN, as was found in the case of individual filaments.

What can a handful of pN buy you? A characteristic mammalian cell has a mass of a few ng (that is, a few 10^{-12} kg), corresponding to a weight of

$$W = m \times g = 10^{-12} \text{ kg} \times 10 \text{ m/s}^2 = 10 \text{ pN}. \tag{3.14}$$

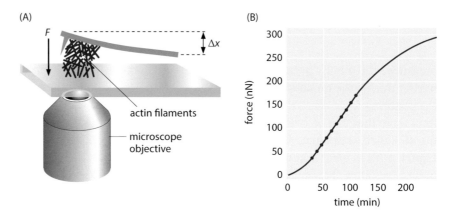

Figure 3-10 Force due to polymerization of a bundle of actin filaments. (A) Schematic of the geometry of force measurement using a calibrated cantilever. (B) Measurement of the buildup of force over time. Note that the number of filaments schematically drawn in (A) is orders of magnitude lower than the actual number that created the actual force measured in (B). (Adapted from Parekh SH, Chaudhuri O, Theriot JA & Fletcher DA [2005] *Nat Cell Biol* 7:1219–1223.)

We can fancifully state that a few filaments can already hold a cell against gravity's pull, just like professional rock climbers can stabilize themselves over a cliff with one hand. Beyond the cellular drama, this simple estimate helps us realize why the force of gravity is usually not of much consequence in the lives of cells. The prevailing forces on components in the cellular environment are much higher than those produced by gravity.

What are the physical limits for detection by cells?

Living organisms have evolved a vast array of technologies for taking stock of conditions in their environment. Some of the most familiar and impressive examples come from our five senses. The "detectors" utilized by many organisms are especially notable for their sensitivity (ability to detect "weak" signals) and dynamic range (ability to detect both very weak and very strong signals). Hair cells in the ear can respond to sounds varying over more than six orders of magnitude in pressure difference between the detectability threshold (as low as 2×10^{-10} atmospheres of sound pressure) and the onset of pain (6×10^{-4} atmospheres of sound pressure). Note that given that atmospheric pressure is equivalent to pressure due to 10 meters of water, a detection threshold of 2×10^{-10} atmospheres would result from the mass of a film of only 10-nm thickness—that is, a few dozen atoms in height. Indeed, the enormous dynamic range of our hearing capacity leads to the use of logarithmic scales (for example, decibels) for describing sound intensity (which is the square of the change in pressure amplitude). The usage of a logarithmic scale is reminiscent of the Richter scale, which permits us to describe the very broad range of energies associated with earthquakes. The usage of the logarithmic scale is also fitting as a result of the Weber–Fechner law, which states that the subjective perception of many different kinds of senses, including hearing, is proportional to the logarithm of the stimulus intensity. Specifically, when a sound is a factor of 10^n more intense than some other sound, we say that that sound is $10 \times n$ decibels more intense. According to this law, we perceive sounds that differ by the same number of decibels as equally different. Some common sound levels, measured in decibel units, are shown in **Figure 3-11**. Given the range from 0 to roughly 130, this implies a dazzling 13 orders of magnitude. Besides this wide dynamic range in intensity, the human ear responds to sounds over a range of 3 orders of magnitude in frequency between roughly 20 Hz and 20,000 Hz, while at the same time being able to detect the difference between 440 Hz and 441 Hz.

Similarly impressively, rod photoreceptor cells can register the arrival of a single photon (BNID 100709). For cones, a value of ≈100 is observed

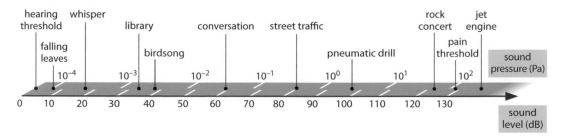

Figure 3-11 Intensities of common sounds in units of pressure and decibels.

(BNID 100710). Here again, the acute sensitivity is complemented by a broad dynamic range, thus permitting us to see not only on bright sunny days but also on moonless starry nights with a 10^9-fold lower illumination intensity. A glance at the night sky in the Northern Hemisphere greets us with a view of the North Star (Polaris). In this case, the average distance between the photons arriving on our retina from that distant light source is roughly a kilometer, demonstrating the extremely feeble light intensity reaching our eyes (as well as how fast the speed of light is).

By studying the observed minimum stimuli detected by cells, we can ask if evolution drove organisms all the way to the limits dictated by physics. To begin to see how the challenge of constructing sensors with high sensitivity and wide dynamic range plays out, we consider the effects of temperature on an idealized tiny, frictionless mass–spring system, as shown in **Figure 3-12,** where the spring is characterized by a spring constant κ. The goal is to measure the force applied on the mass. It is critical to understand how noise influences our ability to make this measurement. The mass will be subjected to constant thermal jiggling as a result of collisions with the molecules of the surrounding environment as well as other internal processes within the spring itself. As an extension to the discussion in the vignette entitled "What is the thermal energy scale and how is it relevant to biology?" (pg. 154), the energy resulting from these collisions equals $1/2\,k_B T$, where k_B is Boltzmann's constant and T is the temperature in kelvin. What this means is that the mass will spontaneously jiggle around its equilibrium position, as shown in Figure 3-12, with the mean squared deflection $\langle x_2 \rangle$ set by the condition that

$$\frac{1}{2}k\langle x_2 \rangle = \frac{1}{2}k_B T. \tag{3.15}$$

As noted above, just like an old-fashioned scale used to measure the weight of fruits or humans, the way we measure the force is by reading out the *displacement* of the mass. Hence, in order for us to measure the force, the displacement must exceed a threshold set by the thermal jiggling. That

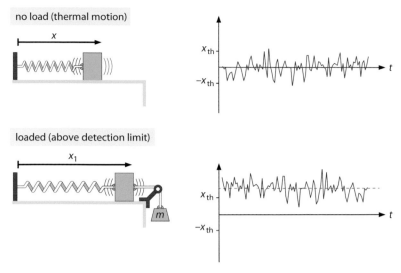

Figure 3-12 Deflection of a mass–spring system. In the top panel, there is no applied force, and the mass moves spontaneously on the frictionless table due to thermal fluctuations. In the lower panel, a force is applied to the mass–spring system by hanging a weight on it. The graphs show the position of the mass as a function of time, revealing both the stochastic and deterministic origins of the motion. As shown in the lower panel, in order to have a detectable signal, the mean displacement needs to be above the amplitude of the thermal motions.

is, we can only say that we have measured the force of interest once the displacement exceeds the displacements that arise spontaneously from thermal fluctuations or

$$\frac{1}{2}kx^2_{\text{measured}} \geq \frac{1}{2}k_\text{B}T. \tag{3.16}$$

Imposing this constraint results in

$$x_{\text{min}} = (k_\text{B}T/k)^{1/2}, \tag{3.17}$$

which gives a force limit of

$$F_{\text{min}} = (kk_\text{B}T)^{1/2}. \tag{3.18}$$

This limit states that we cannot measure smaller forces because the displacements they engender could just as well have come from thermal agitation. One way to overcome these limits is to increase the measurement time (which depends on the spring constant). Many of the most clever tools of modern biophysics, such as the optical trap and atomic-force microscope, are designed both to overcome and exploit these effects.

To give a concrete example, we consider the case of the hair cells of the ear. Each such hair cell features a bundle of roughly 30–300 stereocilia, as shown in **Figure 3-13**. These stereocilia are approximately 10 microns in

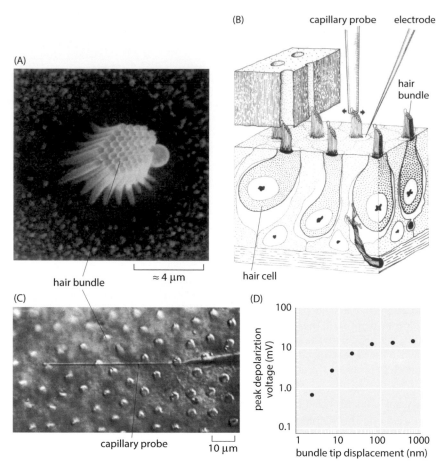

Figure 3-13 Response of hair cells to mechanical stimulation. (A) Bundle of stereocilia in the cochlea of a bullfrog. (B) Schematic of the experiment showing how the hair bundle is manipulated mechanically by the capillary probe and how the electrical response is measured using an electrode. (C) Microscopy image of cochlear hair cells from a turtle and the capillary probe used to perturb them. (D) Voltage as a function of the bundle displacement for the hair cells shown in part (C). (A, adapted from Hudspeth AJ [1989] *Nature* 341:397–404. B, adapted from Hudspeth AJ & Corey DP [1977] *Proc Natl Acad Sci USA* 74:2407–2411. C, and D, adapted from Crawford AC & Fettiplace R [1985] *J Physiol* 364:359–379.)

length (BNID 109301, 109302). These small cellular appendages serve as springs that are responsible for transducing the mechanical stimulus from sound and converting it into electrical signals that can be interpreted by the brain. In response to changes in air pressure, the stereocilia are subjected to displacements that result in the gating of ion channels, which leads, in turn, to further signal transduction. Different stereocilia respond at different frequencies, which makes it possible for us to distinguish the melodies of Beethoven's Fifth Symphony from the cacophony of a car horn. The mechanical properties of the stereocilia are similar to those of the spring that was discussed in the context of Figure 3-12. By pushing on individual stereocilia with a small glass fiber, as shown in Figure 3-13, it is possible to measure the minimal displacements of the stereocilia that can trigger a detectable change in voltage. Rotation of the hair bundle by only 0.01 degree, corresponding to nanometer-scale displacements at the tip, are sufficient to elicit a voltage response of mV scale (BNID 111036, 111038). How do these numbers compare to those expected from thermal jiggling of the stereocilia? Both the observed Brownian motion, as well as simple theoretical estimates using spring models like that described above, reveal thermal motions of the stereocilia tips that are several nanometers in size. However, the hearing threshold appears to correspond to displacements of as little as 0.1 nm, as seen in Figure 3-13. This interesting discrepancy actually points to the fact that the hair bundle is active and amplifies input close to its resonant frequency, as well as the fact that the stereocilia are coupled, effects not considered in the simple estimate.

This same kind of reasoning governs the physical limits for our other senses as well. Namely, there is some intrinsic noise added to the property of the system we are measuring. Hence, to get a "readout" of some input, the resulting output has to be larger than the natural fluctuations of the output variable. For example, the detection and exploitation of energy carried by photons is linked to some of life's most important processes, including photosynthesis and vision. How many photons suffice to result in a change in the physiological state of a cell or organism? In now-classic experiments on vision, the electrical currents from individual photoreceptor cells stimulated by light were measured. **Figure 3-14** shows how a beam of light was applied to individual photoreceptors and how electric current traces from such experiments were measured. The experiments revealed two key insights. First, photoreceptors undergo spontaneous firing, even in the absence of light, revealing precisely the kind of noise that real events (that

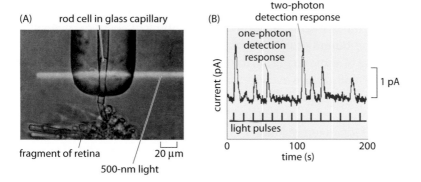

Figure 3-14 Single-photon response of individual photoreceptors. (A) Experimental setup shows a single rod cell from the retina of a toad in a glass capillary and subjected to a beam of light. (B) Current traces as a function of time for a photoreceptor subjected to light pulses in an experiment like that shown in part (A). (A, adapted from Baylor DA, Lamb TD & Yau KW [1979] *J Physiol* 288:589–611. B, adapted from Rieke F & Baylor DA [1998] *Rev Mod Phys* 70:1027–1036.)

is, the arrival of a photon) have to compete against. In particular, these currents are thought to result from the spontaneous thermal isomerization of individual rhodopsin molecules, as discussed in the vignette entitled "How many rhodopsin molecules are in a rod cell?" (pg. 142). This isomerization reaction is normally induced by the arrival of a photon and results in the signaling cascade we perceive as vision. Second, examining the quantized nature of the currents emerging from photoreceptors exposed to very weak light demonstrates that such photoreceptors can respond to the arrival of a single photon. This effect is shown explicitly in Figure 3-14B.

Another class of parameters that cells "measure" with great sensitivity includes the absolute numbers, identities, and gradients of different chemical species. This is a key requirement in the process of development where a gradient of morphogen is translated into a recipe for pattern formation. A similar interpretation of molecular gradients is important for motile cells as they navigate the complicated chemical landscape of their watery environment. These impressive feats are not restricted to large and cognizant multicellular organisms such as humans. Even individual bacteria can be said to have "knowledge" of their environment, as illustrated in the exemplary system of chemotaxis already introduced in the vignette entitled "What are the absolute numbers of signaling proteins" (pg. 135). That "knowledge" leads to purposeful discriminatory power, where even a few molecules of attractant per cell can be detected and amplified and differences in concentrations over a wide dynamic range of about five orders of magnitude can be amplified (BNID 109306, 109305). This enables unicellular behaviors in which individual bacteria will swim up a concentration gradient of chemoattractant. To get a sense of the exquisite sensitivity of these systems, **Estimate 3-3** gives a simple calculation of the concentrations being measured by a bacterium during the chemotaxis process and estimates the changes in occupancy of a surface receptor that is detecting the gradients. In particular, if we think of chemical detection by membrane-bound protein receptors, the presence of a ligand is read out is by virtue of some change in the occupancy of that receptor. As the figure shows, a small change in concentration of ligand leads to a corresponding change in the occupancy of the receptor. For the case of bacterial chemotaxis, a typical gradient detected by bacteria in a microscopy experiment can be reasoned out as follows. If we consider bacteria swimming roughly 1 mm away from a pipette with 1 mM concentration of chemoattractant, the gradient is of order 10^{-2} μM/μm (BNID 111492). Is such a gradient big or small? A single-molecule difference detection threshold can be defined as follows

$$\frac{\Delta c}{\Delta L} = \frac{1\,\text{molecule/bacterium volume}}{\text{length of bacterium}} = \frac{1\,\text{nM}}{1\,\mu\text{m}} = 10^{-3}\,\frac{\mu\text{M}}{\mu\text{m}}. \qquad (3.19)$$

What is the fractional change in occupancy of a receptor in a concentration gradient?

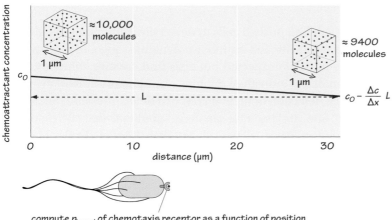

compute p_{bound} of chemotaxis receptor as a function of position

$$p_{bound}(x=0) = \frac{\dfrac{c_0}{K_d}}{1 + \dfrac{c_0}{K_d}} \quad ; \quad p_{bound}(x=30\mu m) = \frac{\dfrac{c_0 - \dfrac{\Delta c}{\Delta x}L}{K_d}}{1 + \dfrac{c_0 - \dfrac{\Delta c}{\Delta x}L}{K_d}}$$

characteristic distance between tumbles

compute fractional change in p_{bound}

$$\frac{\Delta p}{p_0} = \frac{p_{bound}(x=0) - p_{bound}(x=30\mu m)}{p_{bound}(x=0)} \approx \frac{\dfrac{\Delta c}{c_0}}{1 + \dfrac{c_0}{K_d}} \quad \text{for } c_0 = K_d \Longrightarrow \frac{\Delta p}{p_0} = \frac{1}{2}\frac{\Delta c}{c_0}$$

for $c_0 = 10$ µM and $\dfrac{\Delta c}{\Delta x} = 0.02$ µM/µm $\Longrightarrow$ fractional change $= \dfrac{\Delta p}{p_0} \approx \boxed{0.03}$
in occupancy

from measurements

Estimate 3-3

Interestingly, recent work has demonstrated that bacteria can even detect gradients smaller than this (though an attendant insight is that the quantity being "measured" by the cells is actually the gradient of the logarithm of the concentration). As noted above, this small gradient can be measured over a very wide range of absolute concentrations, illustrating both the sensitivity and dynamic range of this process, but also revealing a more nuanced mechanism than the simple occupancy hypothesis described in Estimate 3-3. Like with the hair cell example discussed earlier in the vignette, chemotaxis receptors are adaptive. These same kinds of arguments arise in the context of development, where morphogen gradient interpretation is based upon nucleus-to-nucleus measurements of concentration differences. For example, in the establishment of the anterior–posterior patterning of the fly embryo, neighboring nuclei are "measuring" roughly 500 and 550 molecules per nuclear volume and using that difference to make decisions about developmental fate.

In summary, evolution pushed cells to detect environmental signals with both exquisite sensitivity and impressive dynamic range. In this process, physical limits must be respected but cells find creative solutions. Photoreceptors can detect individual photons, the olfactory system nears the single-molecule detection limit, hair cells can detect pressure differences as small as 10^{-9} atm, and bacteria can detect gradients that correspond to less than one molecule per cell per cell length, a dazzling display of subtle and beautiful mechanisms.

ENERGY CURRENCIES AND BUDGETS

How much energy is released in ATP hydrolysis?

ATP is often referred to as the energy currency of the cell. Hundreds of reactions in the cell, from metabolic transformations to signaling events, are coupled to the hydrolysis (literally meaning "water loosening") of ATP by water. The reaction

$$ATP + H_2O \rightleftharpoons ADP + P_i \qquad (3.20)$$

transforms adenosine triphosphate (ATP) into adenosine diphosphate (ADP) and inorganic phosphate (P_i). The free-energy change associated with this reaction drives a large fraction of cellular reactions, with the membrane potential and reducing power being the other two dominant energy sources. But exactly how much is this energy currency worth, and what does it reveal about the chemical transactions that can be purchased? There is no single answer to this question, because the amount of energy liberated by this hydrolysis reaction depends upon the intracellular conditions, but it is possible to get a feeling for the approximate "value" of this currency by resorting to some simple estimates.

The Gibbs free-energy change (ΔG) due to ATP hydrolysis depends upon the concentrations of the various participants in the reaction, as depicted in **Estimate 3-4**. When the concentrations are farther from their equilibrium values, the absolute value of ΔG is greater. Under "standard" conditions (that is, concentrations of 1 M for all reactants except water, which is

Free energy of ATP hydrolysis under physiological conditions

$ATP + H_2O \rightleftharpoons ADP + P_i$

equilibrium concentrations, $[\ ]_{eq}$, define $\Delta G'^0$ = the standard free energy

$$K'_{eq} = \frac{[ADP]_{eq}/[1M] \times [P_i]_{eq}/[1M]}{[ATP]_{eq}/[1M] \times [H_2O]_{eq}/[55M]} \ ; \ \Delta G'^0 = -RT\ln(K'_{eq}) \approx -30 \text{ to } -40 \ \frac{kJ}{mol}$$

correcting for physiological concentrations, $[\ ]_{phys}$

$$Q' = \frac{[ADP]_{phys}/[1M] \times [P_i]_{phys}/[1M]}{[ATP]_{phys}/[1M]} \ ; \ \Delta G' = \Delta G'^0 + RT\ln Q' \approx -50 \text{ to } -70 \ \frac{kJ}{mol}$$

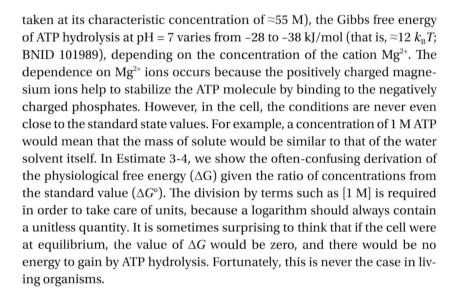

Estimate 3-4

taken at its characteristic concentration of ≈55 M), the Gibbs free energy of ATP hydrolysis at pH = 7 varies from −28 to −38 kJ/mol (that is, ≈12 k_BT; BNID 101989), depending on the concentration of the cation Mg^{2+}. The dependence on Mg^{2+} ions occurs because the positively charged magnesium ions help to stabilize the ATP molecule by binding to the negatively charged phosphates. However, in the cell, the conditions are never even close to the standard state values. For example, a concentration of 1 M ATP would mean that the mass of solute would be similar to that of the water solvent itself. In Estimate 3-4, we show the often-confusing derivation of the physiological free energy (ΔG) given the ratio of concentrations from the standard value (ΔG°). The division by terms such as [1 M] is required in order to take care of units, because a logarithm should always contain a unitless quantity. It is sometimes surprising to think that if the cell were at equilibrium, the value of ΔG would be zero, and there would be no energy to gain by ATP hydrolysis. Fortunately, this is never the case in living organisms.

In practice, the physiological conditions depend on the organism being studied, on the tissue or compartment within the cell under consideration, and on the current energy demands for metabolic and other reactions. For example, in perfused rat liver, the ATP-to-ADP ratio was found to be about 10:1 in the cytosol but 1:10 in the mitochondria under high rates of glycolysis, and under low rates of glycolysis both ratios were much closer to 1 (BNID 111357). Therefore, a range of values for ΔG is expected. The key to understanding this range is to get a sense of how much Q differs from K—that is, how the concentrations differ from standard conditions. The typical intracellular concentrations of all the relevant components (ATP, ADP, and P_i) are in the mM range, much lower than standard conditions. The ratio $[ADP][P_i]/[ATP]$ with concentrations in the mM range is much lower than one, and the reaction will be energetically more favorable than at standard conditions, as shown in **Table 3-2**. The highest value ≈−70 kJ/mol (≈30k_BT) was calculated from values in the human muscle

physiological condition of organism	ATP conc.	ADP conc.	P$_i$ conc.	Inferred ΔG' (kJ/mol)	Inferred ΔG' ($k_B T$)	BNID
standard conditions	1 M	1 M	1 M	−36 to −38	−14 to −15	106580, ionic strength dependent
E. coli aerobic exponential growth on glucose	10 mM	0.6 mM	20 mM	−54	−22	104704
E. coli anaerobic exponential growth on glucose	3 mM	0.4 mM	10 mM	−54	−22	101964
E. coli aerobic exponential growth on glycerol	7 mM	0.7 mM	10 mM	−55	−22	101701
S. cerevisiae aerobic growth on glucose	2 mM	0.3 mM	22 mM	−52	−21	106017
spinach Spinacia oleracea chloroplast stroma in light	2 mM	0.8 mM	10 mM	−51	−21	108113
spinach Spinacia oleracea cytosol + mitochondria in light	3 mM	0.7 mM	10 mM	-52	−22	108113
spinach Spinacia oleracea cytosol + mitochondria in dark	1.5 mM	0.8 mM	10 mM	−50	−21	108113
Homo sapiens—resting muscle	8 mM	9 µM	4 mM	−68	−27	101943
Homo sapiens—muscle recovery from severe exercise	8 mM	7 µM	1 mM	−72	−29	101944

Table 3-2 Free energy for ATP hydrolysis in various organisms and under different physiological conditions. Inferred ΔG' calculations based on a value of ΔG'° of −37.6 kJ/mol. This makes the values in the table consistent among themselves, but creates small deviations from the ΔG' values reported in the primary sources. Such deviations can result from variations in ionic strength, pH, and biases in the measurement method. Values are rounded to one or two significant digits. In spinach, where P$_i$ concentration was not reported, a characteristic value of 10 mM was used (BNID 103984, 103983, 111358, 105540).

of athletes following exertion (BNID 101944). In *E. coli* cells growing on glucose, a value of −47 kJ/mol was reported (≈20 $k_B T$; BNID 101964). To put these numbers in perspective, a molecular motor that exerts a force of roughly 5 pN (BNID 101832) over a 10-nm (BNID 101857) step size does work of order 50 pN nm, requiring slightly more than 10 $k_B T$ of energy, which is well within the range of what a single ATP can deliver.

The calculations of ΔG require an accurate measurement of the relevant intracellular concentrations. Such concentrations are measured *in vivo* in humans by using nuclear magnetic resonance (NMR). The natural form of phosphorus (^{31}P) has magnetic properties, so there is no need to add any external substance. The tissue of interest (for example, muscle) is placed in a strong magnetic field, and shifts in frequency of radio pulses are used to infer the concentration of ATP and P$_i$ directly from the peaks in the NMR spectra. In *E. coli*, the concentrations of ATP can be measured more directly with an ATP bioluminescence assay. A sample of growing bacteria removed from the culture can be assayed using luciferase, a protein from bacteria that live in symbiosis with squids, but that has now joined the toolbox of biologists as a molecular reporter. The luciferase enzyme uses ATP in a reaction that produces light that can be measured using a luminometer, and the ATP concentration can be inferred from the signal strength. So we have cell content as an input and luciferase as a "device" that transforms the amount of ATP into light emission, which serves as the measured output. Using tools such as these, we find that in "real life" ATP is worth twice as much as under "standard" conditions because of the concentrations being more favorable for the forward reaction.

We finish by noting that it is a standing question as to why the adenine nucleotide was singled out to serve as the main energy currency, with GTP and the other nucleotides serving much more minor roles. Is it a case of random choice that later became a "frozen accident" or was there a selective advantage to ATP over GTP, CTP, UTP, and TTP?

What is the energy in transfer of a phosphate group?

ATP hydrolysis is one of the quintessential reactions of the cell and has led some to christen the ATP synthase, which adds phosphate groups onto ADP, as "the world's second most important molecule" (DNA arguably being the first). But phosphate groups have much broader reach than in their role as one of the key energy currencies of the cell. Though the central dogma paints a picture of the great polymer languages as being written to form "sentences" of nucleic acids and proteins as long chains of nucleotides and amino acids, respectively, in fact these languages also use accents. Specifically, the letters making up the alphabets used in these languages are accented by a host of different chemical modifications, some of which involve the addition and removal of charged groups such as phosphates. Just like in the French language, for example, where an accent can completely change the sound and meaning of a word, these molecular accents do the same thing.

What are the functional consequences of these various modifications to nucleic acids and proteins, and how can we understand them in terms of the overall free-energy budget of these molecules? Phosphate groups, for example, are often one of the key carriers of cellular energy. The case of ATP and the energetics associated with its hydrolysis was already discussed in a separate vignette entitled "How much energy is released in ATP hydrolysis?" (pg. 182). In proteins, phosphate groups serve as information carriers. Specifically, a limited set of amino acids can be subject to phosphorylation because only they have the functional groups available that can serve as phosphate tagging sites (–OH in serine, threonine, tyrosine, and rarely, aspartate, and –NH in histidine).

A simple "coarse-grained" picture of the role of such charged groups is that they shift the energetic balance between different allowed states of the molecule of interest. For example, a given protein might have several stable configurations, with one of those states having an overall lower free energy. The addition of a charged group such as a phosphate can then tip the free-energy balance such that now a different conformation has the lowest free energy. As an example, we take the protein Ste5 in yeast, which can be bound to the membrane or unbound. These two states have significant implications for signaling in the process of mating, as well as many other decisions dictated by the MAPK pathway. The propensity to adopt either of these two forms of Ste5 is controlled through phosphorylation. The phosphorylated form of the protein was measured to have a decreased binding energy to the membrane of ≈ 6 kJ/mol (≈ 2 $k_B T$; BNID 105724), which is equivalent to an affinity ratio of ≈ 20 between the phosphorylated and unphosphorylated states. Phosphorylation also decreases the binding affinity energy of the transcription factor Ets1 to DNA by ≈ 1.6 kJ/mol (≈ 0.7 $k_B T$) or about a factor of 2 in the affinity binding constant (BNID 105725).

In shifting from phosphate groups as tags on proteins to their role as energy carriers, it is essential to understand that the amount of energy released when a phosphate group bond dissociates depends on the compound it is attached to. Common metabolites exhibit a big difference in the energy released upon hydrolysis of their phosphate group. For example, it is ≈ 60 kJ/mol (≈ 24 $k_B T$) for the hydrolysis of PEP (phosphoenolpyruvate), but only ≈ 13 kJ/mol (≈ 5 $k_B T$) for glucose-6-phosphate (BNID 105564). In

reaction	Gibbs energy for hydrolysis reaction ΔG_r^0 (kJ/mol)
phosphoanhydride (acid–acid) bond hydrolysis	
$ATP + H_2O \rightarrow ADP + phosphate$	−31 (−13 $k_B T$)
$ADP + H_2O \rightarrow AMP + phosphate$	−31 (−13 $k_B T$)
$ATP + H_2O \rightarrow AMP + PPi$	−38 (−16 $k_B T$)
$PPi + H_2O \rightarrow 2\ phosphate$	−24 (−10 $k_B T$)
phosphoester (alcohol–acid) bond hydrolysis	
glucose 6-phosphate $+ H_2O \rightarrow$ glucose $+$ phosphate	−12 (−5 $k_B T$)
3-phosphoserine $+ H_2O \rightarrow$ serine $+$ phosphate	−10 (−4 $k_B T$)
$AMP + H_2O \rightarrow$ adenosine $+$ phosphate	−14 (−6 $k_B T$)
DHAP $+ H_2O \rightarrow$ dihydroxyacetone $+$ phosphate	−15 (−6 $k_B T$)
fructose 1,6-bisphosphate $+ H_2O \rightarrow$ F6P $+$ phosphate	−16 (−6 $k_B T$)
glyceraldehyde 3-phosphate $+ H_2O \rightarrow$ glyceraldehyde $+$ phosphate	−17 (−7 $k_B T$)
threonine phosphate $+ H_2O \rightarrow$ threonine $+$ phosphate	−19 (−8 $k_B T$)

Table 3-3 Standard Gibbs energy released in the hydrolysis of different types of phosphate bonds. Values are from the Equilibrator website and are based on experimental measurements. Values are rounded to two significant digits.

Table 3-3 we collect information on the energetics of reactions involving phosphate bonds. Data on thermodynamic properties, such as the change in Gibbs energy in biochemical reactions, can be found using the eQuilibrator database (http://equilibrator.weizmann.ac.il/). Such differences are at the heart of the energetic transformations that take place in glycolysis and the TCA cycle, the cell's energy, and carbon highways. What accounts for these differences? Is there an easy rule of thumb that can be applied to predict the energetic content of such groups?

These differences can be partially understood through the changes in bond type, as illustrated in the different scenarios depicted in **Figure 3-15**. Phosphate groups bound to another phosphate group are one type (known as phosphoanhydride bonds), while those phosphate groups that bind to an alcohol are a different type (known as phosphoester bonds). A naïve way to rationalize this is that a carbon surrounded by hydrogens is more "electron rich," and the bond to the overall negative phosphate group is more stable and its hydrolysis less favorable. This contrasts with the case of a phosphate or carboxyl group, where the double bond of the carbon to oxygen makes it "electron poor," and thus the bond to phosphate, which is also "electron poor," is unstable and its hydrolysis is more energetic. A more quantitative explanation is based on the pK_a values of the groups, whereas the fully rigorous explanation requires quantum mechanical analysis.

For example, as the main ingredient of signal transduction, ATP hydrolysis to ADP and inorganic phosphate (breaking a bond between two phosphates) is used in order to phosphorylate amino acids in proteins. In the

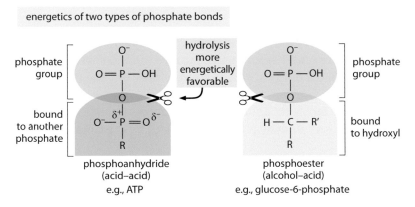

Figure 3-15 The energetics of two types of phosphate bonds. The phosphoanhydride bond is less stable and much further removed from equilibrium and thus much more energetic when hydrolyzed by water. The acid–acid bond could be with another phosphate, as in ATP, or alternatively with a carboxyl (that is, acetyl phosphate), which it is even more energetic than the phosphate–phosphate bond.

most common cases, those of phosphorylation on serine and threonine, the phosphate group reacts with a hydroxyl group (–OH). Other amino acids that can be phosphorylated are tyrosine, histidine, and aspartate, the latter two serving in the important example of two-component signaling systems in prokaryotes. Such transfers are carried out by kinases and are energetically favorable. A phosphatase performs the reverse reaction of severing the phosphate bond in a protein. The action of the phosphatase is thermodynamically favorable (though might require activation), because the phosphate bond on the protein is still far from equilibrium. In biochemist lingo, the transfer of a phosphate group from an ATP (a phosphoanhydride) to an amino acid (a phosphoester) still retains close to half of the free energy of ATP hydrolysis.

In closing, we remind the reader that, as mentioned in the vignette entitled "How much energy is released in ATP hydrolysis?" (pg. 182), the free-energy potential is a function of the distance from equilibrium, which depends on the concentration. We stress the counterintuitive assertion that the energy in a bond depends on the concentrations of the molecules. At equilibrium, the concentrations are such that energies for all transformations are zero, even the so-called "energy-rich" ATP hydrolysis reaction. In fact, there is energy to be had from an ATP molecule only if it is out of equilibrium from its surrounding.

What is the free energy released upon combustion of sugar?

Like humans, bacteria have preferences about what they eat. In a series of beautiful and insightful experiments, Jacques Monod showed that when bacteria are offered different carbon sources, they first use their preferred carbon source before even turning on the genes to use the others. The substrate of choice for bacteria such as *E. coli* is glucose, a molecule known to every biochemistry student as the starting point for the famed reactions of glycolysis.

The free energy released in completely oxidizing glucose to CO_2 by oxygen is $\approx$ –3000 kJ/mol (BNID 103388 and http://equilibrator.weizmann.ac.il/classic_reactions). Expressed in other units, this is $\approx$ –700 kcal/mol, or $\approx$ –1200 $k_B T$, where a kcal is what people often count as Calories (capitalized). As is clear from the schematic showing the range of biological energy scales at the beginning of this chapter, this energy is at the high end of the scale of molecular energies. To get a better idea for how much energy this is, let's think about the delivery of useful work from such reactions. One of the ways of reckoning the potential for useful work embodied in this energy release is by examining the number of ATP molecules that are produced (from ADP and P_i) in the series of reactions tied to combustion of sugar, also known in biochemical jargon as cellular aerobic respiration. The cell's metabolic pathways of glycolysis, the TCA cycle, and the electron transfer chain couple the energy release from the combustion of a single molecule of glucose to the production of roughly 30 ATP molecules (BNID 101778), which is sufficient energy to permit several steps of the molecular motors that drive our muscles or to polymerize a few more amino acids into a nascent polypeptide.

We learn from the labels on our cereal boxes that human daily caloric intake is recommended to consist of 2000 kcal. If supplied only through glucose, that would require about 3 mol of the sugar. From the chemical formula of glucose—namely, $C_6H_{12}O_6$—the molecular weight of this sugar is 180 Da, and thus 3 mol corresponds to ≈ 500 g. So half a kilogram of pure sugar (whether it is glucose, sucrose, or as is often common today, high-fructose corn sugar, so-called HFCS) would supply the required energy of combustion to fuel all the processes undertaken by an "average" person in a single day, though not in a nutritionally recommended fashion.

To get a better sense of the energetic value of all of this glucose, we now consider what would happen if the body did not conduct the heat of combustion of these recommended 2000 kcal into the environment, but rather used that energy to heat the water in our bodies. A calorie is defined as the energy required to increase the temperature of 1 g of water by 1°C (denoted by c below). For a human with a mass (m) of 70 kg, the potential increase in temperature resulting from the energy released in combustion (ΔQ) over a day can be estimated by the relation

$$\Delta T = \Delta Q/(c \times m) = 2 \times 10^6 \text{ cal} / (1 \text{ cal/°C} \times \text{gram})$$
$$\times (70 \times 10^3 \text{ gram}) \approx 30°C, \qquad (3.21)$$

illustrating that the energy associated with our daily diet has a lot of heating capacity.

What is the redox potential of a cell?

Redox potentials are used to characterize the free-energy cost and direction of reactions involving electron transfer, one of the most ubiquitous and important of biochemical reactions. Such reduction–oxidation reactions are characterized by a free-energy change that shares some conceptual features with that used to describe pK_a in acid–base reactions, where proton transfer rather than electron transfer is involved. In this vignette, one of the most abstract in the book, we discuss how the redox potential can be used as a measure of the driving force for a given oxidation–reduction reaction of interest. By way of contrast, unlike the pH, there is no sense in which we can assign a single redox potential to an entire cell.

The redox potential, or more accurately the reduction potential, of a compound refers to its tendency to acquire electrons and thereby to be reduced. Some readers might remember the mnemonic "OILRIG," which reminds us that "oxidation is loss, reduction is gain", where the loss and gain are of electrons. Consider a reaction that involves an electron transfer

$$A_{ox} + ne^- \rightleftharpoons A_{red}, \tag{3.22}$$

where n electrons are taken up by the oxidized form (A_{ox}) to give the reduced form (A_{red}) of compound A. The redox potential difference ΔE between the electron donor and acceptor is related to the associated free-energy change ΔG of the reaction via

$$\Delta G = nF\Delta E, \tag{3.23}$$

where n is the number of electrons transferred and F is Faraday's constant (96,485 J/mol/V or $\approx$100 kJ/mol/V). By inspecting tabulated values of these potentials, it is possible to develop an intuition for the tendency for electron transfer and, hence, of the direction of the reaction.

Though ATP is often claimed to be the energy currency of the cell, in fact, for the energetic balance of the cell, the carriers of reducing power are themselves no less important. The most important example of these carriers is the molecule NADH in its reduced or oxidized (NAD$^+$) forms. We can use the redox potential to connect these two molecular protagonists, in which case we estimate an upper bound on the number of ATP molecules that can be produced from the oxidation of NADH (produced, for example, in the TCA cycle). The NAD$^+$/NADH pair has a redox potential of $E = -0.32$ V, and it is oxidized by oxygen to give water (protons coming from the media) with a redox potential of $E = +0.82$ V. Both are shown in **Figure 3-16** as part of a "redox tower" of key biological half-reactions that can be linked to find the overall redox potential change and, thus, the free energy. For the reaction considered above of NADH oxidation by oxygen, the maximal associated free energy that can be extracted is thus

$$\Delta G = n \times F \times \Delta E = 2 \times 100 \text{ kJ}/(\text{mol} \times \text{V}) \times (0.82-(-0.32)) \text{ V}$$
$$= 230 \text{ kJ/mol} \approx 90 \, k_B T, \tag{3.24}$$

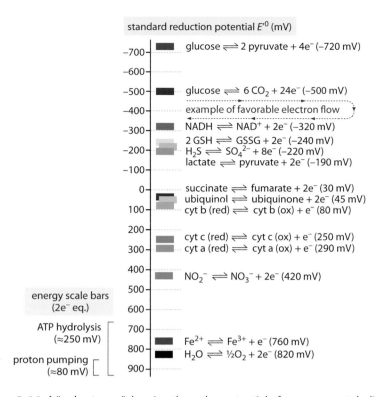

standard reduction potential E'^0 (mV)

−700	glucose $\rightleftharpoons$ 2 pyruvate + 4e$^-$ (−720 mV)
−600	
−500	glucose $\rightleftharpoons$ 6 CO_2 + 24e$^-$ (−500 mV)
−400	example of favorable electron flow
−300	NADH $\rightleftharpoons$ NAD$^+$ + 2e$^-$ (−320 mV)
	2 GSH $\rightleftharpoons$ GSSG + 2e$^-$ (−240 mV)
−200	$H_2S \rightleftharpoons SO_4{}^{2-}$ + 8e$^-$ (−220 mV)
	lactate $\rightleftharpoons$ pyruvate + 2e$^-$ (−190 mV)
−100	
0	succinate $\rightleftharpoons$ fumarate + 2e$^-$ (30 mV)
	ubiquinol $\rightleftharpoons$ ubiquinone + 2e$^-$ (45 mV)
100	cyt b (red) $\rightleftharpoons$ cyt b (ox) + e$^-$ (80 mV)
200	
	cyt c (red) $\rightleftharpoons$ cyt c (ox) + e$^-$ (250 mV)
300	cyt a (red) $\rightleftharpoons$ cyt a (ox) + e$^-$ (290 mV)
400	
	$NO_2{}^- \rightleftharpoons NO_3{}^-$ + 2e$^-$ (420 mV)
500	
600	
700	
800	$Fe^{2+} \rightleftharpoons Fe^{3+}$ + e$^-$ (760 mV)
	$H_2O \rightleftharpoons \tfrac{1}{2}O_2$ + 2e$^-$ (820 mV)
900	

energy scale bars
(2e$^-$ eq.)

ATP hydrolysis
(≈250 mV)

proton pumping
(≈80 mV)

Figure 3-16 A "redox tower" showing the redox potential of common metabolic half-reactions. Metabolic processes can be thought of as moving electrons between molecules, often capturing some of the energy released as the electrons move from high-energy to lower-energy states, as in glycolysis or respiration. Electrons donated by the "half-reactions" on top can be consumed in a half-reaction lower on the tower to complete a thermodynamically favorable reaction. For example, the net process of glycolysis involves the oxidation of glucose to pyruvate, coupled with the reduction of NAD$^+$ to NADH. Since the oxidation of glucose lies at the top of the tower and the reduction of NAD$^+$ is below it, this overall reaction is thermodynamically favorable. Comparing to the ATP hydrolysis scale bar, we can also see that this reaction is sufficiently favorable to generate ATP. Aerobic respiration involves many intermediate electron transfers through the electron transport chain. Several of these transitions are shown, including the oxidation of succinate to fumarate, which is mechanistically coupled to the reduction of ubiquinone to ubiquinol on the inner mitochondrial membranes. Each of these intermediate electron transfers must be thermodynamically favorable on its own in order for respiration to proceed. By comparing to the "ATP hydrolysis scale," we can see that the individual transformations in the electron transport chain are not energetic enough to generate ATP on their own. Yet they are favorable enough to pump a proton across the cell or mitochondrial membrane. This is the energetic basis for chemiosmosis: Cells store quanta of energy too small for ATP synthesis in the proton gradient across a membrane. That energy is later used to generate ATP by converting the H$^+$ gradient into phosphoanhydride bonds on ATP through the ATP synthase. (Courtesy of Avi Flamholz).

where $n = 2$ and $F \approx 100$ kJ/mol/V. Because ATP hydrolysis has a free-energy change of ≈ 50kJ/mol under physiological conditions, we find that 228 kJ/mol suffices to produce a maximum of $228/50 \approx 4.5$ ATPs. In the cell, the oxidation of NADH proceeds through several steps in respiration and results in the transfer of 10 protons across the membrane against the electrochemical potential (BNID 101773). These proton transfers correspond to yet another way of capturing biochemical energy. This energy is then used by the ATPase to produce 2–3 ATPs. We thus find that about one-half of the energy that was released in the transfer of electrons from NADH to oxygen is conserved in ATP. Ensuring that the reaction proceeds in a directional manner to produce ATP rather than consume it requires that some of the energy is "wasted" because the system must be out of equilibrium.

Why should we discuss redox potentials of half-reactions and not free energies of full reactions? The units themselves owe their origins to the ability in the field of electrochemistry to make laboratory measurements of the voltage difference—that is, the potential measured in volts—across two chambers that contain different electron carriers, and to stop the net reaction with a voltage. The usefulness of redox potentials for half-reactions lies in the ability to assemble combinations of different donors and acceptors to assess the thermodynamic feasibility and energy gain of every considered reaction. If you have k possible electron transfer compounds, then $\sim k^2$ possible reactions can be predicted based on only the k redox potentials.

Just as we speak of the pH of a solution, at first guess, we might imagine that it would be possible to speak of an apparently analogous redox potential of the cell. Knowing the concentration of the reduced and oxidized forms of a given reaction pair defines their pool redox potential via the relation

$$E = E_0 - \frac{RT}{nF} \ln \frac{[A_{red}]}{[A_{ox}]}. \tag{3.25}$$

This equation (a so-called Nernst equation) provides the value of the redox potential under concentration conditions typical of the cell as opposed to the standard state conditions (where by definition $[A_{red}] = [A_{ox}]$). As an example, consider the donation of an electron to NAD^+, thereby resulting in the oxidized form NADH. In the mitochondrial matrix, a ratio of 10-fold more of the oxidized form is reported (BNID 100779), as shown in **Table 3-4**. In this case, we find the factor $\frac{RT}{nF} \ln \frac{[A_{red}]}{[A_{ox}]}$ is ≈ 30 mV and thus the redox potential changes from –0.32 V to –0.29 V. To make sure the direction of effect we got is sensible, we notice that with an overabundance of the oxidized form, the tendency to be oxidized by oxygen is somewhat lower, as seen by the fact that the redox potential is now closer than before to that of the oxygen/water electron-exchanging pair (+0.82 V).

condition	[NADH]/ [NAD$^+$]	[NADPH]/ [NADP$^+$]	[NADPH]/[NADP+]/ [NADH]/[NAD+]	BNID
E. coli				
aerobic glucose	0.13	1.3	10	105427
glucose	0.19	6	30	108042
glucose	0.03	60	1900	104679
acetate	0.5	14	30	108042
anaerobic glucose	0.9	2.6	3	108044
mean of various media	0.05	0.8	16	108108
mouse				
embryonic fibroblast	0.4	3.3	8	108105, 108106
rat				
diabetic precataractous lenses	0.008	24	3000	108110, 108109
mitochondrial matrix				
generic "typical conditions"	0.1	100	1000	100779
spinach leaves				
pH 7.2, light	0.0005	3.3	7000	108117, 108115
pH 7.2, dark	0.0007	4.0	6000	108118, 108115

Table 3-4 Concentration ratios of the common electron donor pairs NADH/NAD$^+$ and NADPH/NADP$^+$. As can be seen, NADH/NAD$^+$ is relatively oxidized and NADPH/NADP$^+$ is relatively reduced.

A cell is not at equilibrium, and there is weak coupling between different redox pairs. This situation leads to the establishment of different redox potentials for coexisting redox pairs in the cell. If the fluxes of production and utilization of the reduced and oxidized forms of a redox pair, A_{red} and A_{ox}, and another, B_{red} and B_{ox}, are much larger than their interconversion flux,

$$A_{red} + B_{ox} \rightleftharpoons A_{ox} + B_{red}, \qquad (3.26)$$

then A and B can have very different redox potentials. As a result, it is ill-defined to ask about the overall redox potential of the cell, because it will be different for different components within the cell. By way of contrast, the pH of the cell (or of some compartment in it) is much better defined since water serves as the universal medium that couples the different acid–base reactions and equilibrates what is known as the chemical potential of all species.

For a given redox pair in a given cell compartment, the concentration ratio of the two forms prescribes the redox potential in a well-defined manner. Compounds that exchange electrons quickly will be in relative equilibrium and thus share a similar redox potential. To see how these ideas play out, it is thus most useful to consider a redox pair that partakes in many key cellular reactions and, as a result, is tightly related to the redox state of many compounds. Glutathione in the cytoplasm is such a compound because it takes part in the reduction and oxidation of the highly prevalent thiol bonds (those containing sulfur) in the cysteine amino acids of many proteins. Glutathione is a tripeptide (composed of three amino acids), the central one a cysteine that can be in a reduced (GSH) or oxidized form, where it forms a dimer with a cysteine from another glutathione molecule (denoted GSSG). The half-reaction for glutathione is thus

$$2GSH \rightleftharpoons GSSG + 2e^- + 2H^+. \tag{3.27}$$

The other half-reaction is often a sulfur bond that is "opened up" in a receptive protein, thus being kept in the reduced form owing to the constant action of glutathione. Glutathione is also a dominant player in neutralizing reactive compounds that have a high tendency to snatch electrons and thus oxidize other molecules. Such compounds are made under oxidative stress, as for example when the capacity of the electron transfer reactions of respiration or photosynthesis is reached. Collectively called ROS (reactive oxygen species), they can create havoc in the cell and are implicated in many processes of aging. The dual role of glutathione in keeping proteins folded properly and limiting ROS, as well as its relatively high concentration and electron transfer reactivity, make it the prime proxy for the redox state of the cell. The concentration of glutathione in the cell is $\approx$10 mM (BNID 104679, 104704, 111464), making it the second most abundant metabolite in the cell (after glutamate), ensuring that it plays a dominant role as an electron donor in the redox control of protein function. In other cellular functions, there are other dominant electron pairs. In biosynthetic anabolic reactions, it is the $NADP^+/NADPH$ pair, and in breakdown catabolic reactions, it is $NAD^+/NADH$.

How do we go about measuring redox potentials in living cells? Yet another beneficiary of the fluorescent protein revolution was the subject of redox potentials. A reporter GFP was engineered to be redox sensitive by incorporating cysteine amino acids that affect the fluorescence based on their reduction by the glutathione pool. **Figure 3-17** shows the result of using such a reporter to look at the glutathione redox potential in different compartments of a diatom.

From measurements of the redox state of the glutathione pool in different cellular organelles and under varying conditions, we can infer the ratio of concentrations of the reduced to oxidized forms. Values of the redox potential range from about –170 mV in the ER and in apoptotic cells to about –300 mV in most other organelles and in proliferation cells (BNID 103543, 101823, 111456, 111465). Given that the standard redox potential of glutathione is –240 mV (BNID 111453, 111463), what then is the ratio of reduced to oxidized glutathione? Using the Nernst equation (or

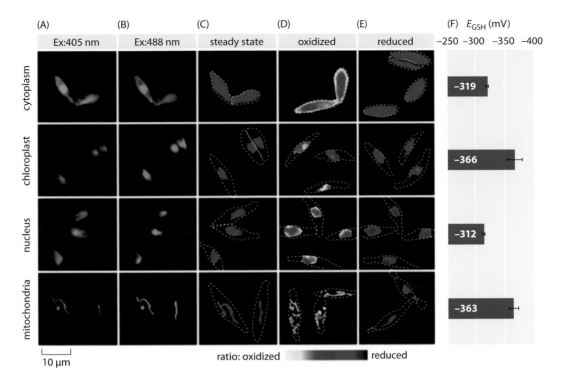

Figure 3-17 Imaging of subcellular redox potential of the glutathione pool in diatom algae *in vivo*. Fluorescence microscopy imaging of *P. tricornutum* cells expressing roGFP2 in various subcellular localizations. Fluorescence images at two excitation wavelengths (A, B) were divided to obtain ratiometric values (C). For calibration, ratiometric images are captured under (D) strong oxidant (150 mM H_2O_2) and (E) reductant (1 mM DTT) conditions. Dashed lines represent the outline of the cells', drawn based on the bright field images. (F) Steady-state redox potential of the glutathione pool, EGSH in mV, was calculated based on the Nernst equation using the oxidation level under given pH values for each organelle. (Adapted from van Creveld SG, Rosenwasser S, Schatz D et al. [2015] *ISME J* 9:385–395.)

equivalently, from the Boltzmann distribution), a 10-fold change in the product/reactant ratio corresponds to an increase of ≈6 kJ/mol in free energy (≈2 k_BT). Given the two electrons transferred in the GSH/GSSG reaction, this concentration ratio change is usually equal to 30mV, though for glutathione, the stoichiometry of two GSH molecules merging to one GSSG covalently bound molecule makes this only an approximation. The 100 mV change reported across conditions reflects a ratio of concentrations between about equal amounts of the reduced and oxidized forms (in apoptotic cells) to over 1000-fold more concentration of the reduced form. Indeed, in most cellular conditions, the oxidized form is only a very small fraction of the overall pool, but it still has physiological implications.

One confusing aspect of redox reactions is that the transfer can take several forms. In one case it is only electrons, as in the reactions carried out by cytochromes in electron transfer chains. In another common case it is a combination of electrons and protons, as in the cofactor $NAD^+/NADH$, where two electrons and one proton (H^+) are transferred. Finally, there are the reactions where the same number of electrons and protons are transferred, when we would naturally be tempted to discuss the transfer of hydrogens. This is, for example, the case for the overall reaction of glucose oxidation, where oxygen is reduced to water. Two hydrogens have thus been transferred, so should we discuss the transfer of electrons, hydrogens, or protons? The definition of the redox potential (given above) focuses only on the electron "state." What about the protons and what happens to them when a chain of electron transfer reactions is encountered where some intermediate compounds contain the hydrogen protons and some do not? The explanation resides in the surrounding water and their pH. The reaction occurs at a given pH, and the reacting compounds are in equilibrium with this pH, so giving off or receiving a proton has no effect on the energetics. The aqueous medium serves as a pool where protons can be "parked" when the transfer reaction is solely of electrons (the analogy borrowed from the very accessible introductory biochemistry book *The Chemistry of Life* by Steven Rose). These parked protons can be borrowed back at subsequent stages, as occurs in the final stage of oxidative respiration, where cytochrome oxidase takes protons from the medium. Because we assume that water is ubiquitous, we do not need to account for protons, except for knowing the prevailing pH, which depicts the tendency to give or receive protons. This is the reason why we discuss electron donors and acceptors rather than hydrogen donors and acceptors.

What is the electric potential difference across membranes?

Many of the most important energy transformations in cells effectively use the membrane as a capacitor, resulting in the storage of energy as a transmembrane potential. Energy-harvesting reactions, such as those involved in photosynthesis and respiration, pump protons across the membrane. On their return back across the membrane, these protons are then harnessed to synthesize ATP and to transport compounds against their concentration gradients. For the mitochondria, this potential difference has a value of roughly 160 mV (BNID 101102, 101103) and for *E. coli* it is about 120 mV (BNID 103386). A pH difference between two compartments that are membrane-bound adds 60 mV per pH unit difference to the overall driving force for proton transport. This sum of electric and concentration difference terms is the so-called proton-motive force, and it is critical for the operation of most membrane-anchored energy transformations (for example, in

compartment	potential difference (mV)	BNID
human red blood cell	−10 to −14	104083
human/rodents resting potential in neurons	−40 to −80	101479,106527,106955, 106956,106104
squid axon membrane	−60	104085
chicken embryo heart muscle	−70	104083
mammalian skeletal muscle	−90	104084
rat liver mitochondria, normal diet	−120 (−170 pmf)	101103
rat liver mitochondria, high fat diet	−140 (−150 pmf)	101103
E. coli fermentive growth on glucose	−110 (−120 pmf)	103386
E. coli spheroplasts growth on aerobic rich media	−130 (−230 pmf)	107128
E. coli aerobic growth on glycerol	−140 (−160 pmf)	103386
S. aureus growth on aerobic rich media	−130 (−210 pmf)	107128
alga Nitella	−140	104083

Table 3-5 Electric potential difference over a range of biological membranes. Negative values indicate that the outer compartment is more positive than the inner compartment. The pmf is the total proton motive force, which includes the effect of pH. When the pH of the media changes, the electric potential of single-celled organisms tends to change such that the pmf remains in the range of −100 to −200 mV.

chloroplasts). A series of representative examples for potential differences in a variety of cellular contexts are listed in **Table 3-5**. To recast these numbers in perhaps more familiar units, recall that the energy scale associated with a potential V is given by qV, where q is the charge moved across that potential. If we take the characteristic 100 mV energy scale of membrane potentials and multiply by the electron charge of 1.6×10^{-19} coulombs, this yields an energy in joules of 1.6×10^{-20} J. If we recall that

$$k_{B}T \approx 4 \text{ pN nm} \approx 4 \times 10^{-21} \text{ J}, \qquad (3.28)$$

then we see that the membrane potential energy scale can be remembered as

$$100 \text{ mV} \approx 4 \, k_{B}T. \qquad (3.29)$$

Though we are accustomed to voltage differences of hundreds of volts or more from our daily experience, these values are actually less impressive than we might think when compared to their microscopic counterparts. The simplest way to see this is to convert these voltages into their corresponding electric fields. Indeed, what we will see is that at microscopic scales, the strengths of the electric fields are extremely high. To estimate these values, we take the characteristic voltage difference and divide it by the length across which it acts. For example, 120 mV across a characteristic

≈4-nm-thick membrane is equivalent to ≈30 kV/mm (BNID 105801), which is a field similar to that of a lightning bolt. Electroporation, used routinely to insert charged DNA into the cell by forming ruptured pores in the cell membrane, occurs at a 300- to 400-mV (BNID 106079) potential difference across the membrane. We thus find that the voltage difference of more than 100 mV across a mitochondrion or an *E. coli* cell is only about a factor of two below the physical limit that would lead to rupture.

How many protons need to be pumped in order to build up these kinds of potential differences? Let's be generous and assume the membrane voltage difference is made fully through proton transport, even though other ionic species are known to make a large contribution. In **Estimate 3-5** we perform a back-of-the-envelope calculation that treats the cell membrane as a parallel-plate capacitor. The areal charge density σ of a parallel-plate capacitor is related to the voltage difference via the relation

$$\sigma = V \varepsilon_r \varepsilon_0 / d, \tag{3.30}$$

where d is the membrane width (≈4 nm) and ε_r and ε_0 are the relative and vacuum permittivity, respectively. The total charge q is

$$q = \sigma A / e, \tag{3.31}$$

where A is the surface area, which for the membrane of *E. coli* is ≈5 μm², and e is the electron charge. The relative permittivity (dielectric constant) of the bilayer is roughly ≈2 (BNID 104080) and plugging in the numbers this leads to about 10^4 protons overall, as shown schematically in Estimate 3-5, and it is consistent with the membrane having a specific capacitance of 1 μF/cm² (BNID 110759). In the vignette entitled "What is the power

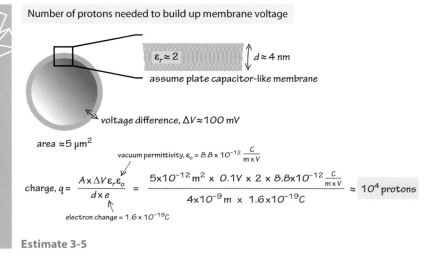

Number of protons needed to build up membrane voltage

$\varepsilon_r \approx 2$ $d \approx 4$ nm

assume plate capacitor-like membrane

voltage difference, $\Delta V \approx 100$ mV

area ≈5 μm²

vacuum permittivity, $\varepsilon_0 = 8.8 \times 10^{-12} \frac{C}{m \times V}$

$$\text{charge, } q = \frac{A \times \Delta V \varepsilon_r \varepsilon_0}{d \times e} = \frac{5 \times 10^{-12} \, m^2 \times 0.1V \times 2 \times 8.8 \times 10^{-12} \frac{C}{m \times V}}{4 \times 10^{-9} m \times 1.6 \times 10^{-19} C} \approx 10^4 \text{ protons}$$

electron change = $1.6 \times 10^{-19} C$

Estimate 3-5

consumption of a cell?" (see below), we noted that the rate of ATP usage by *E. coli* is ≈10^{10} ATP during a cell cycle that can be as short as 1000 s—that is, an expenditure rate of 10^7 ATP/s. With ≈4 protons required to make one ATP, the membrane charge, if not replenished continually, would suffice to produce less than 10^4 ATPs. This potential would be depleted in ≈1 ms under normal load conditions inside the cell.

Another way of viewing these same numbers that yields a surprising insight is to note the ratio of the charges separated across the membrane to the overall charge of ions in the cell. In the opening section on tricks of the trade, we asserted the rule of thumb that in a volume the size of an *E. coli* cell, a concentration of 1 nM is equivalent to one molecule per cell. Thus, an overall ion concentration of ≈100 mM in *E. coli* translates into ≈10^8 charges/cell. On the other hand, in order to achieve the voltage difference across the membrane, the calculation in the previous paragraph shows that it requires 10^4 protons overall—that is, only 1/10,000th of the total ion charge in a bacterial cell. The fraction is even smaller in larger cells, such as neuronal cells, with the charges associated with action potentials being a small fraction of the overall ion concentration in the cell. This shows the property of cells to be close to electroneutral—that is, even though a voltage difference exists, there is only a tiny relative difference in the total ion concentration.

What is the power consumption of a cell?

Cells are out-of-equilibrium structures and require a constant supply of energy to remain in that privileged state. Measuring how much power is required to run a cell or the heat produced as it goes through its normal metabolic operations is experimentally challenging. Beyond the challenges associated with actually measuring cellular power consumption, there are several plausible definitions for a cell's rate of energy usage, making a rigorous discussion of the problem even more demanding. We will explain the meaning and relevance of some of these definitions and then use estimates and reported measurements to explore their order-of-magnitude values. We don't aim for high precision because these values can easily vary by more than an order of magnitude depending upon what growth medium is used, the growth rate, and other environmental factors.

For our first estimate, we consider the rule of thumb that an adult human produces heat at a rate of about 100 W (recall that 100 watts = 100 J/s), similar to a bright incandescent light bulb, which is borne out by noticing how warm a room becomes as more people are packed in. The 100-W value was calculated on the basis of a caloric intake of 2000 kcal per day in the vignette entitled "What is the free energy released upon combustion of sugar?" (pg. 188). Assuming a person has a mass of ≈100 kg (let's forget about our recent post-holiday diet for the moment—this is an order-of-magnitude estimate), we find a power consumption of about 1 W/kg, as depicted in **Estimate 3-6**. This value is about 10^{-15} W/μm^3, where we revert to that useful unit of volume, remembering that a bacterium has a volume of roughly 1 μm^3, a red blood cell has a volume of roughly 100 μm^3, and an adherent mammalian cell has a volume in the range of 1000–10,000 μm^3. The definition of power consumption used here is based on the rate of heat production. We consider other definitions below.

Recent measurements of glucose consumption in primary human fibroblasts make it possible to consider a second estimate for human energy consumption. Quiescent human fibroblasts of unreported volume were found to consume about 1 mmol glucose per gram of protein per hour (BNID 111474). We recall that the total energy released by glucose combustion (where carbon from sugar is merged with oxygen to yield CO_2 and water) is about 3000 kJ/mol, as discussed in the vignette entitled "What is the free energy released upon combustion of sugar?" (pg. 188). The protein content of a characteristic 3000-μm^3 cell volume is about 300 pg, which corresponds to 3×10^9 cells per gram of protein. One cell thus requires

$$(3 \times 10^6 \text{ J/mol glucose}) \times (10^{-3} \text{ mol glucose/}$$
$$(\text{g protein} \times \text{h}) \times (1 \text{ h}/3600 \text{ s}) \times (1 \text{ g protein}/3 \times 10^9 \text{ cell})$$
$$= 3 \times 10^{-10} \text{ W/cell}. \tag{3.32}$$

Energy production rate in human and bacteria

power ≈ 100 W/person

mass ≈ 100 kg/person

$\implies$ $P_{human} \approx 1 \dfrac{W}{kg}$ at rest

oxygen consumption rate during fast exponential growth ≈ 30 $\dfrac{\text{mmol } O_2}{\text{g CDW} \times \text{hour}}$

cell dry weight

$\Delta G_{O_2 \text{ respiration}}$ ≈ −500 kJ/mol

$P_{bacteria} \approx 30 \dfrac{\text{mmol } O_2}{\text{g CDW} \times \text{hour}} \times 500 \dfrac{kJ}{\text{mol } O_2} \times \dfrac{1 \text{ hour}}{3600 \text{ s}} \times 0.3 \times 10^{-12} \dfrac{\text{g CDW}}{\text{cell of 1 } \mu m^3 \text{ volume}} \approx 10^{-12} \dfrac{W}{\text{cell}} \approx 1000 \dfrac{W}{kg}$

wet mass of bacterium ≈ 10^{-15} kg

Estimate 3-6

On a mass basis this is equivalent to

$$3 \times 10^{-10} \text{ W/cell} \times 3 \times 10^{11} \text{ cells/kg} \approx 100 \text{ W/kg} \qquad (3.33)$$

of cell wet weight, which is two orders of magnitude higher than our estimate based on a whole-human-body analysis. Perhaps fibroblasts are more metabolically active than the average human cell. Alternatively, this two order-of-magnitude discrepancy might call into question the accuracy of the reported values. It is hard to tell without more data. Though we are perplexed by this result, it nicely focuses our attention on a concrete scientific question about the energy consumption of fibroblasts grown in the lab versus the "average" cell in the human body, and motivates future experiments and measurements.

It is sometimes more useful to think in ATP units of energy. We can assume ≈20 ATP-molecules are produced per glucose molecule in a combination of respiration and fermentation characteristic of cancerous cells (BNID 111475). We then find that the consumption worked out above translates to about 10^9 ATP/s/mammalian cell of 3000 μm³ volume (BNID 111476). What is knowing this value good for?

Let's think about cell motility. When we watch videos of keratocytes dragging themselves quickly across the microscope field of view on their lamellipodium, as shown in **Estimate 3-7**, it is natural to assume that these

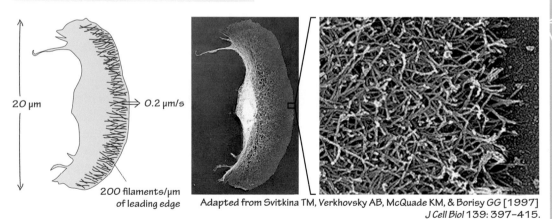

How much ATP is required for actin-driven motility?

20 μm

0.2 μm/s

200 filaments/μm
of leading edge

Adapted from Svitkina TM, Verkhovsky AB, McQuade KM, & Borisy GG [1997]
J Cell Biol 139: 397–415.

$$\text{actin polymerization rate} = 0.2 \text{ μm/s} \times \frac{1000 \text{ nm}}{1 \text{ μm}} \times \frac{2 \text{ monomers}}{5 \text{ nm}} \approx \frac{100 \text{ monomers}}{(\text{s} \times \text{filament})}$$

$$\text{ATP requirement} = 20 \text{ μm} \times \frac{200 \text{ filaments}}{1 \text{ μm}} \times \frac{100 \text{ monomers}}{(\text{s} \times \text{filament})} \times \frac{1 \text{ ATP}}{\text{monomer}} \approx 4 \times 10^5 \text{ ATP/s}$$

Estimate 3-7

processes require a large fraction of the energy available to these cells. But is that really the case? For many eukaryotic cells, motility is driven primarily by dynamic actin polymerization at a steady-state cost of about 1 ATP hydrolysis per polymerizing actin monomer. Labeling actin fluorescently famously showed that actin filaments in moving goldfish epithelial keratocytes polymerize at the same rate that the cell moves—about 0.2 μm/s at room temperature, as depicted in Estimate 3-7 (BNID 111060). Given the actin monomer size, each filament must grow by about 100 monomers/s to support motility, which costs ≈100 ATP per polymerizing filament per second. But how many actin filaments are required to move a cell? As shown in Estimate 3-7, the leading edge of a goldfish keratocyte lamellipodium is about 20 μm long and contains roughly 200 actin filaments per micron of length (BNID 111061), or ≈4000 filaments in total. If actin polymerizes primarily at the leading edge of the lamellipodium, the keratocyte must burn about $4000 \times 100 = 4 \times 10^5$ ATP/s to power its movement. In light of the ATP consumption of a cell calculated above, this value turns out to be a very minor ATP requirement of less than one-tenth of a percent. Having made the effort to calculate these energetic costs, we've refined our understanding of the energy budget of cells.

How do the results described above for humans compare to what we can say about energy consumption in bacteria? An empirical approach is based on keeping track of the rate of oxygen consumption of the cells (which depends on carbon source and growth conditions). For growth on minimal media with glucose, a characteristic value for the oxygen consumption rate is 30 mmol O_2/g dry cell weight/hour (BNID 109687). Performing the necessary unit conversions, noting that oxygen respiration releases 500 kJ/mol O_2 of heat as discussed in the vignette entitled "What is the free energy released upon combustion of sugar?" (pg. 188), we find, as shown in Estimate 3-6, 10^{-12} W/cell = 1000 W/kg. We conclude that under these reference conditions, bacterial consumption of energy per unit biomass is about three orders of magnitude higher than that of a human. A reader curious about similar trends across different organisms is invited to read the next vignette, which is entitled "How does metabolic rate scale with size?" (pg. 204). Similarly, to see how this compares to the energetic requirements of bacterial motility, consult the vignette entitled "What is the frequency of rotary molecular motors?" (pg. 254).

We next analyze rapid bacterial growth in terms of ATP usage. We make use (as done above) of the oxygen requirement of 30 mmol O_2/hour/g dry weight during growth on glucose and now utilize this figure through the so-called P/O ratio, which is the ratio of ATP produced per oxygen atom respired (equal to about 2—that is, 3–5 ATP per 1 O_2 molecule; BNID 111461, 110656, 110628). We thus arrive at about 100 mmol ATP/hour/g dry weight, which translates into ~10^7 ATP/s/bacterial cell. Throughout a cell cycle taking about an hour, this leads to 10^{10}–10^{11} ATP per bacterial cell of 1 μm^3 volume produced. Noting that one hour is also a characteristic doubling time in which each cell produces a new cell of about 10^{10} carbon atoms (BNID 103010), we have a rule of thumb of about 1 ATP per 1 carbon incorporated into biomass during cell growth. How are these numbers useful? Consider the idea of powering an *E. coli* cell using bacterial proteorhodopsin, a membrane protein that sits in the cell membrane of some types of bacteria and pumps protons when exposed to light.[†] One can imagine packing on the order of 10^5 such proteins on the available real estate of a "standard" bacterial membrane (but actual values tend to be much lower, BNID 111296). We can infer that in order to divide once every few hours (say 10^4 s), as expected of bacteria, each of these membrane proteins will have to pump a proton several hundred times per second (at least ~10^{11} protons needed from 10^5 proteins per 10^4 s). These protons will then be used to power the machines that synthesize ATP. If these proteins cannot maintain such a transport rate, or packing such a large number of proteins on the membrane is not possible, powering the cell metabolism is a no go from the start (BNID 111295).

What processes are fueled by all of this energy consumption in cells? Efforts in the 1970s tried to perform "ATP accounting" for bacterial cells—that is, to list all of the processes in cells according to how much ATP they consume. Of the processes that could be clearly quantified (including metabolism and polymerization), protein synthesis from amino acids dominated the budget at fast growth rate and preferred carbon sources. The polymerization of an amino acid into a nascent peptide chain consumes about 4 ATP/amino acid, and with 2–4 million proteins per μm^3 and 300 aa per protein, we are led to about 4×10^9 ATPs spent per μm^3 of cell volume. This should be compared to the value of 10^{10}–10^{11} in the previous paragraph. We conclude that this is a major energy drain, but more surprising is that a large fraction, amounting to about half of the measured energy used (BNID 102605), is not accounted for by any process essential for cell buildup and was generally regarded as lost in the membrane-associated processes of membrane potential buildup and leakage. Revisiting these abandoned efforts at cellular accounting is of great interest, for example, for determining the fraction lost by metabolic futile cycles and by post-translational protein phosphorylations and dephosphorylations.

Trying to perform similar accounting for mammalian cells, the answer again depends on the growth conditions, the relevant tissue, and so on. In **Table 3-6** (BNID 107962), we reproduce the findings for mouse

---

[†] Walter JM, Greenfield D, Bustamante C & Liphardt J (2007) Light-powering *Escherichia coli* with proteorhodopsin. *Proc Natl Acad Sci USA* 104:2408–2412.

tissue	protein synthesis	Na⁺/K⁺ ATPase	Ca²⁺ ATPase	other
liver	20%	5–10%	5%	gluconeogenesis (15–40%), substrate recycling (20%), proton leak (20%), urea synthesis (12%)
kidney	6%	40–70%	—	gluconeogenesis (5%)
heart	3%	1–5%	15–30%	actinomyosin ATPase (40–50%), proton leak (15% max)
brain	5%	50–60%	significant	a single cortical action potential was estimated to require 10^8–10^9 ATP, BNID 111183)
skeletal muscle	17%	5–10%	5%	proton leak (50%), nonmitochondrial (14%)

Table 3-6 Distribution of major oxygen-consuming processes to total oxygen consumption rate of rat tissues in the standard state (BNID 107962). Values are rounded to one significant digit. (Adapted from Rolfe DF & Brown GC [1997] *Physiol Rev* 77:731–758.)

tissues, which shows major contributions from protein synthesis, the Na⁺/K⁺ ATPase (the machine in charge of maintaining the resting electric potential in cells), actinomyosin ATPase (which drives muscle cells), and mitochondrial proton leakage. In neurons it is estimated that actin turnover is responsible for about 50% of the ATP usage (BNID 110642). New bioluminescent probes make it possible to measure the ATP concentration in neurons *in vivo* and connect them to synaptic activity. Such methods promise to give us a new ability for detailed energy censuses in the coming years.

We end by noting that in extreme environments, such as the permafrost of Antarctica, bacteria were found to be viable at depths of 3000 m below ground at temperatures well below 0°C. Due to impurities, the water does not freeze and the metabolic rate is extremely slow—namely, ≈6 orders of magnitude less than under rapid growth (BNID 111454, 111455). This has been called survival metabolism, where cells are dormant, and the energy is thought to be used to repair macromolecule damage.

How does metabolic rate scale with size?

When we arrive at biology from its sister disciplines of physics or engineering, there is a strong temptation to search for consistent quantitative trends and general rules. One such pursuit centers on the power consumption of different organisms—the so-called metabolic energy consumption rate. This example illustrates how scaling arguments work. For many inanimate systems, the energy produced has to be removed through the bounding surface area, and each unit of area allows a constant energy flux. The scaling of surface area A with the radius R goes as

$$A \sim R^2. \tag{3.34}$$

At the same time, the volume V scales as R^3. Assuming constant density, this will also be the scaling of the total mass M. The surface area thus scales as

$$A \sim M^{2/3}. \tag{3.35}$$

How should the energy production (B) per unit mass, B/M, scale? According to our assumption above, the energy is removed through the surface at a constant rate, and thus the total energy produced should be proportional to A—that is, $B \sim A$. Dividing both sides by M and plugging in the scaling of A with M, we finally get

$$B/M \sim A/M \sim M^{2/3}/M \sim M^{-1/3}. \tag{3.36}$$

Does this scaling result based on simple considerations of energy transfer also hold for biological systems?

The metabolic rate of an organism is condition dependent, and thus should be defined if we want to make an honest comparison across organisms. One of the most extreme examples is that bees in flight increase their oxygen consumption and thus their energy consumption by about 1000-fold in comparison to resting conditions (BNID 110031). Similarly, humans taking part in the strenuous Tour de France consume close to 10,000 kcal a day—about five times the normal resting value. It is most common to refer to the resting metabolic rate, which operationally means the animal is not especially active but is well fed. As the alert reader can imagine, it is not easy to ensure rest for all animals—think of an orca (killer whale) as one example. The values themselves are often calculated from the energy consumption rate that is roughly equal to the energy production rate, or in other cases, from the oxygen consumption.

Based on empirical measurements for animals, an observation called Kleiber's law suggests a relationship between the resting metabolic energy requirement per unit mass (B/M) and the total body mass (M) that scales as $M^{-1/4}$. A famous illustration representing this relationship is shown in **Figure 3-18**. Similar to the scaling based on surface area and energy transfer described above, Kleiber's law suggests that heavier animals require less energy per unit mass, but with the value of the scaling

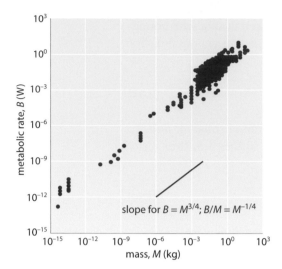

Figure 3-18 Relation of whole organism metabolic rate to body mass. Metabolic rates were temperature standardized to 20°C. (Adapted from Gillooly JF, Brown JH, West GB et al. [2001] *Science* 293:2248–2251.)

exponent being slightly different from the value of –1/3 hypothesized above. The difference between –0.33 and –0.25 is not large, but the law suggests that the data are accurate enough to make such distinctions. Over the years, several models have been put forward to rationalize why the scaling is different from that expected based on surface area. Most prominent are models that discuss the rate of energy supply in hierarchical networks, such as blood vessels in our body, which supply the oxygen required for energy production in respiration. To give a sense of what this scaling would predict, in moving from a human of 100 kg consuming 100 W—that is, 1 W/kg—to a mouse of 10 g (four orders of magnitude) would entail an increase of $(10^{-4})^{-1/4}$ = 10-fold—that is, to 10 W/kg. Jumping as far as a bacterium of mass 10^{-15} kg is 17 orders of magnitude away from a human, which would entail a $(10^{-17})^{-1/4} \sim 10^4$-fold increase or 10,000 W/kg. This is 1–3 orders of magnitude higher than the values discussed in the closely related and complementary vignette entitled "What is the power consumption of a cell?" (pg. 199). But, as can be appreciated in Figure 3-18, the curve that refers to unicellular organisms is displaced in comparison to the curves depicting mammals by about that amount.

The resting energy demand of organisms has recently been compared among more than 3000 different organisms spanning over 20 orders of magnitude in mass (!) and all forms of life. In contrast to the Kleiber law prediction, this recent work found a relatively small range of variation, with the vast majority of organisms having power requirements of 0.3–9 W/kg wet weight, as shown in **Figure 3-19**. Our naïve estimate for a human of 1 W/kg wet weight is somewhere in the middle of this, but the surprising observation is that this range is claimed to also hold for minute bacteria, plant leaves, and across the many diverse branches of the tree of life all the way to elephants. Is this again an indication of Monod's adage that what is true for *E. coli* is true for the elephant? Further evidence for the breaking of Kleiber scaling was provided recently for protists and prokaryotes.[‡] Other recent studies stand behind Kleiber's law and aim to explain it.

[‡] DeLong JP, Okie JG, Moses ME et al. (2010) Shifts in metabolic scaling, production, and efficiency across major evolutionary transitions of life. *Proc Natl Acad Sci USA* 107:12941–12945.

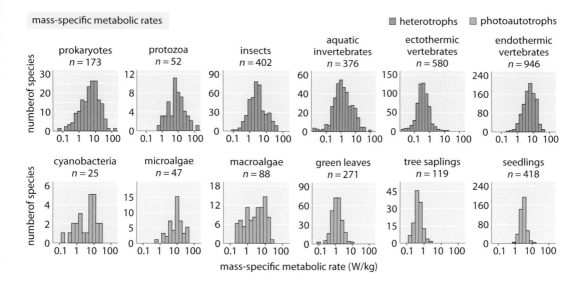

Figure 3-19 Histograms of resting metabolic rates normalized to wet weight. Across many orders of magnitudes of body size and widely differing phylogenetic groups, the rates are very similar at about 0.3–9 W/kg wet weight. (Adapted from Makarieva AM, Gorshkov VG, Li BL et al. [2008] *Proc Natl Acad Sci USA* 105:16994–16999.)

We are not in a position to comment on who is right in this debate, but we are of the opinion that such a bird's-eye view of the energetics of life provides a very useful window on the overarching costs of running the cellular economy.

SPEED

Chapter 4: Rates and Durations

This chapter explores another important quantitative theme in biology—namely, "how fast?" A feeling for the numbers in biology is built around an acquaintance with how large the main players are (that is, the sizes of macromolecules, organelles, cells, and organisms), what concentrations they occur at, and the time scales for the many processes that are carried out by living organisms. Both the hard data and rules of thumb that run through the present chapter and are depicted in **Figure 4-1** can serve as the basis for developing intuition about the rates of a broad spectrum of biological processes.

One of the most obvious features of the living world is its dynamism. If we look down a microscope at low magnification at a sample of pond

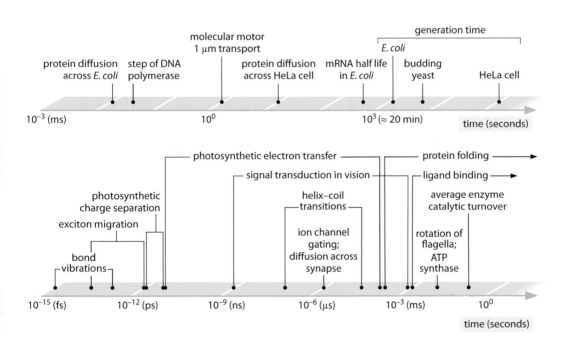

Figure 4-1 Range of characteristic time scales of central biological processes. Upper axis shows the longer time scales, ranging from protein diffusion across a bacterial cell to the generation time of a mammalian cell. The lower axis shows the fast time scales, ranging from bond vibrations to protein folding and catalytic turnover durations.

water or the contents of a termite's gut, we are greeted with a world teeming with activity. If we increase the magnification, we can then resolve the cellular interior, itself the seat of a dazzling variety of processes playing out at all sorts of different time scales. This same dynamic progression continues unabated down to the scale of the ions and molecules that serve as the "ether" of the cellular interior. What sets the time scales for these different processes?

We begin the chapter by considering one of the most important facts of life—namely, that molecules (and even larger assemblies such as viruses or organelles) diffuse. Diffusion is the inevitable motion that results from collisions between the molecule (or particle) of interest and the molecules of the surrounding medium. These diffusive motions mean, for example, that if an ion channel opens and permits the entry of ions into the cellular interior, those ions will leave the vicinity of the channel and spread out into the cellular milieu. Stated simply, these diffusive motions give rise to a net flux from regions of high concentration to regions of low concentration, resulting in an overall homogenization of concentration differences over long time scales.

We then explore the rates associated with the motions of other small molecules in the cell. In addition to the dynamics of the passive motions of molecules within cells, one of the most interesting things that happens to these molecules is that they change their chemical identity through a terrifyingly complex network of reactions. Cells have many tricks for speeding up the rates of many (if not most) of these chemical transformations. The central way in which rates are sped up is through enzymes. Several of our vignettes focus on the time scales associated with enzyme action and what sets these scales.

Once we have settled the diffusive and enzymatic preliminaries, the remainder of the chapter's examples center on the temporal properties of specific processes that are especially important to cell biology. We consider the processes of the central dogma—transcription and translation—and compare their rates. We then analyze other cell processes and tie them together to see the rate at which cells undergo cell division and the cell cycle as another one of the signature time scales in cell and molecular biology. A unifying theme in our depiction is trying to ask what governs the rates and why they are not any faster. All other things being equal, faster rates can enable a smaller investment of cell resources to achieve the same required flux. It has been hypothesized that freeing up resources can lead to faster growth rates and higher fitness, though such hypotheses are fraught with subtleties.

TIME SCALES FOR SMALL MOLECULES

What are the time scales for diffusion in cells?

One of the most pervasive processes that serves as the reference time scale for all other processes in cells is that of diffusion. Molecules are engaged in an incessant, chaotic dance, as characterized in detail by the botanist Robert Brown in his paper with the impressive title, "A Brief Account of Microscopical Observations Made in the Months of June, July, and August, 1827, On the Particles Contained in the Pollen of Plants, and on the General Existence of Active Molecules in Organic and Inorganic Bodies." The subject of this work has been canonized as Brownian motion in honor of Brown's seminal and careful measurements of the movements bearing his name. As he observed, diffusion refers to the random motions undergone by small objects as a result of their collisions with the molecules making up the surrounding medium.

The study of diffusion is one of the great meeting places for nearly all disciplines of modern science. In both chemistry and biology, diffusion is often the dynamical basis for a broad spectrum of different reactions. The mathematical description of such processes has been one of the centerpieces of mathematical physics for nearly two centuries and permits the construction of simple rules of thumb for evaluating the characteristic time scales of diffusive processes. In particular, the concentration of some diffusing species as a function of both position and time is captured mathematically using the so-called diffusion equation. The key parameter in this equation is the diffusion constant, D, with larger diffusion constants indicating a higher rate of jiggling around. The value of D is microscopically governed by the velocity of the molecule and the mean time between collisions. One of the key results that emerges from the mathematical analysis of diffusion problems is that the time scale τ for a particle to travel a distance x is given on the average by

$$\tau \approx x^2/D, \tag{4.1}$$

indicating that the dimensions of the diffusion constant are length^2/time. This rule of thumb shows that the diffusion time increases quadratically with the distance, which has major implications for processes in cell biology, as we now discuss.

How long does it take macromolecules to traverse a given cell? We will perform a crude estimate. As derived in **Estimate 4-1**, the characteristic

Stokes–Einstein relation and the diffusion constant in water

$$R^2 \propto D\tau$$

$$D = \frac{k_B T}{6\pi\eta a} \approx \frac{4 \times 10^{-21}\, N \times m}{6 \times 3 \times 10^{-3}\, \frac{N \times s}{m^2} \times \frac{a}{1\, nm} \times 10^{-9} m} \approx \frac{1}{5} \times \frac{10^{-9} m^2/s}{\frac{a}{1\, nm}} \times \frac{10^{12}\,\mu m^2}{m^2} = \frac{200}{\frac{a}{1\, nm}}\, \frac{\mu m^2}{s}$$

$$\text{viscosity} \approx 10^{-3}\, \frac{N \times s}{m^2}$$

Estimate 4-1

diffusion constant for a molecule the size of a monomeric protein is ≈ 100 $\mu m^2/s$ in water and is about 10-fold smaller, $\approx 10\ \mu m^2/s$, inside a cell, with large variations depending on the cellular context, as shown in **Table 4-1** (larger proteins often show another order of magnitude decrease to ≈ 1 $\mu m^2/s$; BNID 107985). Using the simple rule of thumb introduced above, we find, as shown in **Estimate 4-2,** that it takes roughly 0.01 seconds for a protein to traverse the 1 micron diameter of an *E. coli* cell (BNID 103801). A similar calculation results in a value of about 10 seconds for a protein to traverse a HeLa cell (adhering HeLa cell diameter $\approx 20\ \mu m$; BNID 103788). An axon 1 cm long is about 500 times longer still, and from the diffusion time scaling as the square of the distance, it would take 10^6 seconds or about two weeks for a molecule to travel this distance solely by diffusion. This enormous increase in diffusive time scales as cells approach macroscopic sizes demonstrates the necessity of mechanisms other than diffusion for molecules to travel these long distances. Using a molecular motor moving at a rate of $\approx 1\ \mu m/s$ (BNID 105241), it will take a "physiologically reasonable" 2–3 hours to traverse this same distance. For extremely long neurons, which can reach a meter in length in a human (or 5 m in a giraffe), recent research raises the speculation that neighboring glia cells alleviate much of the diffusional time limits by exporting cell material to the neuron periphery from their nearby position.[*] This can decrease the time for transport by orders of magnitude, but it also requires dealing with transport across the cell membrane.

How much slower is diffusion in the cytoplasm in comparison to water, and what are the underlying causes for this difference? Measurements show that the cellular context affects diffusion rates by a factor that depends

[*] Nave KA (2010) Myelination and the trophic support of long axons. *Nat Rev Neurosci* 11:275–283.

molecule	measured context	diffusion coefficient (μm^2/s)	BNID
H_2O	water	2000	104087, 106703
H_2O	nucleus of chicken erythrocyte	200	104645
H^+ (from H_3O^+ to H_2O)	water	7000	106702
O_2	water	2000	104440
CO_2	water	2000	102625
tRNA ($\approx$20 kDa)	water	100	107933, 107935
protein ($\approx$30 kDa GFP)	water	100	100301
protein ($\approx$30 kDa GFP)	eukaryotic cell (CHO) cytoplasm	30	101997
protein ($\approx$30 kDa GFP)	rat liver mitochondria	30	100300
protein (NLS-EGFP)	cytoplasm of D. melanogaster embryo	20	109209
protein ($\approx$30 kDa)	E. coli cytoplasm	7–8	100193, 107985
protein ($\approx$40 kDa)	E. coli cytoplasm	2–4	107985
protein ($\approx$70–250 kDa)	E. coli cytoplasm	0.4–2	107985
protein ($\approx$140 kDa Tar-YFP)	E. coli membrane	0.2	107985
protein ($\approx$70 kDa LacY-YFP)	E. coli membrane	0.03	107985
fluorescent dye (carboxy-fluorescein)	A. thaliana cell wall	30	105033
fluorescent dye (carboxy-fluorescein)	A. thaliana mature root epidermis	3	105034
transcription factor (LacI)	movement along DNA (1D, in vitro)	0.04 (4×10^5 bp^{2}s^{-1})	102036
morphogen (bicoid-GFP)	cytoplasm of D. melanogaster embryo	7	109199
morphogen (wingless)	wing imaginal disk of D. melanogaster	0.05	101072
mRNA	HeLa nucleus	0.03–0.10	107613
mRNA	various localizations and sizes	0.005–1	110667
ribosome	E. coli	0.04	108596

Table 4-1 A compilation of empirical diffusion constants that shows the dependence on size and cellular context.

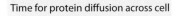

Time for protein diffusion across cell

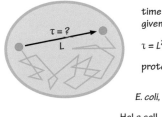

time scale (τ) to traverse distance (L) given diffusion coefficient (D)

$\tau = L^2/6D$

protein in cytoplasm $D \approx 10 \, \dfrac{\mu m^2}{s}$

E. coli, $L \approx 1 \, \mu m \implies \tau \approx 10 \, ms$

HeLa cell, $L \approx 20 \, \mu m \implies \tau \approx 10 \, s$

neuronal cell axon, $L \approx 1 \, cm \implies \tau \approx 10^6 \, s \approx 10 \, days!$

Estimate 4-2

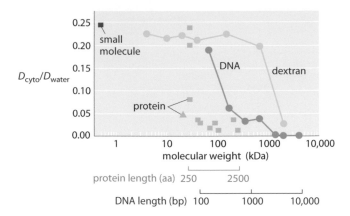

Figure 4-2 The decrease in the diffusion constant in the cytoplasm with respect to water as molecular weight increases. (Adapted from Verkman AS [2002] *Trends Biochem Sci* 27:27–33.)

strongly on the compound's biophysical properties as well as its size. For example, small metabolites might suffer only a fourfold decrease in their diffusive rates, whereas DNA can exhibit a diffusive slowing down in the cell that is tens or hundreds of times slower than in water, as shown in **Figure 4-2**. Causes for these effects have been grouped into categories and explained by analogy to an automobile[†]: the viscosity (like drag due to car speed), binding to intracellular compartments (time spent at stop lights), and collisions with other molecules—also known as molecular crowding (route meandering). Recent analysis[‡] highlights the importance of hydrodynamic interactions—that is, the effects of moving objects in a fluid on other objects, similar to the effect of boats on each other via their wake. Such interactions lead approaching bodies to repel each other, whereas two bodies that are moving away are attracted.

Complications like those described above make the study of diffusion in living cells very challenging. To the extent that these processes can be captured with a single parameter—namely, the diffusion coefficient—we might wonder how these parameters are actually measured. One interesting method turns one of the most annoying features of fluorescent proteins into a strength. When exposed to light, fluorescent molecules are bleached (lose their ability to fluoresce over time). But this becomes a convenience when the bleached region is only part of a cell. After the bleaching event, because of the diffusion of the unbleached molecules from other regions of the cell,

[†] Verkman AS (2002) Solute and macromolecule diffusion in cellular aqueous compartments. *Trends Biochem Sci* 27:27–33.
[‡] Ando T & Skolnick J (2010) Crowding and hydrodynamic interactions likely dominate in vivo macromolecular motion. *Proc Natl Acad Sci USA* 107:18457–18462.

they will fill in the bleached region, thus increasing the fluorescence (the so-called "recovery" phase in fluorescence recovery after photobleaching, or FRAP). This idea is shown schematically in **Figure 4-3**. Using this technique, systematic studies of the dependence of the diffusion coefficient on molecular size for cytoplasmic proteins in *E. coli* have been undertaken, with results as shown in **Figure 4-4**, thus illustrating the power of this method to discriminate the diffusion of different proteins in the bacterial cytoplasm.

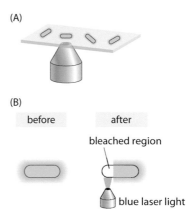

(A)

(B)

before   after

bleached region

blue laser light

Figure 4-3 Fluorescence recovery after photobleaching in bacteria. (A) Schematic of how the FRAP technique works. The laser photobleaches the fluorescent proteins in a selected region. Because of diffusion, proteins that were not bleached come into the bleached region over time. (B) Higher-resolution schematic of the photobleaching process over a selected region within the cell.

How many reactions do enzymes carry out each second?

One oversimplified and anthropomorphic view of a cell is as a big factory for transforming relatively simple incoming streams of molecules like sugars into complex biomass consisting of a mixture of proteins, lipids, nucleic acids, and so on. The elementary processes in this factory are the metabolic transformations of compounds from one to another. The catalytic proteins taking part in these metabolic reactions—the enzymes—are almost invariably the largest fraction of the proteome of the cell (see www.proteomaps.net for a visual impression). They are thus also the largest component of the total cell dry mass. Using roughly one thousand such reactions, *E. coli* cells can grow on a single carbon source such as glucose and some inorganic minerals to build the many molecular constituents of a functioning cell. What is the characteristic time scale (drum beat rhythm or clock rate, if you like) of these transformations?

The biochemical reactions taking place in cells, though thermodynamically favorable, are in most cases very slow if left uncatalyzed. For example, the spontaneous cleavage of a peptide bond would take 400 years at room temperature, and phosphomonoester hydrolysis, routinely breaking up ATP to release energy, would take about a million years in the absence of the enzymes that shuttle that reaction along (BNID 107209). Fortunately, metabolism is carried out by enzymes that often increase rates by

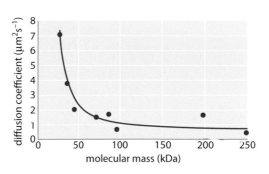

Figure 4-4 Diffusion constant as a function of molecular mass in *E. coli*. The diffusion of proteins within the *E. coli* cytoplasm were measured using the FRAP technique. (Adapted from Kumar M, Mommer MS & Sourjik V [2010] *Biophys J* 98:552–559.)

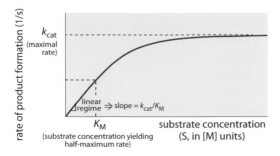

Figure 4-5 The characteristic dependence of enzyme catalysis rate on substrate concentration. Key defining effective parameters, such as k_{cat}, K_M, and their ratio, the second-order rate constant that is equal to the slope at low concentrations, are denoted in the figure.

an astonishing 10 orders of magnitude or more (BNID 105084, 107178). Phenomenologically, it is convenient to characterize enzymes kinetically by a catalytic rate, k_{cat} (also referred to as the turnover number). Simply put, k_{cat} signifies how many reactions an enzyme can possibly make per unit time. This is shown schematically in **Figure 4-5**. Enzyme kinetics is often discussed within the canonical Michaelis–Menten framework, but the so-called hyperbolic shape of the curves that characterize how the rate of product accumulation scales with substrate concentration feature several generic features that transcend the Michaelis–Menten framework itself. For example, at very low substrate concentrations, the rate of the reaction increases linearly with substrate concentration. In addition, at very high concentrations, the enzyme is cranking out as many product molecules as it can every second at a rate k_{cat}, and increasing the substrate concentration will not lead to any rate enhancement.

Rates vary immensely. Record holders are carbonic anhydrase, the enzyme that transforms CO_2 into bicarbonate and back through the reaction,

$$CO_2 + H_2O \rightleftharpoons HCO_3^- + H^+, \tag{4.2}$$

and superoxide dismutase, an enzyme that protects cells against the reactivity of superoxide by transforming it into hydrogen peroxide in the reaction,

$$2O_2^- + 2H^+ \rightleftharpoons H_2O_2 + O_2. \tag{4.3}$$

These enzymes can carry out as many as 10^6–10^7 reactions per second. At the opposite extreme, restriction enzymes limp along while performing only $\approx 10^{-1}$–10^{-2} reactions per second, or about one reaction per minute per enzyme (BNID 101627, 101635). To flesh out the metabolic heartbeat of the cell, we need a sense of the characteristic rates rather than the extremes. **Figure 4-6A** shows the distribution of k_{cat} values for metabolic enzymes based on an extensive compilation of literature sources (BNID 111411). This figure reveals that the median k_{cat} is about 10 s^{-1}, several orders of magnitude lower than the common textbook examples, with the enzymes of central carbon metabolism, which is the cell's metabolic highway, being on the average three times faster with a median of about 30 s^{-1}.

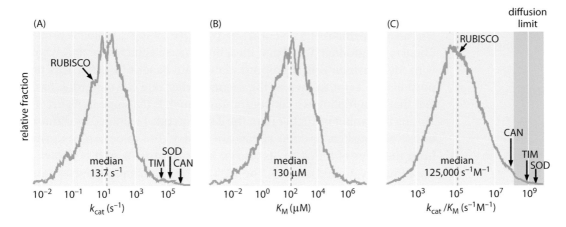

Figure 4-6 Distributions of enzyme kinetic parameters from the literature extracted from the BRENDA database: (A) k_{cat} values ($N = 1942$), (B) K_M values ($N = 5194$), and (C) k_{cat}/K_M values ($N = 1882$). Only values referring to natural substrates were included in the distributions. The location of several well-studied enzymes is highlighted: CAN, carbonic anhydrase; SOD, superoxide dismutase; TIM, triosephosphate isomerase; and Rubisco, ribulose-1,5-bisphosphate carboxylase oxygenase. (Adapted from Bar-Even A, Noor E, Savir Y et al. [2011] *Biochemistry* 50:4402–4410.)

How do we know if an enzyme works close to the maximal rate? From the general shape of the curve that relates enzyme rate to substrate concentration shown in Figure 4-5, there is a level of substrate concentration beyond which the enzyme will achieve more than half of its potential rate. The concentration at which the half-maximal rate is achieved is denoted by K_M. The definition of K_M provides a natural measuring stick used to tell us when concentrations are "low" and when they are "high." When the substrate concentration is well above K_M, the reaction will proceed at close to the maximal rate, k_{cat}. At a substrate concentration [S] = K_M, the reaction will proceed at half of k_{cat}. Enzyme kinetics in reality is usually much more elaborate than the textbook Michaelis–Menten model, with many enzymes exhibiting cooperativity and performing multi-substrate reactions of various mechanisms, resulting in a plethora of functional forms for the rate law. But in most cases, the general shape can be captured by the defining features of a maximal rate and substrate concentration at the point of half-saturation, as indicated schematically in Figure 4-5. Thus, the behavior of real enzymes can be cloaked in the language of Michaelis–Menten, using k_{cat} and K_M, despite the fact that the underlying Michaelis–Menten model itself may not be appropriate.

We have seen that the actual rate depends upon how much substrate is present through the substrate affinity, K_M. What are the characteristic values of K_M for enzymes in the cell? As shown in **Figure 4-6B**, the median K_M value is in the 0.1 mM range. From our rule of thumb (1 nM is about 1 molecule per 1 *E. coli* volume), this is roughly equal to 100,000 substrate molecules per bacterial cell. At low substrate concentration ($[S] << K_M$), we can approximate the reaction rate by $[E_T](k_{cat})[S]/K_M$, which is proportional to the product $[E_T][S]$ that measures the collision rate of the enzyme with the substrate with a proportionality rate factor of k_{cat}/K_M. This proportionality factor, known as the second-order rate constant (due to the fact that it multiplies two concentration terms), is the slope in Figure 4-5. This factor cannot be higher than the collision rate facilitated by diffusion unless electrostatic or other effects are in play. The value can be derived from the rules of diffusion, as shown in **Estimate 4-3**, and is known as the diffusion-limited on rate. The idea of the calculation involves nothing more than working out the diffusive flux of substrate molecules onto our protein "absorber" and then asserting that every arriving molecule

The diffusion-limited on-rate

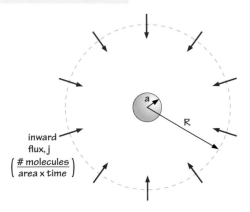

mass conservation says $j(R) \times 4\pi R^2$ is constant for all R, hence $j(R) \propto \dfrac{1}{R^2}$

by Fick's Law, $j \propto -\dfrac{\partial c}{\partial R} \Rightarrow \dfrac{\partial c}{\partial R} = -\dfrac{A}{R^2} \Rightarrow c(R) = \dfrac{A}{R} + B$

apply boundary conditions

$\left.\begin{array}{l} 1) \quad c(\infty) = c_\infty = B \\[2mm] 2) \quad c(a) = \dfrac{A}{a} + c_\infty = 0 \Rightarrow A = -c_\infty a \end{array}\right\} \Rightarrow c(R) = c_\infty\left(1 - \dfrac{a}{R}\right)$

using Fick's Law, $j(a) = -D\left.\dfrac{\partial c}{\partial R}\right|_{R=a} = \dfrac{Dc_\infty}{a}$

reaction rate $\dfrac{dn}{dt} = j(a)\,4\pi a^2 = 4\pi D a c_\infty = k_{on}c_\infty$

protein radius

$k_{on} = 4\pi \times 100\dfrac{\mu m^2}{s} \times 2{\times}10^{-3}\mu m \times 6{\times}10^{23}\dfrac{molecules}{mol} \times \dfrac{1\,L}{10^{15}\mu m^3} \approx 10^9\,s^{-1}M^{-1}$

characteristic diffusion coefficient
of metabolite in cytoplasm

Estimate 4-3

is able to undergo the reaction of interest. For a protein-sized enzyme and a small-molecule substrate, the diffusion-limited rate constant takes the value of roughly 10^9 $s^{-1}M^{-1}$. An enzyme approaching this limit can be described as optimal with respect to its ability to perform a successful transformation on every encounter provided by random diffusion. Few enzymes, with notable exceptions such as the glycolytic enzyme triose isomerase (TIM), merit this status (BNID 103917). How well does the characteristic enzyme do? **Figure 4-6C** shows that the median value is about 10^5 $s^{-1}M^{-1}$, or about 4 orders of magnitude lower than the diffusion limit. This difference can be partially explained by the liberal nature of the diffusion limit that does not depict all the issues related to binding and by noting that for many enzymes there might not be a strong selective pressure to optimize their kinetic properties. Moreover, the rate might be compromised in many cases by the need for recognition and specificity in the interaction.

The value of K_M in conjunction with the diffusion-limited on-rate can be used to estimate the off-rates for bound substrate. The goal of the simple estimate is to find the time scale over which a substrate that is bound to the enzyme will stay bound before it goes back to solution (usually without reacting), the so-called off-rate, k_{off}. The estimate is based upon an ideal limit in which the on-rate is controlled by diffusive encounters with the enzyme characterized by the diffusion-limited on-rate

$$k_{on} \approx 10^9 \, s^{-1}M^{-1}. \tag{4.4}$$

An approximation for k_{off} is the product of this k_{on} and K_M. So, for example, if K_M is a characteristic 10^{-4} M, then this product is 10^5 s^{-1}, so the substrate will unbind in about 10 μs—this is the so-called residence time. For extremely strong binders, where the affinity is, say, 1 nM = 10^{-9} M, the residence time will be 1 s. This gives only a taste of the idealized case; the actual measured values for off-rates (or residence times) are revealed by enzymologists, and this work keeps them busy and confronted with a plethora of surprises. An analogous estimate for the off-rate can be considered for interactions between signaling molecules and for transcription factors binding to DNA, with characteristic time scales from milliseconds to tens of seconds or even longer.

Insight into the rate of interactions at the molecular level can be gleaned from a clever interpretation[§] of the diffusion limit. Say we drop a test substrate molecule into a cytoplasm with a volume equal to that of a bacterial cell. If everything is well mixed and there is no binding, how long will it take for the substrate molecule to collide with one specific protein in the

---

[§] Goodsell, DS (2010) The Machinery of Life. Springer.

cell? The rate of enzyme substrate collisions is dictated by the diffusion limit, which as shown above, is equal to $\approx 10^9 \text{ s}^{-1}\text{M}^{-1}$ times the concentrations. We make use of one of our tricks of the trade, which states that in *E. coli*, a single molecule per cell (say, our substrate) has an effective concentration of about 1nM (that is, 10^{-9} M). The rate of collisions is thus

$$10^9 \text{ s}^{-1}\text{M}^{-1} \times 10^{-9} \text{ M} \approx 1 \text{ s}^{-1}. \tag{4.5}$$

That is, they will meet within a second on average. This allows us to estimate that every substrate molecule collides with each and every protein in the cell on average about once per second. As a concrete example, think of a sugar molecule transported into the cell. Within a second it will have an opportunity to bump into all the different protein molecules in the cell. The high frequency of such molecular encounters is a mental picture worth carrying around when trying to have a grasp of the microscopic world of the cell.

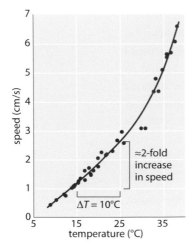

Figure 4-7 Speed of ants as a function of temperature. Measured by the astronomer Harlow Shapley on Mount Wilson above Los Angeles, where he was deeply engaged in measuring the size of our galaxy. The *Liometopumapiculatum* ants he studied on the mountain are active both day and night, thus allowing a larger temperature range to be studied. It was verified that ant body mass had a negligible effect. Similarly, there was no significant difference between incoming and outgoing direction on ant speed. (Adapted from Shapley H [1920] *Proc Natl Acad Sci USA* 6:204–211.)

How does temperature affect rates and affinities?

In the early 1900s, when Harlow Shapley was not busy measuring the size of our galaxy using the telescope on Mount Wilson, he spent his time measuring how fast ants moved and how their speed depended upon the temperature. His observations are shown in **Figure 4-7**, which demonstrates a rapid increase of speed with temperature, with about a twofold increase as the temperature rises from 15°C to 25°C, and another doubling in speed as the temperature rises another 10 degrees from 25°C to 35°C. This relates to an interesting rule of thumb used by enzymologists that states that the catalytic rates of enzymes double when subjected to a 10°C increase in temperature. Though there are many exceptions to this "rule," what is the basis for such an assertion in the first place? A simplified mental picture of enzyme catalysis argues that there is a free-energy barrier that the substrates have to overcome before they can be transformed to products. For a barrier of "height" E_a, where E_a is the Arrhenius energy of activation, the rate scales according to the empirical Arrhenius relationship in which the rate is proportional to $\exp(-E_a/k_B T)$. The theoretical underpinnings of this result come from an appeal

Change in rate for 10°C increase in temperature

$rate \propto e^{-E_a/k_BT}$ (Arrhenius empirical relationship)

$$\Downarrow$$

$$\frac{rate\,(T_2)}{rate\,(T_1)} = \frac{e^{-E_a/k_BT_2}}{e^{-E_a/k_BT_1}} = e^{-\frac{E_a}{k_B}\left(\frac{1}{T_2}-\frac{1}{T_1}\right)} = e^{-\frac{E_a}{k_B}\left(\frac{T_1-T_2}{T_1T_2}\right)} = e^{-\frac{50\,kJ/mol}{8.3\,J/mol\times K}\times\frac{-10\,K}{(300\,K)^2}} \approx e^{0.7} \approx 2$$

$$\Uparrow$$

$E_a \approx 50\ kJ/mol\ (characteristic\ activation\ energy)$

$T_2 = T_1 + 10°C$

$T_1 \approx 300\ K\ (room\ temperature)$

$k_B \approx 1.38 \times 10^{-23}\ J/K \approx 8.3\ J/mol \times K$

Estimate 4-4

to the Boltzmann distribution. If E_a is very large, the barrier is high and the exponential dependence results in a very slow rate. Many reactions have values of E_a of ≈ 50 kJ/mol $\approx 20\ k_BT$ (for example, BNID 107803). In the back-of-the-envelope calculation shown in **Estimate 4-4,** we show how this suggests that a 10-degree (Celsius or kelvin) change around room temperature results in a $\approx$ twofold change in rate.

This rate factor, which can be independently measured for different reactions, is quantified in the literature by a quantity called Q_{10}. Q_{10} reveals the factor by which the rate changes for a 10°C change in temperature. Should an increase in temperature increase or decrease the rate at which some reaction occurs? The Boltzmann distribution states that the number of molecules with sufficient energy to overcome the barrier scales as the exponent of the ratio, $-E_a/k_BT$. At higher temperatures, the ratio is closer to zero, and thus more molecules have the required activation energy, which makes the barrier easier to overcome, resulting in an increase in the reaction rate.

Interestingly, the growth of a whole bacterium also tends to scale with temperature according to a similar functional form (BNID 100919)—that is, the log of the growth rate scales linearly with the inverse temperature below and near the physiological temperature. As an example, growth of *E. coli* increases by ≈ 2.5-fold when the temperature is changed from 17°C to 27°C and then again from 27°C to 37°C. This is often depicted by plotting the growth rate versus $1/T$, as shown in **Figure 4-8**. In this range, one can infer an effective value for E_a of ≈ 60 kJ/mol $\approx 25\ k_BT$. This is called an effective value, because there is no single barrier that the bacterium has to overcome in order to grow and divide, but instead the set of all barriers and processes coalesces into this one effective value.

Though the Arrhenius equation is a staple ingredient of undergraduate education in many disciplines, and it seems like the obvious choice for

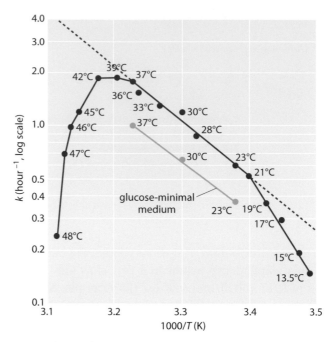

Figure 4-8 Dependence of the growth rate of *E. coli* on temperature. The growth rate is plotted versus the inverse of the temperature (an Arrhenius plot). Note the middle range, where the dependence looks linear in accordance with the Arrhenius rate law. (Adapted from Herendeen SL, VanBogelen RA and Neidhardt FC [1979] *J Bact* 139:185–194.)

characterizing the temperature dependence of biochemical rates, it wasn't always deemed so simple. A menagerie of functional relationships between rate and temperature were suggested over the years and are summarized in **Table 4-2**. As the range of temperatures over which experimental measurements were made only covered a small temperature regime compared to room temperature, most of these different expressions gave similarly

functional form for k	supported by
$k = a \times T^c \times e^{-(b - d \times T^2)/T}$	van't Hoff, 1898; Bodenstein, 1899
$k = a \times T^c \times e^{-b/T}$	Kooij, 1893; Trautz, 1909
$k = a \times e^{-(b - d \times T^2) \times T}$	Schwab, 1883; van't Hoff, 1884; Spohr, 1888; van't Hoff and Reicher, 1889; Buchbock, 1897; Wegscheider, 1899
$k = a \times T^c \times e^{d \times T}$	unknown
$k = a \times e^{-b/T}$	van't Hoff, 1884; Arrhenius, 1889; Kooij, 1893
$k = a \times T^c$	Harcourt and Esson, 1895; Veley, 1908; Harcourt and Esson, 1912
$k = a \times e^{d \times T}$	Berthelot, 1862; Hood, 1885; Spring, 1887; Veley, 1889; Hecht and Conrad, 1889; Pendelbury and Seward, 1889; Tammann, 1897; Remsen and Reid, 1899; Bugarszky, 1904; Perman and Greaves, 1908

Table 4-2 Different expressions suggested for the dependence of the rate of a chemical reaction as a function of temperature. (Adapted from Laidler KJ [1984] *J Chem Educ* 61:494–498.)

good fits. The famed chemist Wilhelm Ostwald stated that temperature dependence "is one of the darkest chapters in chemical mechanics." Many lessons on the balance of models and experiments can be gleaned from following its history as depicted in a careful review.[¶]

What are the rates of membrane transporters?

Cells are buffered from the fluctuating environment that surrounds them by their plasma membranes. These membranes control which molecular species are allowed to cross the membrane and how many molecules are permitted to pass to the cellular interior. Specifically, unless a compound is simultaneously small and uncharged, passage across the plasma membrane is licensed by molecular gatekeepers. Transporting the dazzling complement of molecular building blocks requires a diverse census of membrane proteins that occupy a significant fraction of the membrane real estate, as we now explore and depict schematically in **Estimate 4-5**.

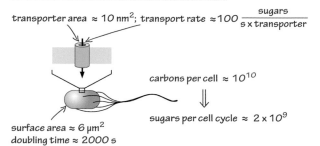

Fraction of membrane taken up by transporters

transporter area $\approx 10 \text{ nm}^2$; transport rate $\approx 100 \dfrac{\text{sugars}}{\text{s} \times \text{transporter}}$

carbons per cell $\approx 10^{10}$

$\Downarrow$

sugars per cell cycle $\approx 2 \times 10^9$

surface area $\approx 6 \text{ μm}^2$
doubling time $\approx 2000 \text{ s}$

number of transporters needed $\approx \dfrac{2 \times 10^9 \frac{\text{sugars}}{\text{cell cycle}}}{2000 \frac{\text{s}}{\text{cell cycle}}} \times \dfrac{1}{100 \frac{\text{sugars}}{\text{s} \times \text{transporter}}} \approx 10^4 \text{ transporters}$

fraction membrane area $\approx \dfrac{10^4 \text{ transporters} \times 10 \text{nm}^2/\text{transporter}}{6 \text{ μm}^2 \times 10^6 \frac{\text{nm}^2}{\text{μm}^2}} \approx 2\%$

Estimate 4-5

[¶] Laidler KJ (1984) The development of the Arrhenius equation. *J Chem Educ* 61:494–498.

The characteristic transport rate for sugar transporters saturated with external substrate, say a glucose transporter, is ≈ 100 s^{-1}. Why should these so-called turnover rates, analogous to the k_{cat} values of enzymes, usually have a range of 30–300 s^{-1} (BNID 102931, 103159, 101737-9) and not be much higher? We can suggest a rationalization for a common subset of transporters. Many transporters are proton-coupled, meaning that they use the proton motive force to drive the transport process, often against a concentration gradient of the sugar substrate. To estimate an upper limit on a proton-coupled transporter turnover rate, we focus on the on-rate of the protons. This is a prerequisite step to the conformational change that will actually perform the transport process. The conformational change might be slower and thus will ensure our estimate is indeed an upper bound. Recall that the proton concentration at pH = 7 is 10^{-7} M and the diffusion-limited on-rate is about 10^9 M^{-1}s^{-1}. This implies that the rate at which protons hit the transporter (k_{on}) is roughly $\sim 10^{-7}$ M $\times$ 10^9 M^{-1}s^{-1} $= 10^2$ s^{-1}, which is the same order of magnitude as the observed turnover rate. This is effectively saying that such a proton-coupled transporter works roughly as fast as it can, given the diffusion-limited rate at which protons that are serving as its energy source arrive. Alas, for the closely related sodium transporters or many ATP-dependent transporters, this logic would give unrealistic limits, with rates on the order of millions per second, thus showing that other kinetic issues are limiting.

The fastest transporter we are aware of is capnophorin—literally meaning "smoke carrier"—a transporter in red blood cells whose physiological role is to transport CO_2 from the lungs (that is, the "smoke" of metabolism). This speed demon chloride–bicarbonate transporter was suggested to reach turnover rates on the order of 100,000 s^{-1} (BNID 111368). Given that the concentration of both of its substrates is in the mM range, we can rationalize the capacity for a 1000-fold increase in rate over the proton-coupled transporters because of the higher concentration that fuels a higher diffusion-limited on-rate. Throughout this vignette, our values originate almost exclusively from studies of glucose and lactose transporters. Surprisingly, we are forced into this situation by a dearth of quantitative information on other transporters.

To get a sense of what measured transporter rates imply about the numbers of membrane proteins, we now estimate how many such proteins are needed for key cellular metabolites. Assume that the carbon source is provided exclusively in the form of glucose or glucose equivalents. Is the maximal division rate dictated by the limited real estate on the surface of the cell membrane to locate glucose carbon transporters? The surface area of an *E. coli* membrane dividing every half an hour is $\approx 6\ \mu m^2$ (BNID 103339, 105026). The structurally determined lactose transporter has an oval shape normal to the membrane with dimensions (long and short axis) of 6 nm $\times$ 3 nm (BNID

102929). Assuming a similar size for the glucose transporter, the area it occupies on the membrane is ≈ 10–20 nm^2 (though a value about fourfold larger for the glucose-like PTS transport system is reported in another species of bacterium). For importing the $\approx 2 \times 10^9$ sugar molecules needed solely to build the cell mass (each consisting of six carbon atoms) within a conservative cell cycle duration of ≈ 2000 s, the fraction of the membrane area required is already $\approx 2\%$, as estimated in Estimate 4-5. Thus, a substantial part of the membrane has to be occupied just to provide the necessary carbon, even under conservative assumptions. Can it be that maximal growth rate (less than 1000-second generation time) is constrained by the ability to transport carbon? Dedicated experiments are required to clarify if there is a limitation on increasing the fraction of transporter much further (say, to 10%). We should also consider that the respiratory system for energizing the cell needs to reside on the membrane in bacteria and that packing idealized oval machines on the membrane's two-dimensional surface cannot reach 100% coverage for geometrical reasons.

Membrane transport is not the only process that might potentially limit the maximal growth rate. Other issues rival the number of available membrane transporters in their role for limiting the maximal growth rates, and probably should be thought of as co-limitations. In the vignette entitled "Which is faster: transcription or translation?" (pg. 231), we discuss the tricks bacterial cells use to achieve fast doubling times with only a single origin of replication. Further, the vignette on "How many ribosomes are in a cell?" (pg. 147) shows that under fast growth rates, ribosome concentration grows linearly with growth rate and that the rate of translation may constrain the maximal growth rate. Indeed, a number of processes can potentially constrain maximal growth rates besides the transport of nutrients across the cell membrane, although the estimates provided here demonstrate the need for careful thought about the management of membrane real estate.

A similar calculation can be performed for budding yeast. The volume, and thus the number of carbons required, is ≈ 50 times (BNID 100427) larger than in *E. coli*, whereas the surface area is ≈ 10 times larger and the fastest generation time is ≈ 5 times longer at ≈ 6000 s (BNID 100270). Thus, the areal fraction required for the transport of carbon building blocks is suggested to be similar. Notice, though, that under maximal growth rate conditions, yeast performs fermentation to supply its energy needs, which dictates a significant additional transport of sugars (actually *E. coli* often does that, as well, and emits carbon as part of overflow metabolism). A measurement shows that under growth rates up to one division per 140 min, approximately half the carbon is lost in fermentation (with an even higher proportion at faster growth rates) (BNID 102324). Thus, the required surface fraction covered by transporters is suggested to be at least double that found in the bacterial setting (i.e., ≈ 4%). This estimate motivates an experimental test: Will the expression of a membrane protein not related to transport decrease the maximal growth rate of yeast and *E. coli* by limiting the available area for transporters?

How many ions pass through an ion channel per second?

Cells regulate their ion concentrations very tightly. Both the identities and quantities of the different ion species within cells play a role in energy storage, protein function, signaling, and a variety of other processes. As with many other key cellular parameters, the ionic disposition within cells is controlled carefully both spatially and temporally. Indeed, whole families of proteins exist (see **Figure 4-9**) whose job is either to open or close pores in the membrane, thus permitting ions (or other species) in and out of the cell, or to actively pump various species, including ions, against their concentration gradients.

Single-molecule studies of the macromolecules of the cell are one of the centerpieces of modern biophysical analysis. These studies had their origins in early work aimed at uncovering the properties of individual ion channels engaged in the transport of ions in the presence of some driving force. There are different classes of driving forces that can gate ion channels. Some

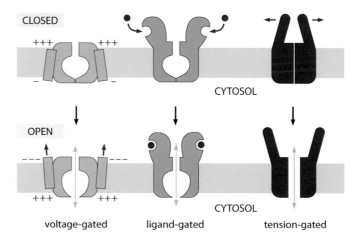

Figure 4-9 Different mechanisms of ion channel gating. The green channel is gated by a transmembrane voltage. The blue channel is gated by ligands that bind the protein and induce a conformational change. The red channel is gated by mechanical forces.

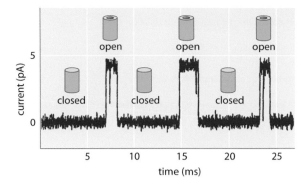

Figure 4-10 Characteristic amplitude of current passing through a channel is a few picoamperes. The channel switches between the closed and open states. When open, the channel permits the passage of ions, which is measured as a current. (Adapted from Phillips R, Kondev J, Theriot J & Garcia H [2012] Physical Biology of the Cell, 2nd ed. Garland Science.)

channels open in response to the presence of some soluble ligand, meaning that the driving force is the concentration of ligands that bind to the channels and change their open probability. In other cases, the driving force is the voltage applied across the membrane that harbors the channels. Finally, we mention the opening through mechanical effects by applying membrane tension. These different gating mechanisms are illustrated in Figure 4-9, which shows schematics of each of the different channel types.

What do the currents measured in single-molecule studies reveal about the dynamics of these channels and the number of ions passing through them? As shown in **Figure 4-10**, the outcome of these kinds of experiments is the observation that the characteristic currents through individual channels are measured in pA. We can convert these current levels into a corresponding number of charges traversing the channel each second, as shown in **Estimate 4-6**. If we recall that an ampere corresponds to a charge flow of 1 coulomb each second and, further, if we use the fact that the charge on a monovalent ion is approximately 1.6×10^{-19} coulombs (that is, 1 electron or 1 proton charge), then we see that a current of about one pA corresponds to roughly 10^7 ions passing through the channel each second. This value is in agreement with measurements (BNID 103163) even if

Characteristic rate of ions passing through channel

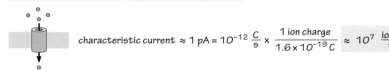

$$\text{characteristic current} \approx 1 \text{ pA} = 10^{-12} \, \frac{C}{s} \times \frac{1 \text{ ion charge}}{1.6 \times 10^{-19} \, C} \approx 10^7 \, \frac{\text{ions}}{s}$$

Estimate 4-6

Estimating the characteristic current flowing through an ion channel

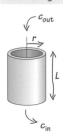

We use Fick's law for the flux: $J_z = -D \frac{\partial c}{\partial z} \approx D \frac{\Delta c}{L}$

A linear profile of concentration satisfies the steady-state diffusion equation:

$c(z) = c_{out} - \frac{(c_{out} - c_{in})}{L} z$

Units of flux are $\frac{number}{area \times s}$; current is $J \times A = J \times \pi r^2 = D \frac{\Delta c}{L} \pi r^2$

$D \approx 1000 \frac{\mu m^2}{s}$ for ions

Membrane thickness, $L \approx 4nm$

Channel radius, $r \approx 1nm$

For ions characteristic $\Delta c \approx 10mM \approx 10^7/\mu m^3$

$$J \times A \approx 1000 \overbrace{\frac{\mu m^2}{s}}^{D} \times \overbrace{10^7 \mu m^{-3}}^{\Delta c} \times \overbrace{3 \times 10^{-6} \mu m^2}^{\pi r^2} / \underbrace{4 \times 10^{-3} \mu m}_{L} \approx 10^7 s^{-1}$$

Estimate 4-7

not with our intuition from daily life that would be hard pressed to imagine 10 million cars passing a bridge that can only hold about four cars at any given moment. We can rationalize the experimentally observed rates by considering the diffusive consequences of a concentration gradient across the membrane and by working out the number of ions we expect to cross the channel each second, as shown in **Estimate 4-7**.

Ions flowing in channels akin to the ionic channels above also drive the flagellar motor of bacteria by coupling the motor to the transport of the protons down their chemiosmotic gradient. The rate of proton transport in these channels is about three orders of magnitude slower, at 10^4 per second (BNID 109822), and is, as a result, one of the channels with lowest conductance.

What is the turnover time of metabolites?

Fast cellular growth rates are associated with proportionally higher utilization rates (fluxes) of precursor metabolites. At the same time, the concentrations of intermediate metabolites need to be kept at levels not

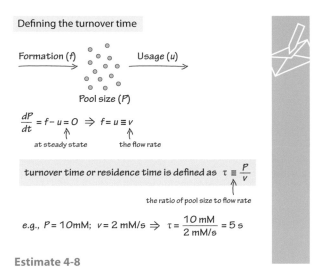

Defining the turnover time

Formation (f) →   ← Usage (u)

Pool size (P)

$$\frac{dP}{dt} = f - u = 0 \;\Rightarrow\; f = u \equiv v$$

at steady state   the flow rate

turnover time or residence time is defined as $\tau \equiv \dfrac{P}{v}$

the ratio of pool size to flow rate

e.g., $P = 10\,mM;\; v = 2\,mM/s \;\Rightarrow\; \tau = \dfrac{10\,mM}{2\,mM/s} = 5\,s$

Estimate 4-8

exceeding a few mM, in order to avoid problems ranging from osmotic pressure imbalance to nonspecific cross reactivity. Achieving these two aims—namely high fluxes at low intermediate concentrations—implies a quick turnover time of the metabolite pool. The turnover time concept is schematically shown in **Estimate 4-8** and is defined to be the mean time over which the pool of a given metabolite will be replaced due to the rates of production and utilization (which are equal in steady state). Indeed, for many key metabolites of central carbon metabolism, the turnover time is on the order of a second, as shown in **Figure 4-11** and **Table 4-3** for the case of the model plant *Arabidopsis Thaliana* (BNID 107358). Similarly, in *E. coli*, the pools of most amino acids were shown to turn over in less than a minute (BNID 101622). The subsecond turnover times in *Arabidopsis* manifest in the startling finding that when aiming to perform a metabolomics experiment that measures the concentration of metabolites, if the researcher briefly passes a hand over the light source when preparing to

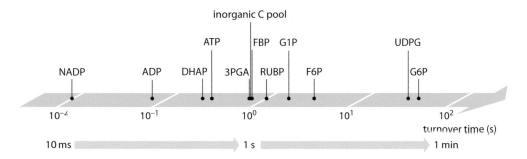

Figure 4-11 The turnover time is defined through the ratio of the pool size to the flux. In steady state, the flux is equal to the formation rate, which also equals the usage rate.

metabolite	turnover time (s)		
	Arabidopsis	*S. cerevisiae*	*E. coli*
NADP	0.01	—	—
ADP	0.07	0.3	0.8
Calvin–Benson cycle intermediates (R5P, S6P, X5P, Ru5P, SBP, RuBP)	0.1–1	—	—
DHAP/G3P	0.2	—	—
ATP	0.3	1.4	2
3PGA	0.7	7	3
inorganic C	0.8	—	—
FBP	0.8	7	1.2
pyruvate	—	1.7	1.5
F6P	3	7	1.2
AMP	—	3	9
UDPG	40	—	—
G6P	40	17	4
glycerol-3-phosphate	—	60	13
TCA cycle (Suc, Fum, Mal)	—	4–30	0.7–9

Table 4-3 Turnover times for key metabolites. For *Arabidopsis*, leaves were measured by mass spectrometry to find turnover times for metabolites in the Calvin cycle, and starch and sucrose synthesis, under light and 485 ppm CO_2. (Adapted from Arrivault S, Guenther M, Ivakov A et al. [2009] *Plant J* 59:826–839. Data for *E. coli* and *S. cerevisiae* are from BNID 109701.)

immerse the plant into the liquid nitrogen, the result will already be different for Calvin–Benson cycle metabolites than if the researcher was careful not to block the light.

Protein synthesis provides another example of fast turnover, where a high rate of polymerization of monomeric amino acids takes place that is mediated by tRNA shuttling. The total number of tRNAs, however, is limited. In *E. coli* growing with a doubling time of 40 minutes, the total number of tRNAs is estimated at ≈200,000 copies per cell (BNID 100066). Given that there are about 30,000 ribosomes (BNID 102015), each working at a rate of polymerization of ≈20 aa per second (BNID 100059), the average turnover time is about $200,000/(20 \text{ s}^{-1} \times 30,000)$ ≈ one-third of a second. This is the timeframe between loading an amino acid onto the tRNA through tRNA synthetase, binding of that tRNA within a ribosome where the amino acid is released and forms the peptide bond, and the replenishment of that tRNA by the loading of a new amino acid. Though this estimate has been carried out in less than one paragraph, careful experiments to actually obtain precise measurements were much harder. Using radioactive pulse-labeling, the numbers from the estimate above were confirmed, resulting in a range of turnover times of 0.1–1 s (BNID 105275) for the turnover of the tRNA pool. In budding yeast, the corresponding numbers are about 2 million tRNAs (BNID 108197), 200,000 ribosomes (BNID 100267, 108197), and a polymerization rate of ≈ 10 aa/s (BNID 107785, 107871), resulting yet again in a turnover time of about 1 s. As an aside, note that the ratio of total tRNA per ribosome tends to be about 10 to 1.

The variety of surprising numbers for turnover times offered throughout this vignette paint a vibrant picture of the chemical hustle and bustle taking place within cells. Though many of our structural descriptions of biology offer a static picture of the molecules of the cell, we see here that whether talking about the molecular components of central metabolic pathways or key players in the central dogma (such as the tRNAs that make protein synthesis possible), these molecules often are transitioning between different states literally in the blink of an eye (taking about 0.1–0.4 s; BNID 100706).

CENTRAL DOGMA

Which is faster: transcription or translation?

Transcription, the synthesis of mRNA from DNA, and translation, the synthesis of protein from mRNA, are the main pillars of the central dogma of molecular biology. How do the speeds of these two processes compare? This question is made all the more interesting as a result of observations like those shown in **Figure 4-12**—namely, the existence of the beautiful "Christmas tree" structures observed in *E. coli* using electron microscopy. These stereotyped structures reflect the simultaneous transcription and translation of the same gene and raise the question of how the relative rates of the two processes compare, making such synchronization of these two disparate processes possible.

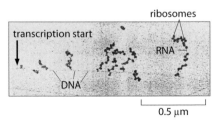

Figure 4-12 Electron microscopy image of simultaneous transcription and translation. The image shows bacterial DNA and its associated mRNA transcripts, each of which is occupied by ribosomes. (Adapted from Miller OL Jr, Hamkalo BA & Thomas CA Jr [1970] *Science* 169:392–395.)

Transcription of RNA in *E. coli* of both mRNA and the stable rRNA and tRNA is carried out by ≈ 1000–10,000 RNA polymerase molecules (BNID 101440), which proceed at a maximal speed of about 40–80 nt/s, as shown in **Table 4-4** (BNID 104900, 104902, 108488). Translation of proteins in *E. coli* is carried out by ≈ 10,000–100,000 ribosomes (BNID 101441) and proceeds at a maximal speed of about 20 aa/s, as shown in **Table 4-5** (BNID 100059, 105067, 108490).

organism	rate (nt/s)	BNID
E. coli	10–100	104900, 104902, 101904, 108488, 108490, 108487, 100060
Monkey cell line	100	105113
H. sapiens	6–70	105566 , 100661, 100662
D. melanogaster	25	111484

Table 4-4 Transcription rate measured across organisms and conditions. All values measured at 37°C, except for *D. melanogaster*, which was measured at 22°C.

organism	rate (aa/s)	BNID
E. coli	10–20	100059 , 105067, 108490, 108487, 100233
S. cerevisiae	3–10	107871
N. crassa	5–8	107872
M. musculus	6	107952

Table 4-5 Translation rate measured across organisms and conditions. All values measured at 37°C, except for *S. cerevisiae* and *N. crassa*, which were measured at 30°C.

Interestingly, since every three base pairs code for one amino acid, the rates of the two processes are nearly matched, as schematically shown in **Estimate 4-9** (see also BNID 108487). If translation were faster than transcription, it would cause the ribosome to "collide" with the RNA polymerase in prokaryotes, where the two processes can happen concurrently. Such co-transcriptional translation has become textbook material through images such as Figure 4-12. But recent single-molecule microscopy shows this occurs relatively rarely, and most translation is not coupled with transcription in *E. coli*.[**] Rather, most translation takes place on mRNA that has already diffused away from the DNA-rich nucleoid region

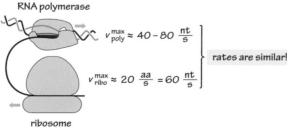

Which is faster: transcription or translation?

RNA polymerase

$v_{poly}^{max} \approx 40 - 80 \ \frac{nt}{s}$

rates are similar!

$v_{ribo}^{max} \approx 20 \ \frac{aa}{s} = 60 \ \frac{nt}{s}$

ribosome

Estimate 4-9

[**] Bakshi S, Siryaporn A, Goulian M et al. (2012) Superresolution imaging of ribosomes and RNA polymerase in live *Escherichia coli* cells. *Mol Microbiol* 85:21–38.

to ribosome-rich cytoplasmic regions. The distribution of ribosomes in the cells is further shown in the vignette entitled "How many ribosomes are in a cell?" (pg. 147). In another twist and turn of the central dogma, it was shown that ribosomes could be important for fast transcription in bacteria.[††] The ribosomes seem to keep RNA polymerase from backtracking and from pauses, which can otherwise be quite common for these machines, thus creating a striking reverse coupling between translation and transcription.

What do the relative rates of transcription and translation mean for the overall time taken from transcription initiation to synthesized protein for a given gene? In bacteria, a 1-kb gene should at a maximal transcription rate take about 1000 nt/80 nt/s ≈ 10s for transcription elongation, with translation elongation at maximal speed being roughly the same. Note that the total time scale is the sum of an elongation time, as above, and the initiation time, which can be longer in some cases. Recently it was observed that increasing the translation rate, by replacing wobble codons with perfect matching codons, results in errors in folding.[‡‡] This suggests a tradeoff, where translation rate is limited by the time needed to allow proper folding of domains in the nascent protein.

How are the rates of these key processes of the central dogma measured? This is an interesting challenge, even with today's advanced technologies. Let's consider how we might attack this problem. One vague idea might be to express a GFP molecule and measure the time until it appears. To see the flaws in such an approach, check out the vignette entitled "What is the maturation time for fluorescent proteins?" (pg. 237) (the short answer is minutes to an hour), which demonstrates a mismatch of time scales between the processes of interest and those of the putative readout. The experimental arsenal available in the 1970s, when the answers were first convincingly obtained, was much more limited. Yet, in a series of clever experiments, using electron microscopy and radioactive labeling, these rates were precisely determined.[§§] As will be shown below, they relied on a subtle quantitative analysis in order to tease out the rates.

Measurements on transcription rates were based upon a trick in which transcription initiation was shut down by using the drug rifampin. Though

[††] Proshkin S, Rahmouni AR, Mironov A & Nudler E (2010) Cooperation between translating ribosomes and RNA polymerase in transcription elongation. *Science* 328:504–508.
[‡‡] Spencer PS, Siller E, Anderson JF & Barral JM (2012) Silent substitutions predictably alter translation elongation rates and protein folding efficiencies. *J Mol Biol* 422:328–335.
[§§] Miller OL Jr, Hamkalo BA & Thomas CA Jr (1970) Visualization of bacterial genes in action. *Science* 169:392–395; Young RY & Bremer H (1975) Analysis of enzyme induction in bacteria. *Biochem J* 152:243–254.

no new transcription events can begin, those that are already under way continue unabated—that is, rifampin inhibits the initiation of transcription, but not the elongation of RNA transcripts. As a result, this drug treatment effectively begins the running of a stopwatch that records the length of time since the last transcription process began. By fixing the cells and stopping the transcription process at different times after the drug treatment and then performing electron microscopy, resulting in images like that shown in **Figure 4-13**, it was possible to measure the length of RNA polymerase-free DNA. By taking into account the elapsed time since drug treatment, the rate at which these polymerases were moving was inferred.

The measurement of translation rates similarly depended upon finding an appropriate stopwatch, but this time for the protein synthesis process. The crux of the method is the following: Start adding labeled amino acid at time zero and follow ("chase," as it is often called) the fraction of labeled protein of mass m, as defined by looking at a specific band on a gel. Immediately after the pulse of labeled amino acids, we start to see proteins of mass m

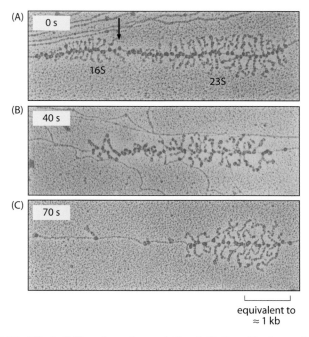

equivalent to
≈ 1 kb

Figure 4-13 Effect of rifampin on transcription initiation. Electron micrographs of *E. coli* rRNA operons: (A) before adding rifampin, (B) 40 s after addition of rifampin, and (C) 70 s after exposure. After drug treatment, no new transcripts are initiated, but those already initiated are carrying on elongation. In (A) the arrow indicates the site where RNaseIII cleaves the nascent RNA molecule, thus producing 16S and 23S ribosomal subunits. (Adapted from Gotta SL, Miller OL Jr & French SL [1991] *J Bacteriol* 173:6647–6649. With permission from the American Society for Microbiology.)

with radioactive-labeled amino acids on their ends. With time, the fraction of a given protein mass that is labeled will increase because the chains have a larger proportion of their length labeled. After a time τ_m, depending on the transcript length, the whole chain will be labeled, because these are proteins that began their translation at time zero when the label was added. At this time, we observe a change in the accumulation dynamics (when appropriately normalized to the overall labeling in the cell). From the time that elapsed, τ_m, and by knowing how many amino acids are in a polypeptide chain of mass m, it is possible to derive an estimate for the translation rate. There are uncertainties associated with doing this that are minimized by performing this for different protein masses, m, and calculating a regression line over all the values obtained. For a full understanding of the method, the reader will benefit from the original study by Young and Bremer.[¶¶] It remains as a reliable value for *E. coli* translation rate to this day. We are unaware of newer methods that give better results.

What are the corresponding rates of transcription and translation in eukaryotes? As shown in Tables 4-4 and 4-5, transcription in mammalian cells consists of elongation at rates similar to those measured in *E. coli* (50–100 nt/s; BNID 105566, 105113, 100662). It is suggested that these stretches of rapid transcription are interspersed with pauses, thus leading to an average rate that is about an order of magnitude slower (≈ 6 nt/s; BNID 100661), but some reports do not observe such slowing down (BNID 105565). Recent *in vivo* measurements in fly embryos have provided a beautiful real-time picture of the transcription process by using fluorescence to watch the first appearance of mRNA, as shown in **Figure 4-14**. Recently another approach utilizing the power of sequencing inferred the distribution of transcription elongation rates in a HeLa cell line, as shown in **Figure 4-15**, obtaining a range of 30–100 nts/s with a median rate of 60 nts/s (BNID 111027). Remember that in eukaryotes, transcription and translation are spatially segregated, with transcription taking place in the nucleus and translation taking place in the cytoplasm. Introns are excised from transcripts prior to translation, taking about 5–10 min on average for this process of mRNA splicing (BNID 105568). Though our focus here was on transcript elongation, in some cases the rate-limiting process seems to be the initiation of transcription. This is the process in which the RNA polymerase complex is assembled and the two DNA strands are separated to form a bubble that enables transcription.

What about the eukaryotic rates of translation? In budding yeast, the rate is about twofold slower than that in bacteria (3–10 aa/s; BNID 107871), but note that the "physiological" temperature at which it is measured is 30°C,

---

[¶¶] Young RY & Bremer H (1976) Polypeptide-chain-elongation rate in *Escherichia coli* B/r as a function of growth rate. *Biochem J* 160:185–194.

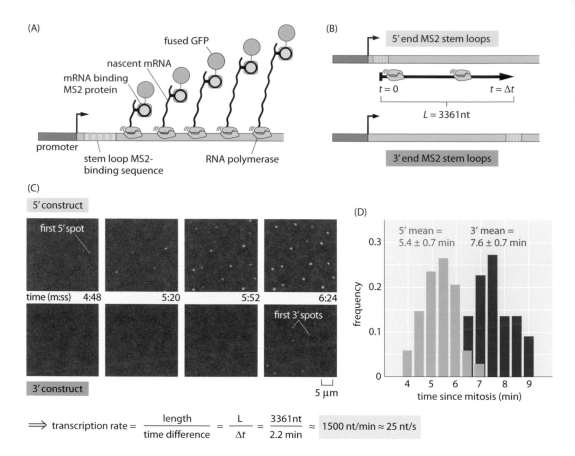

$$\Longrightarrow \text{transcription rate} = \frac{\text{length}}{\text{time difference}} = \frac{L}{\Delta t} = \frac{3361 \text{nt}}{2.2 \text{ min}} \approx \boxed{1500 \text{ nt/min} \approx 25 \text{ nt/s}}$$

Figure 4-14 Dynamics of transcription in the fly embryo. (A) Schematic of the experiment showing how a loop in the nascent RNA molecule serves as a binding site for a viral protein that has been fused to GFP. (B) Depending upon whether the RNA loops are placed on the 5′ or 3′ end of the mRNA molecule, the time it takes to begin seeing GFP puncta will be different. The delay time is equal to the length of the transcribed region divided by the speed of the polymerase. (C) Microscopy images showing the appearance of puncta associated with the transcription process for both constructs shown in (B). (D) Distribution of times of first appearance for the two constructs yielding a delay time of 2.2 min, from which a transcription rate of 25 nt/s is inferred. Measurements were performed at room temperature of 22°C. (Adapted from Garcia HG, Tikhonov M, Lin A, & Gregor T [2013] *Curr Biol* 23:2140–2145.)

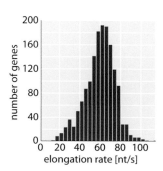

whereas for *E. coli*, measurements are at 37°C. As discussed in the vignette entitled "How does temperature affect rates and affinities?" (pg. 220), the slower rate is what we would expect based on the general dependence of a factor of 2–3 per 10°C (the Q_{10} value; BNID 100919). Using the method of ribosome profiling based on high-throughput sequencing and schematically

Figure 4-15 Distribution of measured transcription elongation rates inferred from relieving transcription inhibition and sequencing all transcripts at later time points. (Adapted from Fuchs G, Voichek Y, Benjamin S et al. [2014] *Genome Biol* 15:R69–R78.)

(A)

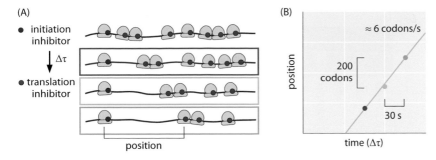

(B)

Figure 4-16 Inferring the rate of translation by the ribosome in mouse embryonic stem cells using ribosome profiling. (A) Inhibiting translation initiation followed by inhibition of elongation creates a pattern of ribosome stalling that depends on the time differences and rates of translation. Using modern sequencing techniques, this can be quantified genome wide and the translation rate accurately measured for each transcript. (B) Measurement of the translation rate using the methodology indicated schematically in (A). (Adapted from Ingolia NT, Lareau LF & Weissman JS [2011] *Cell* 147:789–802.)

depicted in **Figure 4-16**, the translation rate in mouse embryonic stem cells was surveyed for many different transcripts. It was found that the rate was quite constant across proteins and was about 6 aa/s (BNID 107952). After several decades of intense investigation and ever more elaborate techniques at our disposal, we seem to have arrived at the point where the quantitative description of the different steps of the central dogma can be integrated to reveal its intricate temporal dependencies.

What is the maturation time for fluorescent proteins?

Fluorescent proteins have become a dominant tool for the exploration of the dynamics and localization of the macromolecular contents of living cells. Given how pervasive the palette of different fluorescent proteins, with their many colors and properties has become (shown in **Figure 4-17**), it is incredible that we have really only seen a decade of concerted effort with these revolutionary tools. Indeed, it is difficult to imagine any part of biology that has not been touched in some way or another (and often deeply) by the use of fluorescent reporter proteins.

However, as a tool for exploring the many facets of cellular dynamics, fluorescent proteins have both advantages and disadvantages. Once a

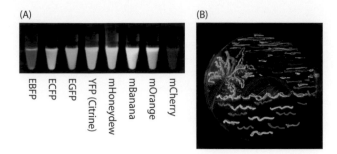

Figure 4-17 Illustration of some of the palette of fluorescent proteins that have revolutionized cell biology. (A) Fluorescent proteins spanning a range of excitation and emission wavelengths. (B) A petri dish with bacteria harboring eight different colors of fluorescent proteins that are used to "paint" an idyllic beach scene. (Adapted from Tsien RY [2010] *Integr Biol* 2:77–93.)

fluorescent protein is expressed, it has to go through several stages until it becomes functional, as shown in **Figure 4-18**. These processes are together called maturation. Until completion of the maturation process, the protein, even though already synthesized, is not fluorescent. To study dynamics, it is most useful if there is a separation of time scales between the reporter maturation process (which preferably should take place on "fast" time scales) and the dynamics of the process of real interest (which should be much slower than the maturation time). The first stage in the maturation process (not depicted) is the most intuitive and refers to the protein folding itself, which is relatively fast and should take less than a minute, assuming there is no aggregation. The next stage is a torsional rearrangement (Figure 4-18B–C) of what can be thought of as the active site of the fluorophore—the amino acids where the conjugated electrons that will fluoresce are located. The next step, known as cyclization (where

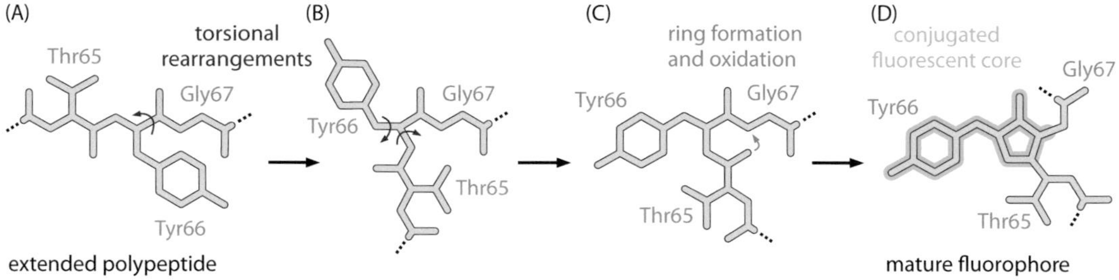

Figure 4-18 Schematic diagram of the chromophore formation in maturing enhanced green fluorescent protein (EGFP). (A) The prematuration EGFP fluorophore tripeptide amino acid sequence (Thr65-Tyr66-Gly67) stretched into a linear configuration. The first step in maturation is a series of torsional adjustments (B) and (C). These torsional adjustments allow a nucleophilic attack that results in formation of a ring system (the cyclization step). (D) Fluorescence occurs following oxidation of the tyrosine by molecular oxygen. The final conjugated and fluorescent core atoms are shaded. (Adapted from Olenych SG, Claxton NS, Ottenberg GK & Davidson MW [2007] *Curr Protoc Cell Biol* 36:21.5.1–21.5.34.)

a ring is formed between two amino acids; Figure 4-18C–D), is longer but still fast in comparison to the final and rate-limiting step of oxidation. In this final oxidation step, molecular oxygen grabs electrons from the fluorophore, creating the final system of conjugated bonds. All these steps are a prerequisite to making the active site fluoresce.

There are only a limited number of reliable measurements of the maturation time as summarized in **Table 4-6**, though the values are still far from being completely agreed upon. One approach to measure fluorophore maturation is by switching the cells from anaerobic growth, where the fluorophore protein is expressed but cannot perform the slowest step of oxidation, to aerobic conditions, and by watching the rate of fluorescent signal formation. More commonly, inducible promoters or cycloheximide-induced translation arrest are used. Nagai et al. (BNID 103780) measure a time scale of less than 5 min for the maturation of YFP and

fluorophore	maturation time (min)	cell type	BNID
ECFP	50	S. cerevisiae	106883
GFP wildtype	50	in vitro	106892
sfGFP	6	E. coli	110546
GFPmut3	7	E. coli	102972
GFPmut3	7	in vitro	107004
EGFP	60	E. coli	107001
EGFP	14	in vitro	107000
Emerald	12	in vitro	106893
GFPem	5	in vitro	106887
EYFP	40	S. cerevisiae	102974
EYFP	20	in vitro	106891
Venus	40	in vitro	106890
mCherry	15	E. coli	106877
mCherry	40	S. cerevisiae	110551
mCherry	40–100	E. coli	111423
mCherry	17 + 30	S. cerevisiae	110552
mStrawberry	50	E. coli	106880
tdTomato	60	E. coli	106876
mPlum	100	H. sapiens, B cell line	106878

Table 4-6 Common fluorescent protein maturation times. Because different approaches and conditions still give quite different values, we have to be careful in studies where the maturation time can affect the conclusions. In mCherry, there are indications of two time scales, the first leading to fluorescence at a different wavelength regime (BNID 110552). Values are rounded to one significant digit. (Adapted from Iizuka R, Yamagishi-Shirasaki M & Funatsu T [2011] Anal Biochem 414(2):173–178. For definitions of fluorophores via mutations relative to WT, see Shaner NC, Steinbach PA & Tsien RY [2005] Nat Methods 12:905–909.)

7 min for the corresponding maturation of GFP in *E. coli*. By way of contrast, Gordon et al. (BNID 102974) report a time scale of ≈ 40 min for the maturation of YFP and a very slow ≈ 50 min for the maturation of CFP, though part of the difference can be explained by the fact that, in this case, the measurements were carried out in yeast at 25°C. The measurements were done by inducing expression, and after 30 min, by inhibiting translation using cycloheximide. The dynamics of continued fluorophore accumulation, even after no new proteins were synthesized, was used to infer the maturation time scale. Note that for many of the processes that occur during a cell cycle, such as expression of genes in response to environmental cues, the maturation time can be a substantial fraction of the time scale of the process of interest. If a marathon runner stops for a drink in the middle of a race, this will hardly affect the overall time of the racer's performance. On the other hand, if the runner stops to have a massage, this will materially affect the time scale at which the racer completes the race. By analogy with the runner stopping at a restaurant, the maturation time can seriously plague our ability to accurately monitor the dynamics of a variety of cellular processes.

Chromophore maturation effectively follows first-order kinetics in most studies performed. As a result, this implies that we will find a small fraction of functional fluorophore much earlier than the maturation time. Still, to have the majority of the population active, the characteristic time scale is roughly the maturation time itself. This effect results in a built-in delay in the reporting system and should be heeded when estimating response times based on fluorescent reporters. Similarly, if translation were being stopped (say, by the use of a ribosome inhibitor such as cyclohexamide), we would still have a period of time where some proteins that were translated before the inhibition were coming "online" and were adding to the signal. This again should be taken into account when estimating degradation times.

Another dynamical feature of these proteins that can make them tricky for precisely characterizing cellular dynamics is the existence of photobleaching. This process has a characteristic time scale of tens of seconds using standard levels of illumination and magnification. This value means that after a continuous exposure to illumination for several tens of seconds, the fluorescent intensity will have decayed to $1/e$ of its original value. Though sometimes a nuisance, recently this apparent disadvantage has been used as a trick both in the context of FRAP, which allows inference about diffusion rates, and in super-resolution microscopy techniques, where the bleaching of individual fluorophores makes it possible to localize these proteins with nanometer-scale resolution.

Differences in maturation times of different fluorophores were recently turned into a way to measure rates of degradation and translocation without the need for time course measurements.[***] The protein of interest was fused not to one, but to two fluorescent tags, a fast-maturing GFP (a so-called superfolder) and a slower-maturing mCherry. The ratio of intensities was measured, and this can serve as a built-in timer. If the protein of interest were short lived, the slowly maturing tag would often not have time enough to fluoresce before the protein is degraded, and its intensity ratio to the quickly maturing tag would be low. At the other extreme, if the protein were long lived, there would be ample time for the more slowly maturing tag to fluoresce, and its ratio to the fast-dividing tag would be high. The ratio of intensities thus serves as a timer that was used, for example, to show that daughter cells tend to get the old copies of some protein complexes, such as spindle pole bodies and nuclear pore complexes, while the mothers retain the newly formed copies.

How fast do proteasomes degrade proteins?

One of the ways in which the protein content of the cell is controlled is by the regulated degradation of its proteins. The main macromolecular machine in charge of degradation is the proteasome. It can be thought of as the "evil" twin of the ribosome. The size and shape of this barrel-shaped machine is seen in **Figure 4-19**. What fraction of the proteome is made up of these machines? In HeLa cells, about 1% of the total bulk protein was reported to be proteasomes (BNID 108028, 108717). This is far less than the investment in the proteins that build the ribosomes, which can be as high as one-third of the proteome in fast-growing bacteria and often 5–10% in other cells (http://www.proteomaps.net/)—still a much larger fraction than that taken up by proteasomes. In blood cells, the fraction of proteasomes out of the proteome varies between 0.01 and 0.3% for different cell types (BNID 108041). The half-life of these machines is found to be about five days (BNID 108031). The degradation rate associated with proteasome-mediated degradation is currently based on *in vitro* measurements. These rates exhibit a great deal of variability, with rates coming in with values from ≈ 0.05 through ≈ 0.2 to ≈ 5 "characteristic" peptide chains

[***] Khmelinskii A, Keller PJ, Bartosik A et al. (2012) Tandem fluorescent protein timers for *in vivo* analysis of protein dynamics. *Nat Biotechnol* 30:708–714.

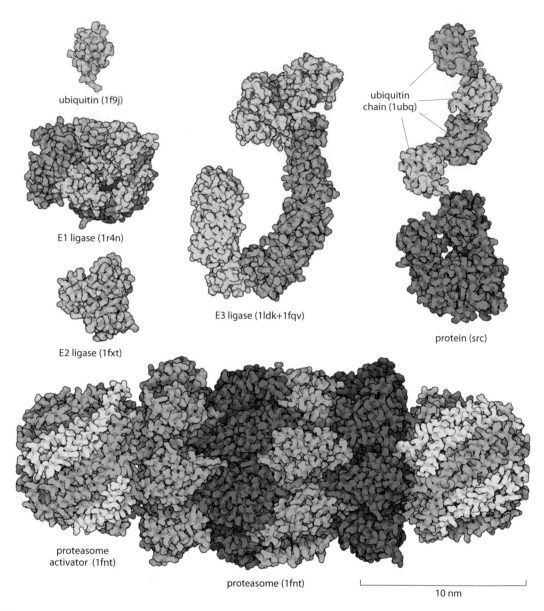

ubiquitin (1f9j)

E1 ligase (1r4n)

E2 ligase (1fxt)

E3 ligase (1ldk+1fqv)

ubiquitin
chain (1ubq)

protein (src)

proteasome
activator (1fnt)

proteasome (1fnt)

10 nm

Figure 4-19 Proteins involved in the ubiquitin-proteasome pathway for protein degradation. Key molecules in the degradation process range from the ubiquitin molecular tag that marks a protein for degradation to the ligases that put these molecular tags on their protein targets. Once proteins are targeted for degradation, the proteasome actively carries out this degradation. The depicted proteasome is based on the structure determined for budding yeast. (Courtesy of David Goodsell.)

per minute (BNID 108032, 109854). Given this wide range of values, we are faced with the key question of whether there is any reason to favor one of these numbers as a "characteristic" value over the others, at least for the rates observed in cell lines studied in the lab. The rate of degradation by the proteasome can vary as a function of the protein substrate, thus limiting the reliability of a "characteristic" average value. Based on relatively meager information, we can try a sanity check. For example, we can ask whether there are enough molecular machines for degrading a significant fraction of the proteome at each of these rates. As will be seen below, in carrying out this sanity check, which is one of the main mantras of the entire book, one of these results is more plausible than the others.

Assume that the proteome consists overall of N_{aa} amino acids, as shown in **Estimate 4-10**. For example, if there are 3 million proteins in the relevant HeLa cell of 3000 μm^3, and the average length is 400 amino acids per protein, then N_{aa} is 4×10^{12} aa. As we shall see, though, the exact value of N_{aa} will cancel in our estimate. Assuming $\approx 1\%$ of the proteome is proteasomes, we have $0.01 N_{aa}$ amino acids present in those machines. The average molecular weight of a proteasome is $\approx 2.4 \times 10^6$ Da (BNID 104915)—that is, about 20,000 amino acids. So, there are about

$$(0.01 \times N_{aa} \text{ aa})/(20{,}000 \text{ aa/proteasome}) \approx 0.5 \times 10^{-6} N_{aa} \text{ proteasomes} \quad (4.6)$$

in the cell (that is, on the order of a million proteasomes in the Hela cell considered above). Taking the higher rate of proteasome degradation from above of 5 protein/min ≈ 0.1 protein/s, we find that on an amino acid basis, this degradation rate is equivalent to ≈ 40 aa/s. The protein, though, is degraded by the proteasome to chunks of 2–30 amino acids

---

What is the turnover time of proteins through active degradation?

we denote the number of amino acids per cell by N_{aa}

(e.g., for HeLa cell, $N_{aa} \approx 3 \times 10^6 \dfrac{proteins}{\mu m^3} \times 3000 \dfrac{\mu m^3}{cell} \times 400 \dfrac{aa}{protein} \approx 4 \times 10^{12} \dfrac{aa}{cell}$)

$\approx 1\%$ of proteome mass is proteasomes

number of proteasomes $\approx \dfrac{0.01 \times N_{aa} \text{ aa}}{20{,}000 \text{ aa/proteasome}} \approx 0.5 \times 10^{-6} N_{aa}$ proteasomes/cell

molecular mass of proteasome ≈ 2.4 MDa $\approx 20{,}000$ aa

proteasome deg. rate ≈ 5 proteins/min ≈ 0.1 protein/s ≈ 40 aa/s

total deg. rate $\approx 40 \dfrac{aa}{s \times proteasome} \times 0.5 \times 10^{-6} N_{aa} \dfrac{proteasomes}{cell} \approx 20 \times 10^{-6} N_{aa} \dfrac{aa}{s \times cell}$

turnover time $\approx \dfrac{\text{number of aa per cell}}{\text{total deg rate}} \sim \dfrac{N_{aa} \text{ aa}}{N_{aa} \times 20 \times 10^{-6} \text{aa/s}} \approx 0.5 \times 10^5 s \approx 1 \text{ day}$

i.e., it would take all the proteasomes working at full speed about one day to degrade all of the proteome.

Estimate 4-10

each (BNID 108111), which are only later further degraded by peptidases, so the aa/s unit is only an effective value for easy calculation and comparison and not the actual biophysical process taking place. Note that the rate of protein polymerization by the ribosome of ≈ 10 aa/s, as discussed in the vignette entitled "Which is faster: transcription or translation?" (pg. 231), is not very far from this rate of degradation by the proteasome. The two machine complexes also share a similar molecular weight. Focusing back on our sanity check, we thus have an overall degradation rate of

$$(40 \text{ s}^{-1}) \times (0.5 \times 10^{-6} \, N_{aa} \text{ aa}) = 20 \times 10^{-6} \, N_{aa} \text{ aa/s}. \tag{4.7}$$

So, the turnover time, which is the total number of amino acids divided by the overall degradation rate, is about

$$\frac{N_a \text{ aa}}{20 \times 10^{-6} \, N_a \text{ aa/s}} \approx 0.5 \times 10^5 \text{ s}, \tag{4.8}$$

or about a day. This time scale is about the same as the characteristic cell cycle time for a happily dividing cell line. As is seen in Estimate 4-10, the value of N_{aa} is of no importance for this estimate. This is in agreement with the observations detailed in the vignette entitled "How fast do RNAs and proteins degrade?" (pg. 244) that for cell lines, an average protein degradation rate of 1–2 days was measured (BNID 109937). Had we taken the lower limit value on the degradation rate, we would have found a turnover time of about a month, way longer than the measured value for fast-dividing cells, but probably more relevant for cells in our body that turn over slowly, indicating an inclination to trust the rate of five peptide chains per minute as the more reliable measurement for fast-dividing cells. This is but one example of how a simple calculation can help us perform a sanity check of contrasting measured values.

How fast do RNAs and proteins degrade?

The central dogma focuses on the production of the great nucleic acid and protein polymers of biology. However, the control and maintenance of the functions of the cell depend upon more than just the synthesis of new molecules. Degradation is another key process in the lives of the macromolecules of the cell and is itself tightly controlled. Indeed, in the simplest model of mRNA production, the dynamics of the average level of mRNA is given by

$$\frac{d\bar{m}}{dt} = r - \gamma \bar{m} \tag{4.9}$$

where r is the rate of mRNA production and γ is the rate constant that dictates mRNA decay. The steady-state value of the mRNA is given by

$$\bar{m} = \frac{r}{\gamma},$$ (4.10)

thus showing that to first approximation, it is the balance of the processes of production and decay that controls the steady-state levels of these molecules. If our equation is for the copy number of molecules per cell, there will be an abrupt change in the number each time the cells divide, since the total mRNA and protein content is partitioned between the two daughter cells. If, instead, our equation is thought of in the language of concentrations, we do not have to face this problem, because as the cell grows, so too does the number of molecules, in which case the concentration varies smoothly. The growth effect on concentration can be absorbed into the rate constant for degradation to take account of the dilution. This is a common mathematically elegant solution, but it is not immediately intuitive, and so we will try to clarify it below. But first, what are the characteristic values for mRNA and protein degradation times?

The lifetime of mRNA molecules is usually short in comparison with the fundamental time scale of cell biology defined by the time between cell divisions. As shown in **Figure 4-20A**, for *E. coli*, the majority of mRNA molecules have lifetimes between 3 and 8 min. The experiments leading to these results were performed by inhibiting transcription through the use of the drug rifampicin, which interacts with the RNA polymerase, and then by querying the cells for their mRNA levels in two-minute intervals after drug treatment. In particular, the RNA levels were quantified by

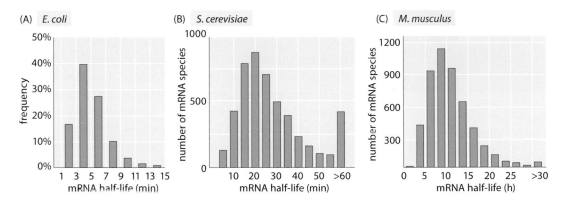

Figure 4-20 Measured half-lives of mRNAs in *E. coli*, budding yeast, and mouse NIH3T3 fibroblasts. (A, adapted from Bernstein JA, Khodursky AB, Lin PH et al. [2002] *Proc Natl Acad Sci USA* 99:9697–9702. B, adapted from Wang Y, Liu CL, Storey JD et al. [2002] *Proc Natl Acad Sci USA* 99:5860–5865. C, adapted from Schwanhäusser B, Busse D, Li N et al. [2011] *Nature* 473:337–342.)

hybridizing with complementary DNAs on a microarray and measuring the relative levels of fluorescence at different time points. These degradation times are only several times longer than the minimal time required for transcriptional and translational elongation, as discussed in the vignette entitled "Which is faster: transcription or translation?" (pg. 231). This reflects the fleeting existence of some mRNA messages.

Given such genome-wide data, various hypotheses can be explored for the mechanistic underpinnings of the observed lifetimes. For example, is there a correlation between the abundance of certain messages and their decay rate? Are there secondary-structure motifs or sequence motifs that confer differences in the decay rates? One of the big surprises of the measurements leading to Figure 4-20A is that none of the conventional wisdom on the origins of mRNA lifetime was found to be consistent with the data, which revealed no clear correlation with secondary structure, message abundance, or growth rate.

How far does Monod's statement that "what is true for *E. coli* is true for the elephant" (depicted by Monod in **Figure 4-21**) take us in our assessment of mRNA lifetimes in other organisms? The short answer is, "not very."

Figure 4-21 Jacques Monod's L'éléphant et l'Escherichia Coli, décembre 1972. "Tout ce qui est vrai pour le Colibacille est vrai pour l'éléphant", or in English, "What is true for *E. coli* is true for the elephant." (Courtesy of the Institut Pasteur Archives.)

Whereas the median mRNA degradation lifetime is roughly 5 min in *E. coli,* the mean lifetime is ≈ 20 min in the case of yeast (see Figure 4-20B) and 600 min (BNID 106869) in human cells. Interestingly, a clear scaling is observed, with the cell cycle times for these three cell types of roughly 30 min (*E. coli*), 90 min (budding yeast), and 3000 min (human), under the fast exponential growth rates that the cells of interest were cultivated in for these experiments. As a rule of thumb, these results suggest that the mRNA degradation time scale in these cases is thus about one-fifth of the fast exponential cell cycle time.

Messenger RNA is not the only target of degradation. Protein molecules are themselves also the targets of specific destruction, though their lifetimes tend to be longer than the mRNAs that lead to their synthesis, as discussed below. Because of these long lifetimes, under fast growth rates the number of copies of a particular protein per cell is reduced, not because of an active degradation process, but simply because the cell doubles all its other constituents and divides into two daughters, leaving each of the daughters with half as many copies of the protein of interest as were present in the mother cell. To understand the dilution effect, imagine that all protein synthesis for a given protein has been turned off, while the cell keeps on doubling its volume and shortly thereafter divides. In terms of absolute values, if the number of copies of our protein of interest before division were N, afterwards it would be $N/2$. In terms of concentrations, if it started with a concentration c, during the cell cycle it would be diluted to $c/2$ by the doubling of the volume. This mechanism is especially relevant in the context of bacteria, where the protein lifetimes are often dominated by the cell division time. As a result, the total protein loss rate α (the term carrying the same meaning as γ for mRNA) is the sum of a part due to active degradation and a part due to the dilution that occurs when cells divide. We can write the total removal rate in the form

$$\alpha = \alpha_{\text{active}} + \alpha_{\text{dilution}}. \tag{4.11}$$

The statement that protein lifetimes in rapidly growing bacteria are longer than the cell cycle itself is supported by measurements already from the 1960s, where radioactive labeling was used as a way to measure rates. In this case, the degradation of labeled proteins was monitored by looking at the accumulation of radioactive amino acids in a rapidly exchanged perfusate. Only 2–7% of the proteome was estimated to be actively degraded, with a half-life of about one hour (BNID 108404). More recently, studies showed specific cases of rapid degradation, including some sigma factors, transcription factors, and cold-shock proteins, yet the general statement that dilution is the dominant protein-loss mechanism in bacteria remains valid.

Just like with the genome-wide studies of mRNA lifetimes described above, protein lifetimes have been subjected to similar scrutiny. Surprisingly, we could not find genome-wide information in the literature on the degradation times for proteins in *E. coli*, but in budding yeast, a translation–inhibition drug (cycloheximide) was used to inhibit macromolecular synthesis, and then protein content was quantified at later time points using Western blots. The Western blot technique is a scheme in which the proteins of interest are fished out by specific binding to some part of the protein (for example, by antibodies) and the amount of protein is read off of the intensity of a reporter that has been calibrated against a standard. Inhibiting translation might cause artifacts, but with that caveat in mind, the measured lifetimes shown in **Figure 4-22A** using the method of translation–inhibition reveal the longer lifetimes of proteins in comparison with their mRNA counterparts, with a mean lifetime of roughly 40 min (BNID 104151). Issues with the precision of these results still calls for the development of new methods for constructing such surveys.

Using modern fluorescence techniques, it has become possible to measure degradation rates of human proteins *in vivo* without the need to lyse the cells. The long removal times observed in human cells are shown

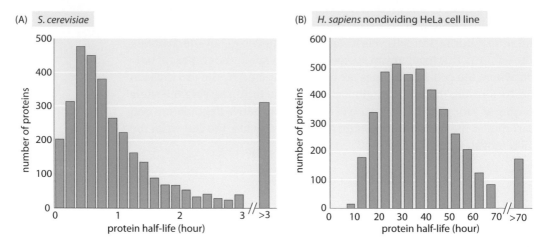

Figure 4-22 Measured half-lives of proteins in budding yeast and a HeLa human cancer cell line. The yeast experiment used the translation inhibitor cycloheximide, which disrupts normal cell physiology. The median half-life of the 4100 proteins measured in the nondividing HeLa cell is 36 h. (A, adapted from Belle A, Tanay A, Bitincka L et al. [2006] *Proc Natl Acad Sci USA* 103:13004–13009. B, adapted from Cambridge S, Gnad F, Nguyen C et al. [2011] *J Proteome Res* 10:5275–5284.)

in **Figure 4-22B**. The measurements were done by fusing the protein of interest to a fluorescent protein. Then, by splitting the population into two groups, one of which is photobleached and the other of which is not, and watching the re-emergence of fluorescence in the photobleached population, it is possible to directly measure the degradation time. As shown in **Figure 4-23**, for human cells there is an interesting interplay between active degradation and protein removal by dilution. Active degradation half-lives were seen to be broadly distributed, with the fastest observed turnover being less than an hour and the slowest showing only negligible active degradation in the few days of time-lapse microscopy. These results can be contrasted with a prediction based on the N-end rule, which states that the amino acid at the N-terminus of the protein has a strong effect on the active degradation performed through the ubiquitination system. For example, in mammalian systems, it predicts that arginine, glutamate, and glutamine will lead to degradation within about an hour, while valine, methionine, and glycine will be stable for tens of hours.

In trying to characterize the lifetimes of the most stable proteins, mice were given isotopically labeled food for a short period at an early age and then analyzed a year later. The results showed that most proteins turn over within a few days, but a few show remarkable stability. Histone half-lives were measured at ≈ 200 days; even more tantalizing, the nuclear pore consists of a protein scaffold with half-life >1 year, while all the surrounding components were replenished much faster.

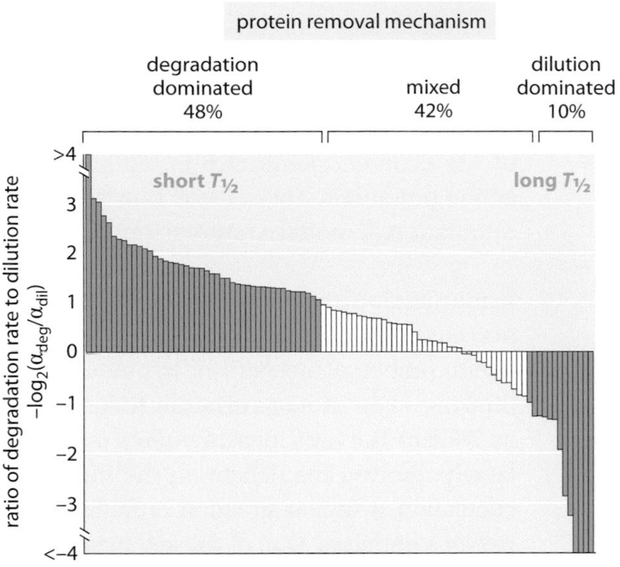

Figure 4-23 Protein degradation rates in human cells. Distribution of 100 proteins from a H1299 human cell line, comparing the rate of degradation to dilution to find which removal mechanism is dominant for each of the proteins. The overall removal rate α ranges between 0.03 and 0.82 hour^{-1} with an average of 0.1 $\pm$ 0.09 hour^{-1}. This is equivalent to a half-life of $\approx$7 h via the relationship half-life, $T_{1/2}$ = ln(2)α. (Adapted from Eden E, Geva-Zatorsky N, Issaeva I et al. [2011] *Science* 331:764–768.)

CELLULAR DYNAMICS

How fast are electrical signals propagated in cells?

Nerve cells are among the most recognizable of human cell types, noted not only for their enormous size relative to many of their cellular counterparts, but also for their unique shapes, as revealed by their sinuous and elongated structures. Already in the early days of microscopy, biological pioneers found these cells a fascinating object of study, with van Leeuwenhoek musing, "Often and not without pleasure, I have observed the structure of the nerves to be composed of very slender vessels of an indescribable fineness, running length-wise to form the nerve." See **Figure 4-24** for several examples of the drawings made by van Leeuwenhoek as a result of his observations with the early microscope. The mystery of nerve cells went beyond their intriguing morphology as a result of their connection with electrical conduction and muscle action. In famed experiments like those shown in **Figure 4-25**, Luigi Galvani

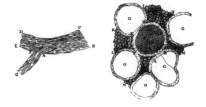

Figure 4-24 Antoine van Leeuwenhoek's 1719 drawings of nerve cells in a letter to a friend. The drawing on the left shows a longitudinal view of nerves and the drawing on the right shows a cross-sectional view of a central nerve surrounded by five others (labeled with "G"). (Adapted from Lopez-Munoz F, Boya J & Alamo C [2006] *Brain Res Bull* 70:391–405. With permission from Elsevier Inc.)

discovered that muscles in dead frogs could be stimulated to twitch by the application of an electrical shock. This work set the stage for several centuries of work on animal electricity, culminating in our modern notions of the cellular membrane potential and propagating action potentials. These ideas now serve as the cellular foundation of modern neuroscience.

In the middle of the nineteenth century, the mechanism of nervous impulses was still hotly contested, with wildly different competing mechanistic hypotheses, similar to early thinking on the motions of bodily fluids such as blood. Just as William Harvey's measurements on the flow of blood largely resolved the debate on the mechanism of blood circulation, a similar situation unfolded in the context of nervous impulses. One of the key measurements that set the path towards the modern understanding of electrical communication in nerve cells was the measurement by Hermann von Helmholtz of the speed of propagation of such impulses. The apparatus he used to make such measurements is shown in **Figure 4-26**. Helmholtz tells us, "I have found that there is a measurable period of time during which the effect of a stimulus consisting of a momentary electrical current applied to the iliac plexus of a frog is transmitted to the calf muscles

Figure 4-25 The experiments of Luigi Galvani on the electrical stimulation of muscle twitching. Using a dead frog, Galvani discovered that he could use an electrical current to induce muscle twitching, lending credence to the idea that nervous impulses are electrical. (Adapted from Galvani A [2010] De Viribus Electricitatis In Motu Musculari,1792 Latin Edition. Kessinger Publishing.)

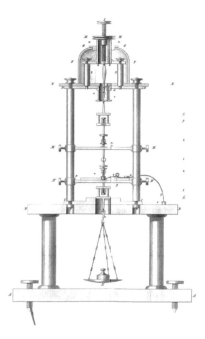

Figure 4-26 The measurements of Hermann von Helmholtz on the propagation of nervous impulses. Schematic of the apparatus used by Helmholtz in his measurements. The stimulated nerve was used to lift the weight shown at the bottom of the apparatus. (Adapted from Helmholtz H [1850] *Archiv für Anatomie, Physiologie und wissenschaftliche Medicin* MPIWG:2AT3G7QD. Courtesy of the Max Planck Institute for the History of Science.)

at the entrance of the crural nerve. In the case of large frogs with nerves 50-60 mm in length, this period of time amounted to 0.0014 to 0.0020 of a second." If we use his values of 50 mm as the distance of propagation and 1.5 ms as the propagation time, this leads to an estimate of 30 m/s for the propagation velocity of the nerve impulse. This value compares very favorably with the modern values of 7–40 m/s for frogs, depending on axonal diameter (BNID 110597, 110594). Helmholtz's measurement of the velocity of nervous impulses was inextricably linked to mechanism. Specifically, it helped dispel earlier notions, where the propagation of nervous impulses had been attributed to all sorts of mystical properties, including some that posited instantaneous communication between different parts of the same cell. Without the measurement of a finite velocity, the ideas on how it worked remained muddled.

In the time since, numerous measurements have confirmed and extended the early insights of Helmholtz, with a broad range of propagation speeds ranging from less than 1 m/s all the way to over 100 m/s. The fastest took

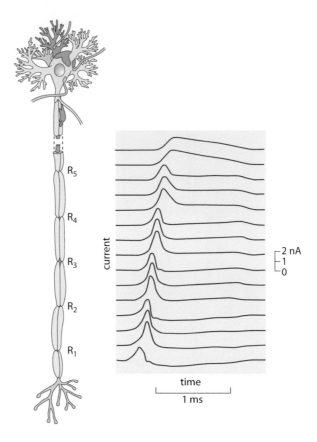

Figure 4-27 Measurement of the propagation of a nervous impulse. The cell on the left shows an axon with 5 nodes of Ranvier labeled R1–R5. (Adapted from Huxley AF & Stampfli R [1949] *J Physiol* 108:315–339. With permission from John Wiley & Sons.)

place in a shrimp giant fiber (BNID 110502, 110597), with a value over half of the speed of sound. **Figure 4-27** shows the results of one of these classic studies. Determining the mechanistic underpinnings of variability in the speed of nervous impulses has been one of the preoccupations of modern neurophysiology and has resulted in insights into how both the size and anatomy of a given neuron dictate the action potential propagation speed. An important insight that attended more detailed investigations of the conduction of nervous impulses was the realization that the propagation speed depends both on the cellular anatomy of the neuron in question, such as whether it has a myelin sheath (increasing the propagation speed several fold), and also on the thickness of the nerve (propagation speed being proportional to the diameter in myelinated neurons and proportional to the area in unmyelinated neurons).

Early work on impulse conduction along peripheral fibers by Joseph Erlanger and Herbert Gasser, for which they shared the Nobel Prize in Physiology or Medicine in 1942, demonstrated remarkable relationships between the conduction velocity of the axons and the type of neuron and thus the information that they conveyed. The largest motor fibers (13–20 μm, conducting at velocities of 80–120 m/s) innervate the extrafusal fibers of the skeletal muscles, and smaller motor fibers (5–8 μm, conducting at 4–24 m/s) innervate intrafusal muscle fibers. The largest sensory fibers (13–20 μm) innervate muscle spindles and Golgi tendon organs, both conveying unconscious proprioceptive information. The next largest sensory fibers (6–12 μm) convey information from mechanoreceptors in the skin, and the smallest myelinated fibers (1–5 μm) convey information from free nerve endings in the skin, as well as pain and cold receptors.

One of the beautiful outcomes of recent fluorescent methods is the invention of genetically encoded voltage reporters. These molecular probes have differing fluorescence depending upon the membrane voltage. An impressive usage of these methods has been to watch in real time the propagation of nerve impulses. Specifically, the readout of the passing of such an impulse comes from the transient change in fluorescence along the neuron. An example of this method is shown in **Figure 4-28**. We note these alternative methods because of the strict importance we attach to the ability to measure the same quantity in multiple ways, especially to make sure that they yield consistent values.

Can we connect the reported action potential speeds in humans of 10–100 m/s (BNID 110594) to our human response limits? From the moment of hearing the firing of the starter's pistol in a 100-m dash to activating the muscles in the feet, at least one meter of impulse propagation had to take place. This dictates a latency of 10–100 ms even before taking into account all other latencies, such as the processing happening in the brain and the propagation time due to the finite speed of sound in air. Indeed, in a 100-m dash race, the best athletes have response times of roughly 120 ms (BNID 111450), and anything below 100 ms is actually disqualified as a false start according to the binding rules of the International Association of Athletics Federations. If the propagation speed of nerve impulses were significantly slower, running races, as well as soccer games or indeed most sports events, would be much less interesting to watch. Speaking of watching, we know that when shown frames at the standard rate of 24 Hz, the brain sees or interprets the movement as continuous. It is interesting to speculate what the frame rate would have to be if our action potentials were moving at, say, 1000 m/s.

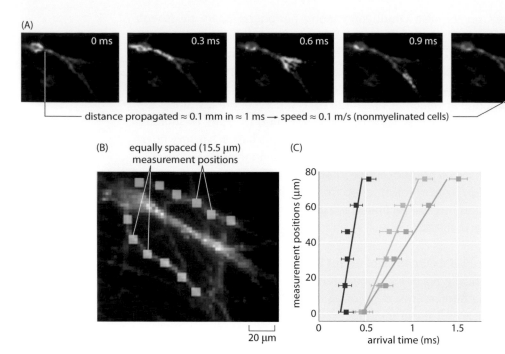

(A)

0 ms   0.3 ms   0.6 ms   0.9 ms   1.2 ms

distance propagated ≈ 0.1 mm in ≈ 1 ms → speed ≈ 0.1 m/s (nonmyelinated cells)   20 μm

(B)   equally spaced (15.5 μm)
       measurement positions

20 μm

(C)

arrival time (ms)

measurement positions (μm)

Figure 4-28 Optical measurement of action potential speed. (A) Series of images of the fluorescence in the cell as a function of time. (B) A series of equally spaced (15 μm) measurement points along three different processes are used to measure the arrival time of a propagating action potential. (C) Arrival times for the three processes shown in (B). For example, for the action potential propagating along the fiber labeled with red boxes, the signal arrives with a time delay of roughly 0.05 ms from one measurement point to the next. The action potential speed can be read off of the graph in the usual way by dividing distance traveled by time elapsed. Due to technical limitations, these are unmyelinated neuronal cells, and thus the propagation speed is much slower than *in vivo*. (Courtesy of Daniel Hochbaum and Adam Cohen.)

What is the frequency of rotary molecular motors?

Wheels are a remarkable human invention that revolutionized our mobility. Interestingly, rotary motion is not exclusively the province of humans, though it has sometimes been argued that human ingenuity outpaced nature on the grounds that "nature did not invent the wheel." Different rationalizations for the putative absence of wheels have been put forward to explain this apparently surprising observation, ranging from developmental and anatomical constraints to the absence of a selective advantage for such wheels. However, in fact, cells of all kinds exploit rotary motion, whether in the form of the ATP synthase machines that generate ATP or the motors that power flagella to propel cells forward.

Perhaps the most well-known example of molecular rotary motion is that of the flagella that drive the swimming behavior of *E. coli* and other

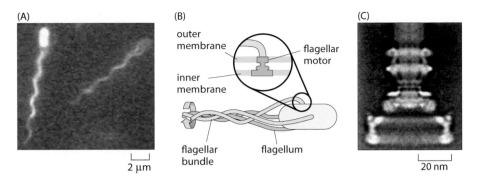

(A)

(B)

outer
membrane

flagellar
motor

inner
membrane

flagellar   flagellum
bundle

2 µm

(C)

20 nm

Figure 4-29 Flagellar-based motility in *E. coli*. (A) Two *E. coli* cells and their bundle of fluorescently labeled flagella. (B) Schematic of the bundling of flagella that drives bacterial motility. The inset shows how the rotary motor is embedded in the cell membrane. (C) Electron microscopy image of the rotary motor. (A, and C, adapted from Berg HC [2000] *Phys Today* 53:24–29. Courtesy of David DeRosier.)

bacteria. This model system is also a foundational example of cell signaling, where dissection of the signal transduction in chemotaxis has made it possible to give a quantitative explanation of "behavior" in molecular terms. The motion of these bacteria is driven by the rotation of one to 10 flagella (**Figure 4-29A**; BNID 100100), which are propelled by an exquisite rotary motor (**Figure 4-29B**). The free energy that drives these motors is provided by protons moving down a transmembrane electrochemical potential gradient. The flagellar motor rotates at 100 turns per second under normal motility speed and can reach a maximal speed of around 300 turns per second (BNID 103813, 109337), a rate that surpasses the rapid turbine blades of modern jet engines.

The rotation of the flagellar motor is energized by the cell's membrane serving as a circuit element known as a capacitor. Pumps that continuously pump protons out of the cell ensure that this energy source is not drained by maintaining an imbalance in the electrochemical potential across the membrane. One interesting question raised by this process is how much power does motility require and how efficient is the motor? **Estimate 4-11** provides a schematic of the conceptual framework we use to make an estimate of the efficiency of these motors. About 1200 protons were measured to flow through the motor per revolution (BNID 109759). This number is roughly consistent with our knowledge that each motor is composed of about 11 stator complexes (with four copies of MotA protein and two copies of MotB protein each; BNID 109768) and was measured to take 26 steps per rotation (BNID 110614). Each complex at each step requires about 2–4 protons. So with ≈ 1200 H$^+$/rotation, and, say, four flagella rotating at ≈ 100 Hz, we get a proton consumption rate of 5×10^5 H$^+$/s. Each proton transfer releases about 0.15 eV or 0.2×10^{-19} J (see the vignette on "What is the electric

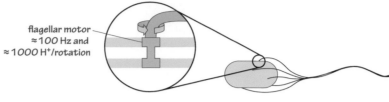

What is the energy demand for flagella-based rotation?

flagellar motor
≈ 100 Hz and
≈ 1000 H⁺/rotation

$$\text{protons needed} \approx 1000 \; \frac{H^+}{\text{rotation} \times \text{flagella}} \; \times \; 100 \; \frac{\text{rotations}}{s} \; \times \; 4 \; \text{flagella} \approx 4 \times 10^5 \; \frac{H^+}{s}$$

$$\text{proton motive force} \approx 150 \; mV \rightarrow \text{energy per proton} \approx 0.15 \, V \times 1.6 \times 10^{-19} \, J/V \approx 0.2 \times 10^{-19} \, J$$

$$\text{power expended} \approx 4 \times 10^5 \; \frac{H^+}{s} \; \times \; 0.2 \times 10^{-19} \; \frac{J}{H^+} \approx 10^{-14} \, W \approx 10 \, W/\text{kg cells}$$

Estimate 4-11

potential difference across membranes?" pg. 196) and so 5×10^5 H⁺/s release about 10^{-14} W. The power required for driving a sphere the size of an *E. coli* at a velocity of ≈ 30 µm/s (BNID 109419) against the force of viscosity can be calculated based on the Einstein–Stokes equation, as elegantly derived in the classic book on random walks in biology by Howard Berg.[†††] This theoretical value for the minimal needs for motility is 10^{-17} W, and so we find that the "efficiency" is about $10^{-17}/10^{-14} = 0.1\%$. Edward Purcell showed that with a helical flagellum, we cannot have an efficiency higher than 1%. So this mode of motility is not very energy efficient, but it works all the same, which is not an easy feat, as can be appreciated by reading one of the all-time favorite papers of physicists on biology—namely, "Life at low Reynolds number."[‡‡‡] Should bacteria care about the efficiency of the process of cellular motility? Consulting the vignette on the power consumption of a bacteria (pg. 199), we remind ourselves that at fast growth rates a bacterium uses about 10^{-12} W, which makes the motility cost about 1% of the total energetic budget. When the cell is starved, the maintenance energy is about 10^{-14} W, and so the motility energy requirements are expected to be a significant fraction of the total.

The diversity of life has become one of our typical refrains, and rotary motion is no exception. Beyond the *E. coli* paradigm are all sorts of other interesting and bizarre examples of rotary motion. One such example is presented by the periplasmic flagella of the spirochaete *Treponema primitia*, which is characterized by the spinning of its flagellum *within* the cell, resulting in a corkscrew motion of the cell, as shown in **Figure 4-30**. Like

[†††] Berg HC (1993) Random Walks in Biology, Princeton University Press.
[‡‡‡] Purcell EM (1977) Life at low Reynolds number. *Am J Phys* 45:3-11.

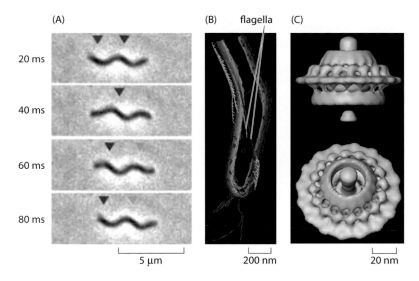

(A)   (B) flagella   (C)

20 ms

40 ms

60 ms

80 ms

5 µm   200 nm   20 nm

Figure 4-30 Spirochaete motility powered by periplasmic flagella. (A) A swimming *T. primitia* cell shown at various times during a swimming trajectory. (B) Cryo-electron microscopy image reconstruction of the internal flagella. (C) The molecular motor that powers rotation of the flagellum. (A, and B, adapted from Murphy GE, Matson EG, Leadbetter JR et al. [2008] *Mol Microbiol* 67:1184–1195. Adapted from Murphy GE, Leadbetter JR & Jensen GJ [2006] *Nature* 442:1062–1064.)

its *E. coli* counterpart, the rotary motor that controls this flagellum can turn as fast as 300 Hz (BNID 103813) and has a structure with many of the same key features of an exterior flagellum, as depicted in Figure 4-30.

Rotary motion is a part of many cell's "gadgets" in contexts beyond motility. Indeed, some have ventured that the world's second most important molecule is the ATP synthase protein complex, which is responsible for the enormous ATP biosynthetic flux central to organisms ranging from bacteria to humans. The study of the dynamics of this rotary motor culminated in one of the most beautiful single-molecule experiments, in which the rotations of individual synthases were followed in real time by attaching actin filaments to the top of the motor, as shown in **Figure 4-31**. In the *in vivo* setting, these complexes also rotate at ≈300 turns per second (at 37°C; BNID 104890).

Interestingly, the ATP synthase process, as well as many others, use the same power source as the flagellar motor—namely, they rely on the transmembrane voltage created by pumps. In times of need, when these pumps cannot function, the ATP synthase rotor can reverse direction in order to ensure the capacitor keeps its charge. It then breaks down ATP and moves protons up the chemical gradient, thus replenishing the original driving force. So this machine is actually a dual-purpose rotor.

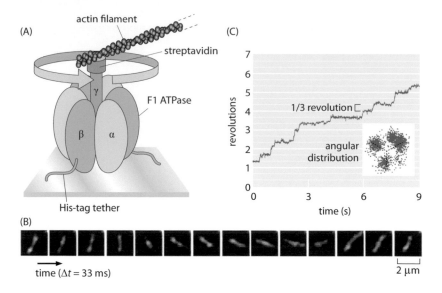

Figure 4-31 Single-molecule observation of a rotary motor using actin filaments to reveal the motor rotation. (A) The F1 portion of ATP synthase is tethered to a glass slide. The rotation of the complex is monitored by attaching a fluorescently labeled actin filament. (B) Fluorescence images of the F1 shaft as it turns. (C) At low ATP concentrations, the rotation occurs in three evenly spaced angular substeps. The graph shows the angular revolution for a single actin filament over a period of a few seconds, and the inset shows the positions of the filament end over a longer movie. (A, and B, adapted from Noji H, Ysuda R, Yoshida M & Kinosita K Jr [1997] *Nature* 386:299–302. With permission from Macmillan Publishers Ltd. C, adapted from Yasuda R, Noji H, Kinosita K Jr & Yoshida M [1998] *Cell* 93:1117–1124.)

What are the rates of cytoskeleton assembly and disassembly?

What is it that makes the polymers of the cytoskeleton so different from the polymers that make up the plastic bags and containers that fill our stores and the nylon in the clothes we wear? Above all, it is their fascinating and counterintuitive dynamics that makes cytoskeletal filaments such as actin and microtubules so distinct from the polymers of the industrial age. To get an idea of these complex dynamics, we need only consider the defining act of individual cells as they divide to become two new daughters. The microtubules in the mitotic spindle of dividing cells are engaged in a constant dance as they grow and shrink over and over again (see **Figure 4-32**). Similarly, the actin at the leading edge of motile cells also engages in an incessant parade of nucleation, branching, growth, and depolymerization. In this vignette, we take stock of the rates associated with the assembly

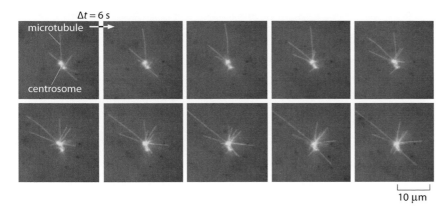

Figure 4-32 Snapshots of the dynamics of microtubules. This series of snapshots comes from a *Xenopus* egg extract, which makes it possible to reconstitute microtubule dynamics *in vitro*. The time interval between images is 6 s. Note that individual filaments both grow and shrink with a characteristic half-life of a little less than a minute. (Adapted from Tournebize R, Andersen SSL, Verde F et al. [1997] *EMBO J* 16:5537–5549.)

and disassembly of these biological polymers, with the numbers discussed here serving to provide insights into contexts ranging from the timing of metaphase in the cell cycle to the speed with which motile cells can move across surfaces.

If we were to look down a microscope at fluorescently labeled micro-tubules, we would see that these filaments perform a bizarre series of growth and shrinkage events, as shown in the snapshots from a video microscopy study of the dynamic instability in Figure 4-32. From a quantitative perspective, we can monitor the length of microtubules as a function of time. Snapshots from a video, like those shown here, lead us to recognize four key parameters that characterize microtubule dynamics: the growth and shrinkage rates themselves, as well as the rates at which the microtubules transition between growth and shrinkage phases. As seen in the data in **Figure 4-33**, such time courses allow us to immediately read off approximate values for the *in vitro* rates that characterize both the growth and shrinkage of these polymers. To be concrete, we note that for the data shown in Figure 4-33, the microtubule grows roughly 8 μm over a time of approximately 4 min, thereby corresponding to a growth rate of 2 μm/min ≈ 30 nm/s. These numbers remind us of the timing of the cell cycle, where mitosis takes several tens of minutes, consistent with this 2 μm/min polymerization rate, where we see that to move chromosomes over distances of several tens of microns should take tens of minutes. Further, if we recall from the vignette entitled "How big are the cell's filaments?" (pg. 57) that the size of a monomer is of order 5 nm, this means that the growth

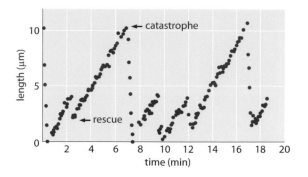

Figure 4-33 Microtubule length vs. time. The length of microtubules as a function of time reveals periods of growth punctuated by catastrophes in which the filaments rapidly depolymerize. (Adapted from Fygenson DK, Braun E & Libchaber A [1994] *Phys Rev* E50:1579–1588.)

rate is roughly 5–10 monomers added per second. Since a microtubule is comprised of 13 protofilaments, we need to multiply this 5–10 monomers per second by a factor of 13, resulting in a net addition rate of roughly 100 monomers per second at the growing end of a microtubule.

What about the *in vivo* rates of microtubule dynamics? Recent experiments on the dynamics of chromosome segregation by the mitotic spindle have been carried out in extracts of eggs from the frog *Xenopus laevis*. The idea of this experiment is that by cutting the microtubules using a laser, as shown in **Figure 4-34**, we can watch and measure the resulting dynamics as the newly formed plus ends shrink by depolymerization. The experiment is revealed in Figure 4-34, which shows an example of the dynamics after the spindle is cut. The measured rate of depolymerization is 35 ± 2 µm/min. This rate corresponds to roughly 500 nm/s. If we recall that each monomer is roughly 5 nm in size, this means that one protofilament on a given microtubule is losing roughly 100 monomers every second. Since there are 13 protofilaments per microtubule, the total loss rate from a shortening microtubule is roughly 1000 monomers per second.

Just as microtubules exhibit the dynamic instability that leads to periods of growth and shrinkage, actin filaments too are subject to an array of interesting dynamics. The simplest way to characterize the important character of the dynamics of polymerization and depolymerization in actin is to note that there is a structural asymmetry between the two ends of the filament. This dictates that the rate of monomer addition and loss on the two ends is different, a fact that is central to their intriguing dynamics. One of the earliest efforts to parameterize the different rate constants on the two ends was a tour de force study using electron

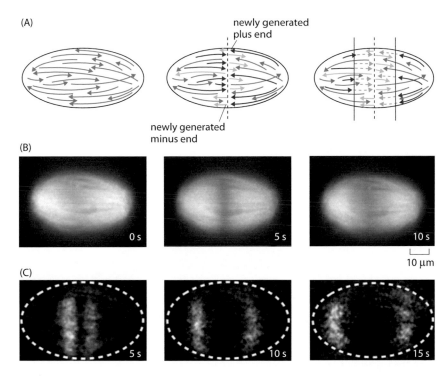

Figure 4-34 Measuring the rate of microtubule depolymerization in the mitotic spindle. (A) Schematic of the microtubules before and after the cut. The newly formed plus ends are then subject to depolymerization, which can be visualized fluorescently. (B) Fluorescent images of the spindle before and after the laser cutting. (C) Loss of fluorescent intensity at various times after the cut, thus revealing the depolymerization dynamics. (Adapted from Brugués J, Nuzzo V, Mazur E & Needleman DJ [2012] *Cell* 149:554–564.)

microscopy, where the lengths of actin filaments were measured as a function of time after incubation at various actin concentrations. The resulting data from that experiment are shown in **Figure 4-35**, which reveals the striking asymmetry in rate constants on the two ends (barbed and pointed), and further, how these rates depend upon whether the actin is bound to ATP or ADP.

Actin is a key participant in cell motility, as we have already seen in several other vignettes. How do the *in vitro* rates described above compare to what is seen in living cells? Sophisticated image analysis tools have made it possible to watch the dynamics of all of the many filaments within a motile cell simultaneously, as shown in **Figure 4-36**. Note that the results of this *in vivo* study using fluorescence microscopy are remarkably consistent with the *in vitro* rates reported in Figure 4-35. Specifically, if we look at the growth rate in Figure 4-35 for barbed ends and extrapolate the change in

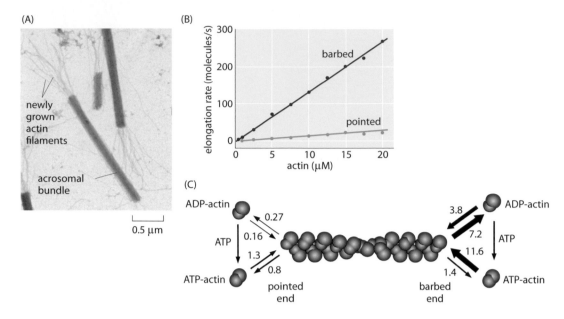

Figure 4-35 Measuring the rate constants for actin filament polymerization. (A) Electron microscopy image showing the structures used to determine the polymerization rates at both the barbed and pointed ends. (B) Elongation rate for the barbed and pointed ends as a function of actin concentration. (C) Rate constants for both ATP and ADP actin. On-rates have units of $\mu M^{-1}\,s^{-1}$, while off-rates have units of s^{-1}. Note the large asymmetry in rates between the barbed and pointed ends. (A, courtesy of Matt Footer. B, and C, adapted from Pollard TD [1986] *J Cell Biol* 103:2747–2754. Courtesy of TD Pollard.)

length over a minute time scale, we see that the growth rates are tens of μm/min, consonant with the fluorescence studies and a happy self-consistency between widely different methods.

Cytoskeletal filaments are also key players in the dynamics of bacteria. One of the most interesting case studies is the ParM system, which is responsible for the segregation of bacterial plasmids prior to cell division. In schematic form, it is thought that the way these polymers work is that each extremity of the growing polymer is attached to a plasmid, and as the polymer grows across the cell, it pushes the two plasmids to the different future daughter cells. As seen in **Figure 4-37**, the rate of ParM polymerization *in vitro* can be estimated by noting that the length increases by several microns over a minute time scale, resulting in a growth rate of several μm/min. Structurally, ParM is essentially indistinguishable from actin, though its role in the segregation of DNA, as well as the fact that it exhibits a dynamic instability, shown in Figure 4-37B, makes it occupy a conceptual middle ground between actin and microtubules.

(A)

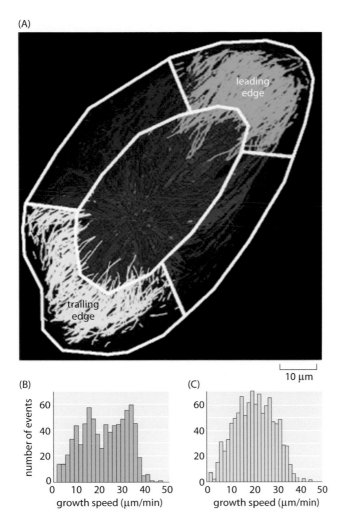

leading
edge

trailing
edge

10 µm

(B)

number of events

60

40

20

0
  0   10   20   30   40   50
       growth speed (µm/min)

(C)

60

40

20

0
  0   10   20   30   40   50
       growth speed (µm/min)

Figure 4-36 Growth rates of actin filaments *in vivo*. (A) Growth rates in different regions of a human endothelial cell. Partitioning of the cell into different "growth zones" for each of which the speed is measured. (B) Growth speeds at the leading edge of the cell. (C) Growth speeds at the trailing edge of the cell. (Adapted from Applegate KT, Besson S, Matov A et al. [2011] *J Struct Biol* 176:168–184.)

How fast do molecular motors move on cytoskeletal filaments?

Molecular motors are central to a vast array of different processes, including cell crawling, cell division, chromosome segregation, intracellular trafficking, and so forth. These active processes are driven by motors of many different types moving about on both actin and microtubule filaments.

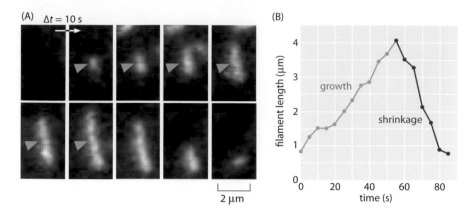

Figure 4-37 Dynamics of ParM. (A) At the beginning of the filmstrip, the filament of ParM is growing. The red arrows indicate the terminal ends of the filament and provide a fiducial marker for evaluating filament shrinkage. (B) The graph shows the length of a filament as a function of time. (Adapted from Garner EC, Campbell CS & Mullins RD [2004] *Science* 306:1021–1025.)

We noted in the vignette entitled "What are the time scales for diffusion in cells?" (pg. 211) (that is, those bigger than several tens of microns) that diffusion times become exorbitant and cells need to resort to motor-mediated, directed transport, paid for at the cost of ATP hydrolysis. The presence of these motors makes it possible for cargos of many types, including transport vesicles and even organelles, to be directed to various places throughout the cell.

Motors moving on cytoskeletal filaments can be classified into three types: myosins, kinesins, and dyneins, as shown in **Figure 4-38**. Though the diversity of these motors mirrors that of life more generally, we can attempt to classify them broadly into those that move on actin filaments and those that move on microtubules and according to the directionality of their motion. Both actin and microtubules have asymmetric filaments characterized by a plus-end and a minus-end. The motion of motors can, in turn, be characterized by the directions of their movement (that is, they can be plus-end or minus-end directed).

Once these motors have engaged their cargo, how fast can they move? For a single motor, as opposed to the collective motion of many motors that can engage a cargo simultaneously, measurements have been made using single-molecule techniques *in vitro* as well as *in vivo*. **Figure 4-39** shows examples of the kinds of microscopic observations that make these measurements possible. In now-classic optical-tweezers experiments, individual motors are tethered to a much larger bead that is then trapped using laser light. This trapping makes it possible to characterize the motor's velocity as a function of the resistive force applied by the trapped bead. In

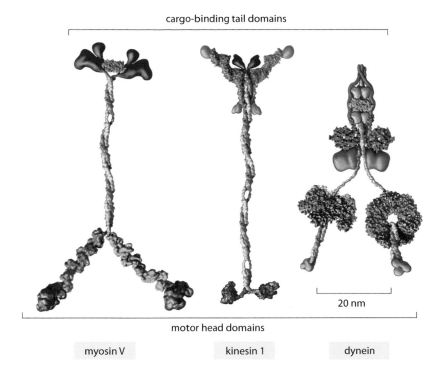

cargo-binding tail domains

20 nm

motor head domains

| myosin V | kinesin 1 | dynein |

Figure 4-38 Key classes of translational motor. A myosin V molecule is one of about 20 different types of myosins that move on actin filaments. Kinesin 1 is also a member of a large family of related molecular motor proteins, but these move on microtubules rather than on actin. Although myosins and kinesins have different substrates, the detailed structures of their motor heads are quite similar, and they are thought to be derived from a single common molecular ancestor. Cytoplasmic dynein represents a different class of microtubule-based motors that appears to be unrelated to kinesin or myosin. (Adapted from Vale RD [2003] *Cell* 112:467–480.)

many ways, the *in vivo* measurement of motor velocities is more conceptually straightforward since it involves essentially video microscopy of the motion of cargo within cells, as shown in Figure 4-39B.

Broadly speaking, translational motors move at rates somewhere between several tenths of a micron to several microns per second, with some notable and very interesting outliers, which broaden the distribution of motor speeds considerably. For example, conventional kinesins have an *in vitro* speed of 800 nm/s (BNID 101506) and an *in vivo* speed of 2000 nm/s. This directed motion is made up of individual steps of length 8 nm (BNID 101857), thus requiring about 100 steps per second to achieve such speeds *in vitro*. Though we are talking about the average response, the stochastic variations in these parameters are of great interest as well. After a characteristic duration of 100 steps, the motor is

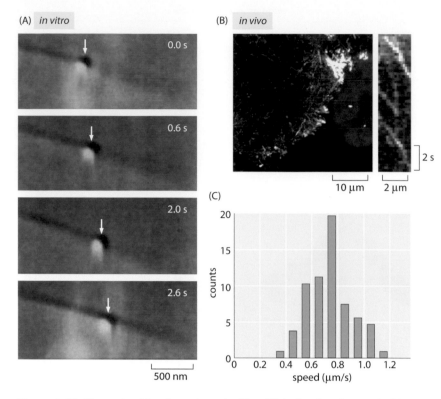

Figure 4-39 Measuring kinesin motor velocities. (A) A glass bead coated with kinesin motors was brought in contact with a microtubule using an optical trap. Both the microtubule and the bead can be seen using DIC microscopy, and the optical trap is visible as a slightly shiny spot around the bead. When the trap is shut off, the bead begins to move down the microtubule processively over several seconds. (B) Fluorescently labeled *in vivo* measurements of kinesin molecules fused to GFP. The kymograph shown on the right shows that the motors move roughly 2 microns in roughly 4 s. (C) Histogram of motor speeds from the measurements of 10 cells like those made in (B). (A, adapted from Block SM, Goldstein LSB & Schnapp BJ [1990] *Nature* 348:348–352. B, and C, adapted from Tanenbaum ME, Gilbert LA, Qi LS et al. [2014] *Cell* 159:635–646.)

released from the microtubule (BNID 103552). In every step, one ATP molecule gets hydrolyzed, releasing about 20 $k_B T$ of free energy. The force this can exert over the 8-nm step length is about 5 pN (assuming 50% efficiency; BNID 103008). A parametric spec sheet for these microscopic transport machines is given in **Table 4-7**.

How big of an object can be moved through a viscous environment at a 1 μm/s velocity with this amount of force? Stokes' law governs the relation of force (F) to velocity (V) in a fluid of viscosity (η) through the relation

$$F = 6\pi\eta RV,\qquad(4.12)$$

motor	function	speed *in vivo* (nm/s)	rate (ATPase, s^{-1}, *in vitro*)	mode of action
		myosins		
myosin XI	cytoplasmic streaming in algae	60,000	not determined	unknown
myosin II	fast skeletal muscle	6000	20	large arrays (10^4–10^9)
myosin IB	amoeboid motility, hair cell adaptation	200 (*in vitro*)	6	small arrays (10–10^3)
myosin II	smooth muscle contraction	200	1.2	large arrays (10^4–10^9)
myosin V	vesicle transport	200	5	alone or in small numbers (<10)
myosin VI	vesicle transport?	–60 (*in vitro*)	0.8	unknown
		dyneins		
axonemal	sperm and cilial motility	–7000	10	large arrays (10^4–10^9)
cytoplasmic	retrograde axonal transport, mitosis, transport in flagella	–1000	2	alone or in small numbers (<10)
		kinesins		
Fla10/KinII	transport in flagella, axons, melanocytes	2000	not determined	small arrays (10–10^3)
conventional	anterograde axonal transport	1800	40	alone or in small numbers (<10)
Nkin	secretory vesicle transport	800	80	alone or in small numbers (<10)
Unc104/KIF	transport of synaptic vesicle precursors and mitochondria	700	110	alone or in small numbers (<10)
Bimc/Eg5	mitosis and meiosis	18	2	small arrays (10–10^3)
Ncd	mitosis and meiosis	–90 (*in vitro*)	1	small arrays (10–10^3)

Table 4-7 Summary of experimental data on the dynamics of translational molecular motors. Based on BNID 106501, 101506. Values were rounded to one significant digit. Negative speeds indicate movement towards the minus end of the filament.

where R is the radius of the object moving through the fluid. Plugging in the value for water—namely $\eta = 0.1$ Pa s—we find a characteristic size of $R = 2$ μm. This is about the upper limit on the size of an organelle, whereas most transport vesicles are significantly smaller than this bound. The value of the viscosity we used is that for water; in the highly crowded cellular interior, the viscosity is higher but only by a factor of about 2–3 (BNID 105903, 103392).

Together, diffusion and motor-mediated active transport constitute two of the dominant mechanisms governing the lively comings and goings of molecules within cells. For active transport, evolution has resulted in a huge array of molecular motors with all sorts of elaborations that make it possible for them to move in different directions on different kinds of filaments while pulling along cargos that are themselves of a great diversity. Further, these motors are engaged in all sorts of dynamic activities within cells that do not relate to the transport of cargo at all, but rather, endow cells with their dynamism when separating chromosomes, moving around, or separating in two.

The biophysical study of molecular motors helps clarify a seemingly magical sleight of hand—namely, how does the hydrolysis of phosphate bonds of diameter smaller than 1 nm get spatially amplified to entail a movement of your hand over a distance on the order of centimeters? The step of the myosin motor transforms the <1 nm phosphate bond severing to a movement two orders of magnitude longer of about 36 nm across a half period of the actin filament. This same action happening in a concerted direction in 10^4–10^5 sarcomeres per muscle amplifies the movement to the level of millimeters. Finally, the anatomy of the arm and its muscles give the final leveraging to the domain of centimeters. With many biophysicists clarifying each of these steps in ever more rigorous detail, the micro to macro magic is demystified, as described in the book, *Mechanisms of Motor Proteins and the Cytoskeleton* by Jonathan Howard.

How fast do cells move?

Cell movements are one of the signature features of the living world. Whether we observe the many and varied movements of microbes in a drop of water, the crawling of *Dictyostelium* cells to form fruiting bodies, or the synchronized cell movements during gastrulation in the developing embryo, each of these processes paints a lively picture of cells in incessant motion. Fascination with cellular movements is as old as the microscope itself. In 1683, Leeuwenhoek wrote to the Royal Society about his observations with his primitive microscope (in a letter to the Royal Society of September 17, 1683) on the plaque between his own teeth: "a little white matter, which is as thick as if 'twere batter." He repeated these observations on two ladies (probably his wife and daughter) and on two old men who had never cleaned their teeth in their lives. Looking at these samples with his microscope, Leeuwenhoek reported how in his own mouth, "I then most always saw, with great wonder, that in the said matter there were many very little living animalcules, very prettily a-moving. The biggest sort . . . had a very strong and swift motion, and shot through the water (or spittle) like a pike does through the water. The second sort . . . oft-times spun round like a top . . . and these were far more in number." These excerpts beautifully illustrate both our attention to and wonder at the microscopic movement of cells.

As noted by van Leeuwenhoek himself, there are many different types of cell movements. Many microorganisms (and larger organisms, too!) make their way hither and yon by swimming, as classically exemplified by *E. coli* and *Paramecium*. Another classic mechanism is the subject of one of the most famous series of time-lapse images in all of biology, where David Rogers captured the motion of a neutrophil crawling along a surface in hot pursuit of a bacterium. Yet another mode of bacterial motility is known as gliding and refers to a form of motion that is not yet fully understood.

Such cell movements are not at all the exclusive prerogative of single-celled organisms, with all sorts of cell movements at the heart of developmental processes giving multicellular organisms their shape. One impressive example of such movements is revealed in the developing nervous system in which neurons undergo a kind of pathfinding, where protrusions from certain neurons grow outward, say, from the brain to the eye.

One of the best ways to put all of these movements of cells of different scales in perspective is to evaluate how many body lengths a given organism moves every second. In swimmer Michael Phelps' performances in several Olympics, he traveled 100 m in roughly 50 s, meaning that he was moving at roughly 1 body length per second. The sailfish *Istiophorus platypterus* swims at a speed of roughly 110 km/h ≈ 30 m/s, corresponding in this case to roughly 15 body lengths per second. When undergoing its chemotactic wanderings, an *E. coli* cell has a mean speed of roughly 30 µm/s, meaning that, like the sailfish, it travels roughly 15 times its 2 µm body lengths every second. Amoeba such as *Dictyostelium* move at a rate of 10 µm/min or 1 body length per minute, very similar to the speeds seen in the motion of the neutrophil chasing down its prey, as shown in the famed Rogers video. A collection of cell speeds is presented in **Table 4-8**.

Taking the analogy of the Olympic race to a new level, a world cell lines race was recently performed that competed crawling cell lines from labs around the world on a racecourse made of micro-fabricated lanes. **Figure 4-40** shows an overlay of the fastest cells in the competition. The winner was a human embryonic mesenchymal stem cell, which showed the fastest migration speed recorded at 5.2 µm/min. Comparison to Table 4-8 shows that this event, limited to crawling cell lines, is actually at a much slower pace than a possible microbial swimming event.

What is the limit on the crawling speed of cells? Why should crawling be slower than swimming? The molecular basis is quite different, because crawling depends on actin polymerization, whereas the swimming bacterium exploits flagellar rotation, for example. Actin polymerization-based motility is key for the development of protrusions in polarized eukaryotic cells as well as for bacteria (such as *Listeria*) that move around inside cells by hijacking the host cell cytoskeleton.

organism	speed	speed in body lengths (bl) per time	BNID and comments
bacteria and archaea			
Ovobacter propellens	1000 µm/s	200 bl/s	111235
Thiovulum majus	600 µm/s	90 bl/s	107652, 111231, , cell length ≈7 µm
Methanocaldococcus jannaschii	400 µm/s	200 bl/s	107649, measured at ≈80°C
Bdellovibrio bacteriovorus	160 µm/s	160 bl/s	101969, has to catch other bacteria it preys on
Vibrio cholerae	40–100 µm/s	20–50 bl/s	108083, sodium ion motor, one polar flagellum
Caulobacter crescentus	40 µm/s	20 bl/s	108085, proton motor, one polar flagellum
Spirochete Brachyspira hyodysenteriae	40 µm/s	8 bl/s	104904, assuming 5 µm cell length
E. coli	16–30 µm/s	8–15 bl/s	101793, 106819, 108082, proton motor, 4–8 lateral flagella
S. typhimurium	30 µm/s	15 bl/s	106818
Synechococcus	5–25 µm/s	2–10 bl/s	109314, mysterious propulsion by one third of wild isolates
Myxococcus Xanthus motility system S	>20 µm/min	>10 bl/min	106811
Myxococcus Xanthus motility systemA	2–4 µm/min	1–2 bl/min	106811
Listeria monocytogenes	6 µm/min	3 bl/min	106823 *in vitro* motility assays
Halobacterium halobium	2–3 µm/min	1 bl/min	111147
eukaryotes			
Ciliate Paramecium tetraurelia	100–1000 µm/s	1–5 bl/s	108087, ciliated, assuming 200 µm cell length
Tetrahymena thermophila	200–400 µm/s	4–8 bl/s	111429, 111435, 111436, ciliated
Gyrodinium dorsum	300 µm/s	10 bl/s	111432, flagellated
green algae *Chlamydomonas Reinhardtii*	50–150 µm/s	5–15 bl/s	108086, 111430
fish keratocytes - wound healing fibroblasts of the cornea	10–50 µm/min	0.7–3 bl/min	106807, 106817
Amoeba Dictyostelium discoideum	10 µm/min	≈1 bl/min	106825
human neutrophil	9 µm/min	≈1 bl/min	106809
glioma cells	50 µm/h	4 bl/h	106810
mouse fibroblastoid L929 cells	30 µm/h	2 bl/h	106808
human H69 small cell lung cancer cell	16 µm/h	1 bl/h	106815

Table 4-8 Cell speeds of different cells given in µm per time unit and as body lengths per time unit. Assume a bacterial length of ≈2 µm and a eukaryotic cell length of ≈15 µm unless otherwise stated. Speeds depend on temperature, experimental conditions, etc. Values given here are those reported in the literature. Most measurements are based on time-lapse microscopy.

What can be said about the sources for the diversity in speeds? Some of the fastest bacteria are at high temperatures, where rates of nearly everything tend to be higher, or in organisms that have to depend on their speediness to make a living, such as in the case of *Bdellovibrio bacteriovorus*, which has to be faster than the bacteria it preys on. The record holder, *Ovobacter*

time = 0 min

human BM.C.SC
human Bre.E.Fib.2
human Bre.E.Fib.6
mouse Emb.C.Tra.1
mouse Emb.C.Tra.2
human Emb.C.Pri
mouse SG1.E.Sar
mouse Emb.C.Tra.5
human Skin.EPri
mouse Hip.N.Tra.2

time = 70 min

human BM.C.SC
human Bre.E.Fib.2
human Bre.E.Fib.6
mouse Emb.C.Tra.1
mouse Emb.C.Tra.2
human Emb.C.Pri
mouse SG1.E.Sar
mouse Emb.C.Tra.5
human Skin.EPri
mouse Hip.N.Tra.2

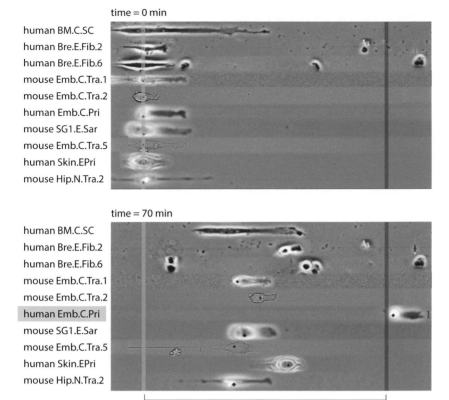

350 µm

Figure 4-40 Finals of the World Cell Race. The 10 fastest cell lines are displayed competing over a 350 µm micro-fabricated sprinting lane. Each of the cells was found to be the fastest among its cell type. Each cell type was recorded in a separate well, and movies were combined to show one lane per cell type. The time difference between the two images is about an hour. The winner is highlighted in brown. (Courtesy of Matthieu Piel.)

propellens, moves at an astonishing 1 mm/s, armed with about 400 flagella on its 5 µm cell (BNID 111233, 111232, 111235). The pressure to run swiftly is less clear. The functional significance of different swimming speeds for bacteria is usually discussed in terms of the ability of bacteria to achieve chemotaxis, where they perform a biased random walk using their flagella to move to environments of higher nutrient concentrations. Different lines of evidence suggest that motility might have important parts to play in the dense communities of bacteria, where the survival and growth often depend on more intricate issues of communication, cooperation, and relative location, all affected by motility.

LIFE CYCLE OF CELLS

How long does it take cells to copy their genomes?

Genomes and the management of the vast array of information they contain are one of the signature features that make living matter so different from its inanimate counterpart. From the moment of the inception of the modern view of DNA structure, Watson and Crick made it clear that one of the most compelling features of DNA's double-helix structure was that it suggested a mechanism for its own replication. But what sets the time scale for the replication process itself and how do the mechanisms and associated rates differ from one organism to the next? Does the time required to complete replication ever impose a limitation on the growth rate of the organism?

An elegant way to directly measure the replication rate is through the use of a single-molecule technique in which the progress of the replication machinery is monitored by using a microscope to watch the motion of a tiny bead attached to the DNA template, as shown in **Figure 4-41**. By

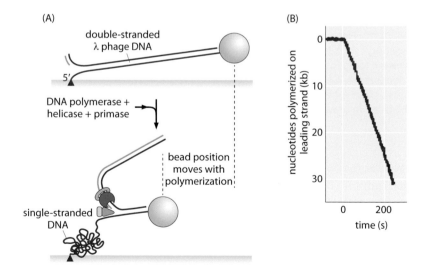

Figure 4-41 Schematic of a single-molecule experiment used to measure the rate of replication. (A) The progress of the replication process during leading-strand synthesis converts double-stranded into single-stranded (plus another double-stranded) DNA. (B) Because the "spring constant" of single-stranded DNA is larger than that for double-stranded DNA, the stretched tether recoils, resulting in the bead position time course shown. (Adapted from Lee JB, Hite RK, Hamdan SM et al. [2006] *Nature* 439:621–624.)

permitting only leading-strand synthesis, the replication process results in the conversion of double-stranded DNA into one double-stranded fragment and a second single-stranded fragment on the uncopied strand. The trick in this method is that it exploits the difference in entropic elasticity of the single-stranded and double-stranded fragments. As a result, with increasing replication, more of the template is converted into the single-stranded form, which as seen in Figure 4-41, serves as a much stronger entropic spring than the double-stranded fragment whose persistence length is orders of magnitude larger. The spring moves the bead at the same rate as the polymerase proceeds forward, serving as a readout of the underlying replication dynamics. These measurements resulted in an *in vitro* replication rate of 220 ± 80 nucleotides/s (BNID 103995) for the replication machinery from a T7 bacterial virus. With a genome size of ≈ 40,000 bp and without taking into account initiation and similar processes that might complicate directly importing these *in vitro* insights to the *in vivo* setting, we can estimate that it will require at least 40,000 bp/220 bp/s ≈ 200 s or about 3 min to replicate the compact viral genome.

Given these insights into the replication rate, how do they stack up against the known division times of different cell types? *E. coli* has a genome of roughly 5 million bp (BNID 100269). Replication rates are observed to be several hundred bp/s (BNID 104120, 109251). Further, replication in these bacteria takes place with two replication forks heading in opposite directions around the circular bacterial chromosome. As shown in **Estimate 4-12**, the replication rates imply that it should take the two replisomes at least 2500 s (≈ 40 min) to replicate the genome, a number that is much longer than the minimal division time of ≈ 20 min (BNID 103514). This interesting estimate delivers a paradox that is resolved by the observation that *E. coli* under ideal growth conditions employ nested replication forks like those seen in Estimate 4-12, which begin

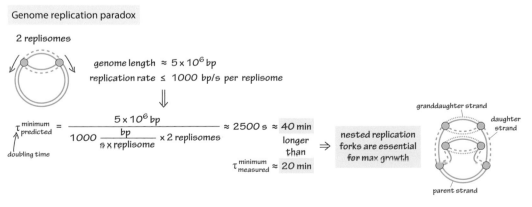

Estimate 4-12

replicating the granddaughter and great-granddaughter cells' genomes while the daughter cells are still themselves engaged in replication. At fast growth rates, more than six origins of replication and more than 10 replication forks coexist in a single cell (BNID 102356), as deduced from elegant models on the co-dependence of the generation time, genome replication time, and the numbers of replication forks and origins. Recently, single-molecule microscopy revealed that the most common stoichiometry of the replication machinery, the replisome, consists of three DNA polymerases per replisome, in contrast to the naïve picture of two DNA polymerases (BNID 107686). It seems that the third polymerase can sometimes be engaged in the lagging strand replication, together with another polymerase, or in other cases to be awaiting engagement in the replication process.

Eukaryotic genomes are usually much larger than those of their prokaryotic cousins, and as a result, the replication process must depend upon more than a single origin of replication. The number of origins leading to replication is a subject of active research. Recently, microarrays and deep sequencing have been used to find peaks of DNA content in S phase that indicate putative origins. Estimates for the total number of origins still vary widely—for example, in mouse they range from as low as 1000 to as many as 100,000, while for *Drosophila* the estimate is about 10,000 (BNID 107654, 109283). Each origin is associated with a replisome that proceeds at a rate of 4–40 bp/s, or roughly 1 kb/min (BNID 104930, 104935, 104936, 104937). A classic view of the replication process has been offered by electron microscopy images, such as that presented in **Figure 4-42**, which shows a collection of replication forks associated with the copying of the *Drosophila* genome. From the rate of replication and the observed distance between replication forks, we can see that a complete replication cycle can proceed much more quickly than if there were only one replication origin. This is a necessity, given the rapid genome replication in the early stage of *D. melanogaster* development, where the embryo replicates its ≈120-million-bp genome (BNID 100199) at a dizzying pace of once every ≈8 min (BNID 101971). In humans

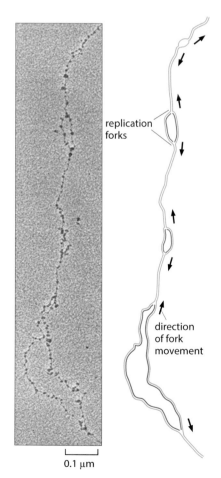

replication forks

direction of fork movement

0.1 μm

Figure 4-42 Replication forks on the DNA of *D. melanogaster*. Replication forks move away in both directions from replication origins. (Electron micrograph courtesy of Victoria Foe.)

(**Figure 4-43**) the S phase in many cell types is on the order of 10 hours (BNID 103742, 103741, 102204). What selective force should push the DNA replication time to be relatively short in cells that have very long overall cell cycle times? And for fast growing cells why not make the replication time even shorter by employing more origins of replication?

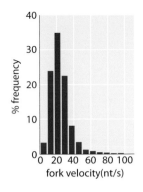

Figure 4-43 Histogram of fork velocities for human primary keratinocytes (mean = 1.5 kb/min; N = 5460). (Adapted from Conti C, Saccà B, Herrick J et al. [2007] *Mol Biol Cell* 18:3059–3067.)

How long do the different stages of the cell cycle take?

Replication is one of the hallmark features of living matter. The set of processes known as the cell cycle, which is undertaken as one cell becomes two, has been a dominant research theme in the molecular era, with applications that extend far and wide, including to the study of diseases such as cancer, which is sometimes characterized as a disease of the cell cycle gone awry. Cell cycles are interesting, both for the ways they are similar from one cell type to the next, and for the ways they are different. To bring the subject in relief, we consider the cell cycles in a variety of different organisms, including a model prokaryote, mammalian cells in tissue culture, and a fruit fly during embryonic development. Specifically, what individual steps are undertaken for one cell to divide into two and how long do these steps take?

Arguably the best-characterized prokaryotic cell cycle is that of the model organism *Caulobacter crescentus*. One of the appealing features of this bacterium is that it has an asymmetric cell division, which enables researchers to bind one of the two progeny to a microscope cover slip while the other daughter drifts away, thus enabling further study without obstructions. This has given rise to careful depictions of the ≈ 150-min cell cycle (BNID 104921), as shown in **Figure 4-44**. The main components of the cell cycle are G1 (the first growth phase, ≈ 30 min; BNID 104922), where at least some minimal amount of cell size increase needs to take place, S phase (synthesis, ≈ 80 min; BNID 104923), where the DNA gets replicated, and G2 (the second growth phase, ≈ 25 min; BNID 104924), where chromosome segregation unfolds, thereby leading to cell division (this final phase lasts ≈ 15 min). *Caulobacter crescentus* provides an interesting example of the way in which certain organisms get promoted to "model organism" status because they have some feature that renders them particularly opportune for the question of interest. In this case, the cell cycle progression goes hand in hand with the

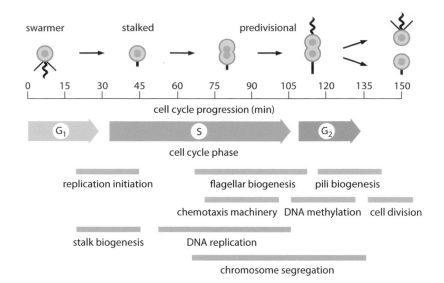

Figure 4-44 The 150-min cell cycle of *Caulobacter* is shown, highlighting some of the key morphological and metabolic events that take place during cell division. M phase is not indicated, because in *Caulobacter* there is no true mitotic apparatus that gets assembled as in eukaryotes. Much of chromosome segregation in *Caulobacter* (and other bacteria) occurs concomitantly with DNA replication. The final steps of chromosome segregation, and especially decatenation of the two circular chromosomes, occurs during G2 phase. (Adapted from Laub MT, McAdams HH, Feldblyum T et al. [2000] *Science* 290:2144–2148.)

differentiation process, giving readily visualized identifiable stages and thereby making them preferable to cell cycle biologists over, say, the model bacterium *E. coli*.

The behavior of mammalian cells in tissue culture has served as the basis for much of what we know about the cell cycle in higher eukaryotes. The eukaryotic cell cycle can be broadly separated into two stages—namely, interphase, that part of the cell cycle when the materials of the cell are being duplicated, and mitosis (M), the set of physical processes that attend chromosome segregation and subsequent cell division. The rates of processes in the cell cycle are mostly built up from many of the molecular events, such as the polymerization of DNA and cytoskeletal filaments whose rates we have already considered. For the characteristic cell cycle time of 20 h in a HeLa cell, almost half is devoted to G1 (BNID 108483), and close to another half is S phase (BNID 108485), whereas G2 and M are much faster at about 2–3 h and 1 h, respectively (BNID 109225, 109226). The stage most variable in duration is G1. In less favorable growth conditions, when the cell cycle duration increases, this is the stage that is mostly affected, possibly due to the time it takes until some regulatory size checkpoint is reached. Though different types of evidence point to the existence of such a checkpoint, it is currently very poorly understood. Historically, stages in the cell cycle have usually been inferred using fixed cells, but recently, genetically encoded biosensors that change localization at different stages of the cell cycle have made it possible to get live-cell temporal information on cell cycle progression and arrest.

How does the length of the cell cycle compare to the time it takes a cell to synthesize its new genome? A decoupling between the genome length and the doubling time exists in eukaryotes due to the usage of multiple DNA replication start sites. For mammalian cells, it has been observed that for many tissues with widely varying overall cell cycle times, the duration of the S phase, where DNA replication occurs, is remarkably constant. For mouse tissues, such as those found in the colon or tongue, the S phase varied in a small range from 6.9 to 7.5 h (BNID 111491). Even when comparing several epithelial tissues across human, rat, mouse, and hamster, S phase was between 6 and 8 h (BNID 107375). These measurements were carried out in the 1960s by performing a kind of pulse-chase experiment with the radioactively labeled nucleotide thymidine. During the short pulse, the radioactive compound was incorporated only into the genome of cells in S phase. By measuring the duration of the appearance and then the disappearance of labeled cells in M phase, we can infer how long S phase lasted. The fact that the duration of S phase is relatively constant in such cells is used to this day to estimate the duration of the cell cycle from a knowledge of only the fraction of cells at a given snapshot in time that are in S phase. For example, if one-third of the cells are seen in S phase, which lasts about 7 h, then the cell cycle time is inferred to be about 7 h/(1/3) ≈ 20 h. Today these kinds of measurements are mostly performed using BrdU as the marker for S phase. We are unaware of a satisfactory explanation for the origin of this relatively constant replication time and how it is related to the rate of DNA polymerase and the density of replication initiation sites along the genome.

The diversity of cell cycles is shown in **Figure 4-45** and depicts several model organisms and the durations and positioning of the different stages of their cell cycles. An extreme example occurs in the mesmerizing process of embryonic development of the fruit fly *Drosophila melanogaster*. In this case, the situation is different from conventional cell divisions, since rather than synthesizing new cytoplasmic materials, mass is essentially conserved, except for the replication of the genetic material. This happens in a very synchronous manner for about 10 generations, and a replication cycle of the thousands of cells in the embryo, say between cycle 10 and 11, happens in about 8 min, as shown in Figure 4-45 (BNID 103004, 103005, 110370). This is faster than the replication times for any bacteria, even though the genome is ≈ 120 million bp long (BNID 100199). This is a striking example of the ability of cells to adapt their temporal dynamics.

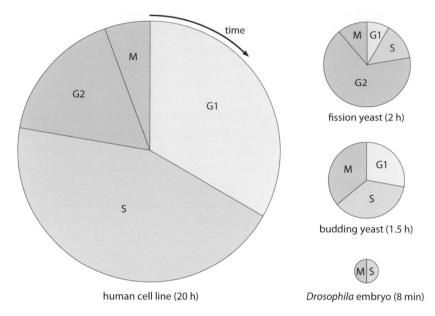

Figure 4-45 Cell cycle times for different cell types. Each pie chart shows the fraction of the cell cycle devoted to each of the primary stages of the cell cycle. The area of each chart is proportional to the overall cell cycle duration. Cell cycle durations reflect minimal doubling times under ideal conditions. (Adapted from Morgan D [2007] The Cell Cycle: Principles of Control. Sinauer Associates.)

How quickly do different cells in the body replace themselves?

The question of cell renewal is one that all of us have intuitive daily experience with. We all notice that our hair falls out regularly, yet we don't get bald (at least not until males reach a certain age!). Similarly, we have all had the experience of cutting ourselves only to see how new cells replaced their damaged predecessors. And we donate blood or give blood samples without gradually draining our circulatory system. All of these examples point to a replacement rate of cells that is characteristic of different tissues and in different conditions, but that makes it abundantly clear that for many cell types renewal is a part of their story. To be more concrete, our skin cells are known to constantly be shed and then renewed. Red blood cells make their repetitive journey through our bloodstream with a lifetime of about four months (BNID 107875, 102526). We can connect this lifetime to the fact calculated in the vignette entitled "How many cells are there in an organism?" (pg. 314) that there are about 3×10^{13} red blood cells to infer that about 100 million new red blood cells are being formed in our body every minute! Replacement of our cells also occurs in most

cell type	turnover time	BNID
small intestine epithelium	2–4 days	107812, 109231
stomach	2–9 days	101940
blood neutrophils	1–5 days	101940
white blood cells eosinophils	2–5 days	109901, 109902
gastrointestinal colon crypt cells	3–4 days	107812
cervix	6 days	110321
lungs alveoli	8 days	101940
tongue taste buds (rat)	10 days	111427
platelets	10 days	111407,111408
bone osteoclasts	2 weeks	109906
intestine paneth cells	20 days	107812
skin epidermis cells	10–30 days	109214, 109215
pancreas beta cells (rat)	20–50 days	109228
blood B cells	1 month	111516
trachea	1–2 months	101940
hematopoietic stem cells	2 months	109232
sperm (male gametes)	2 months	110319, 110320
bone osteoblasts	3 months	109907
red blood cells	4 months	101706, 107875
liver hepatocyte cells	0.5–1 year	109233
fat cells	8 years	103455
cardiomyocytes	0.5–10% per year	107076, 107077, 107078
central nervous system	life-time	101940
skeleton	10% per year	109908
lens cells	life-time	109840
oocytes (female gametes)	life-time	111451

Table 4-9 Cell renewal rates in different tissues of the human body. Values are rounded to one significant digit. Giving context through daily life replacement processes, we note that hair elongates at about 1 cm per month (BNID 109909), while fingernails grow at about 0.3 cm per month (BNID 109900), which is about the same speed as the continental spreading in plate tectonics that increases the distance between North America and Europe (BNID 110286).

of the other tissues in our body, though the cells in the lenses of our eyes and most neurons of our central nervous system are thought to be special counter examples. A collection of the replacement rates of different cells in our body is given in **Table 4-9**.

How can the replacement rates of the cells in various tissues in our body be measured? For rapidly renewing tissues, common labeling tricks can be useful, as with the nucleotide analog BrdU. But what about the very slow tissues that take years or a lifetime? In a fascinating example of scientific serendipity, Cold War nuclear tests have come to the aid of scientists as a result of the fact that they changed the atmospheric concentrations of the isotope carbon-14 around the globe. These experiments are effectively

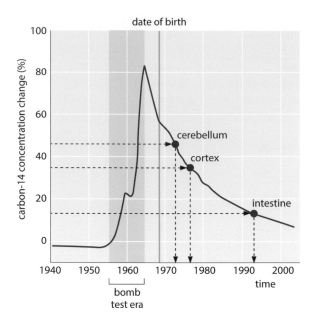

Figure 4-46 Inferring tissue turnover time from natural stable isotope labeling. The global ^{14}C levels in the environment are shown in red. A large addition of ^{14}C in 1955–1963 is the result of nuclear bomb tests. Cell age in different adult human organs is inferred from analysis of ^{14}C levels in genomic DNA measured in 2003–2004 from the cerebellum, occipital cortex, and small intestine. The birth year of the individual is indicated by a vertical line. Stable isotope levels reveal the differing turnover rates of cells in different tissues. (Adapted from Spalding KL, Bhardwaj RD, Buchholz BA et al. [2005] *Cell* 122:133–143.)

pulse-chase experiments but on a global scale. Carbon-14 has a half-life of 5730 years, and thus even though radioactive, the fraction that decays within the lifetime of an individual is negligible, and this time scale should not worry us. The "labeled" carbon in the atmosphere is turned into CO_2 and later into our food through carbon fixation by plants. In our bodies, this carbon gets incorporated into the DNA of every nascent cell and the relative abundance of carbon-14 remains stable, because the DNA is not replaced through the cell's lifetime. By measuring the fraction of carbon-14 in a tissue, it is possible to infer the year in which the DNA was replicated, as depicted in **Figure 4-46**. The carbon-14 time course in the atmosphere initially spiked due to bomb tests and then subsequently decreased as it got absorbed in the much larger pools of organic matter on the continents and the inorganic pool in the ocean. As can be seen in Figure 4-46, the time scale for the exponential decay of the carbon-14 in the atmosphere is about 10 years. The measured dynamics of the atmospheric carbon-14 content is the basis for inferring the rates of tissue renewal in the human body and yielded insights into other obscure questions, such as how long sea urchins live and the origins of coral reefs.

Using these dating methods, it has been inferred that fat cells (adipocytes) replace at a rate of $8 \pm 6\%$ per year (BNID 103455). This results in the replacement of half of the body's adipocytes in ≈ 8 years. A surprise arrived when heart muscle cells were analyzed. The long-held dogma in the cardiac biology community was that these cells did not replace themselves. This paradigm was in line with the implications of heart attacks, where scar tissue is formed instead of healthy muscle cells. Yet it was found that replacement does occur, albeit at a slow rate. Estimates vary from 0.5% per year (BNID 107076) to as high as 30% per year (BNID 107078), depending on age and gender (BNID 107077). A debate is currently taking place over the very different rates observed, but this peculiar scientific side effect of Cold War tensions is providing a fascinating window into the interesting question of the life history of the cells that make up multicellular organisms.

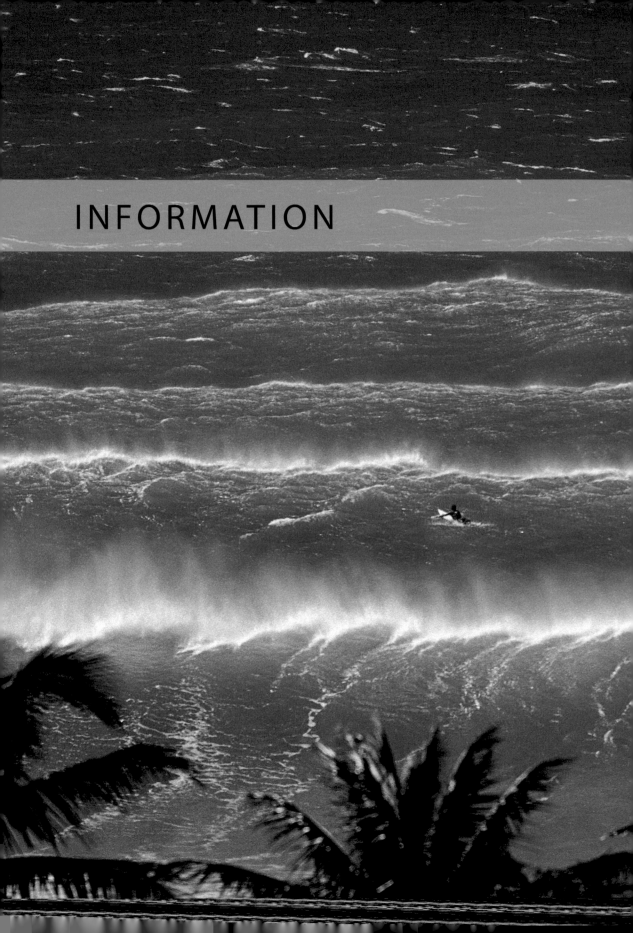

INFORMATION

Chapter 5: Information and Errors

What is it that makes living matter so different from its inanimate counterpart? Stated simply, living matter carries within it the blueprint for its own construction. The storehouse of information contained both in genomes and in the post-translational modifications of proteins leads to an ability to pass information along from one generation to the next with staggering fidelity. Genomes preside over the management of the molecules of the cell in ways that forbid them from becoming an inactive soup of chemicals whose potential for further reactions has been exhausted. This feat is all the more impressive given that, on evolutionary time scales, this information content changes as a result of adaptations and genetic drift.

The vignettes presented in this chapter all focus either directly or indirectly on quantifying the management of the information content in cells. The scale of information storage in biological components is depicted in **Figure 5-1** and compared to human designed information storage devices. The juxtaposition of biological and human information storage is both surprising and enlightening. To get a sense of the astonishing information density of biological systems, consider an estimate made by one of our students in a class entitled "Cell Biology by the Numbers." What this student found is that if one imagines the information storage density of the influenza virus scaled up to the size of a modern disk-on-key device, it would account for several exabytes of data (10^{18})—equivalent to the global internet traffic over a few days.

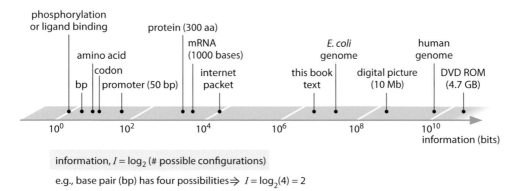

Figure 5-1 Information content of biological entities and some human designed information storage devices. Information is quantified through binary bits, where a base pair that has four possibilities is 2 bits, etc.

In this chapter, we begin by examining genomes themselves. How big are they and how many genes do they harbor? We will see that there is a huge eight orders of magnitude (or even more) difference in sizes between the largest and smallest genomes, though the number of genes they contain shows much less variability. The next set of questions that we broach in considering information management within cells centers broadly on the question of biological fidelity. In the Middle Ages, the promulgation of sacred texts took place through the patient action of scribes whose job it was to copy the contents of these books. Like any copying process, these reproductions were subject to mistakes, and it is the biological analog of such mistakes that will concern us here. We mainly examine the error rate associated with the processes of the central dogma. How many mistakes are made each time a genome is copied? When new proteins are synthesized, how often is the wrong amino acid added onto the nascent polypeptide chain?

Future work should expand the scope of the discussion presented here to ask about substrate recognition more broadly. Many proteins have their activity regulated by the addition and removal of charged groups such as phosphates through the action of kinases. But what prevents these kinases from adding a phosphate group on the wrong substrate, and how often are such mistakes made? It is of great interest to better understand the discriminatory powers that are exploited in selecting residues for phosphorylation.

All told, information management is one of the great themes of biology, and the task of this chapter is to provide a quantitative view of some of these questions.

GENOME

How big are genomes?

Genomes are now being sequenced at such a rapid rate that it is becoming routine. As a result, there is a growing interest in trying to understand the meaning of the information that is stored and encoded in these genomes and to understand their differences and what these differences say about the evolution of life on Earth. Further, it is now even becoming possible to compare genomes between different individuals of the same species, which serves as a starting point for understanding the genetic contributions to their observed phenotypes. For example, in humans, so-called genome-wide association studies associate variations in genetic makeup with the susceptibility to diseases such as diabetes and cancer.

Naïvely, the first question we might ask in trying to take stock of the information content of genomes would be, "How large are they?". Early thinking held that the genome size across the whole tree of life should be directly related to the number of genes it contains. This was strikingly refuted by the similarity in the number of protein-coding genes in genomes of very different sizes— one of the unexpected results of sequencing many different genomes from organisms far and wide. For example, as shown in **Table 5-1**, *Caenorhabditis elegans* (a nematode) has a very similar number of protein-coding genes to that of human or mouse ($\approx$ 20,000), even though their genomes vary in size by over 20-fold. As shown in **Figure 5-2**, the range of genome sizes runs from 0.16 Mbp for the endosymbiotic *Candidatus ruddii* to $\approx$150 Gbp (BNID 110278) for the enormous genome of the plant *Paris Japonica,* thus revealing a million-fold difference in genome size. An often-cited claim for a world record genome size—namely, 670 Gbp for the amoeba *Polychaos dubium*—is considered dubious because it used 1960s methods that analyzed the whole cell rather than single nuclei. Because of this approach, the result could be muddled by including contributions from mitochondrial DNA, possible multiple nuclei, and anything the amoeba recently engulfed (BNID 104470). At the other extreme of small genome sizes, viral genomes are in a class of their own, where sizes are usually considerably smaller than the smallest bacterial genome, with many of the most-feared RNA viruses having genomes that are less than 10 kb in length.

What is the physical size of these DNA molecules? Converting the length as measured in base pair units to physical length of the fully stretched-out DNA molecule can be carried out by noting that the distance between bases along the DNA strand is $\approx$0.3 nm (BNID 100667). For the human genome, with its length of $\approx$3Gbp, this conversion tells us that each of our more than 10^{13} cells harbors more than a meter of DNA. Remarkably, each cell in our body has to compress this one meter's worth of DNA into a nuclear volume with a radius of only a few microns. There is actually double trouble because our cells are diploid, meaning that each nucleus has to pack in roughly two meters, worth of DNA. To carry out this extreme compaction requires architectural proteins, such as histones, and much dexterity in reading the stored information during transcription. Similarly, in bacteria, every operon (such as the Lac operon), if it were stretched in a straight line, would by itself traverse the whole length of the bacterium.

Figure 5-2 and Table 5-1 give examples of different genome sizes with the ambition of illustrating some of the useful and well-known model organisms, some of the key outliers characterized by genomes that are either extraordinarily small or large, and examples that are particularly exotic. For some of the largest genomes, such as the record holder of the animal kingdom, the marbled lungfish, sequencing is not yet available. Older methods of measuring DNA in bulk refer to the genome size through

organism	genome size (base pairs)	protein-coding genes	number of chromosomes
model organisms			
model bacteria *E. coli*	4.6 Mbp	4300	1
budding yeast *S. cerevisiae*	12 Mbp	6600	16
fission yeast *S. pombe*	13 Mbp	4800	3
amoeba *D. discoideum*	34 Mbp	13,000	6
nematode *C. elegans*	100 Mbp	20,000	12 (2n)
fruit fly *D. melanogaster*	140 Mbp	14,000	8 (2n)
model plant *A. thaliana*	140 Mbp	27,000	10 (2n)
moss *P. patens*	510 Mbp	28,000	27
mouse *M. musculus*	2.8 Gbp	20,000	40 (2n)
human *H. sapiens*	3.2 Gbp	21,000	46 (2n)
viruses			
hepatitis D virus (smallest known animal RNA virus)	1.7 Kb	1	ssRNA
HIV-1	9.7 kbp	9	2 ssRNA (2n)
influenza A	14 kbp	11	8 ssRNA
bacteriophage λ	49 kbp	66	1 dsDNA
Pandoravirus salinus (largest known viral genome)	2.8 Mbp	2500	1 dsDNA
organelles			
mitochondria – *H. sapiens*	16.8 kbp	13 (+22 tRNA +2 rRNA)	1
mitochondria – *S. cerevisiae*	86 kbp	8	1
chloroplast – *A. thaliana*	150 kbp	100	1
bacteria			
C. ruddii (smallest genome of an endosymbiont bacteria)	160 kbp	182	1
M. genitalium (smallest genome of a free-living bacteria)	580 kbp	470	1
H. pylori	1.7 Mbp	1600	1
Cyanobacteria S. elongatus	2.7 Mbp	3000	1
methicillin-resistant *S. aureus* (MRSA)	2.9 Mbp	2700	1
B. subtilis	4.3 Mbp	4100	1
S. cellulosum (largest known bacterial genome)	13 Mbp	9400	1
archaea			
Nanoarchaeum equitans (smallest parasitic archaeal genome)	490 kbp	550	1
Thermoplasma acidophilum (flourishes in pH<1)	1.6 Mbp	1500	1
Methanocaldococcus (Methanococcus) jannaschii (from ocean bottom hydrothermal vents; pressure >200 atm)	1.7 Mbp	1700	1
Pyrococcus furiosus (optimal temp 100°C)	1.9 Mbp	2000	1
eukaryotes – multicellular			
pufferfish *Fugu rubripes* (smallest known vertebrate genome)	400 Mbp	19,000	22
poplar *P. trichocarpa* (first tree genome sequenced)	500 Mbp	46,000	19
corn *Z. mays*	2.3 Gbp	33,000	20 (2n)
dog *C. familiaris*	2.4 Gbp	19,000	40
chimpanzee *P. troglodytes*	3.3 Gbp	19,000	48 (2n)
wheat *T. aestivum* (hexaploid)	16.8 Gbp	95,000	42 (2n = 6x)
marbled lungfish *P. aethiopicus* (largest known animal genome)	130 Gbp	unknown	34 (2n)
herb plant *Paris japonica* (largest known genome)	150 Gbp	unknown	40 (2n)

Table 5-1 Genomic census for a variety of selected organisms. The table features the genome size, current best estimate for number of protein-coding genes, and number of chromosomes. Genomes often also include extra chromosomal elements such as plasmids, which might not be indicated in the genome size and number of chromosomes. The number of genes is constantly under revision. The numbers given here reflect the number of protein-coding genes. tRNA and noncoding RNAs are not accounted for since many of them are still to be discovered. Bacterial strains often show significant variations in genome size and number of genes among strains. Values were rounded to two significant digits. See full references in BNID 111493.

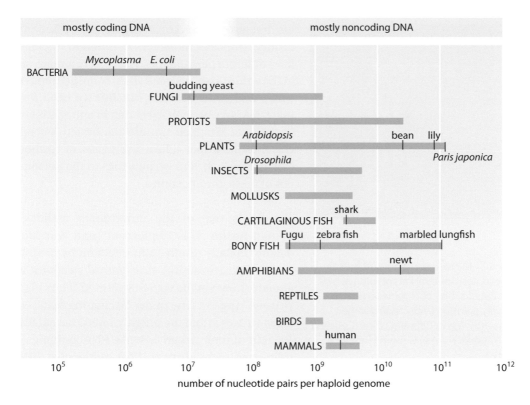

Figure 5-2 Genome sizes of different organisms.

the C-value, which represents the amount of DNA and, thus, the genome length without regard to its specific sequence. The next vignettes now take up the question of how many chromosomes and genes are present in these various genomes and whether there are any useful rules of thumb for predicting the gene number on the basis of genome size.

How many chromosomes are found in different organisms?

Living matter is programmed by its genome, the iconic DNA molecule that carries not only the instructions needed to make new copies of that very same organism through the many RNAs and proteins that run its daily

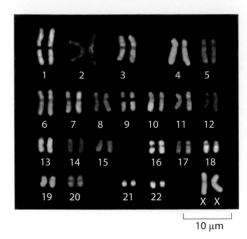

10 µm

Figure 5-3 Microscopy images of human chromosomes. Spectral karyotyping allows for the visualization of chromosomes by effectively painting each chromosome fluorescently with a different color. (Courtesy of the National Human Genome Research Institute.)

life, but also a record of an organism's evolutionary history. The DNA molecules of different organisms have different personalities. As we have already seen in the write-up on genome sizes, some genomes are small, and some are very big. But we can't forget that DNA is a physical object that in animals is compacted and wrapped up into the famed X-shaped chromosomes that adorn the pages of textbooks (see **Figure 5-3**), and we now turn to the personalities of these chromosomes.

The flu is one of the unfortunate realities of human health. This unpleasant (and sometimes deadly) malady results from infection by the influenza virus, a beautiful virus whose structure was already shown in the vignette entitled "How big are viruses?" (pg. 5). One of the fascinating features of these viruses is that the roughly 14,000 bases (BNID 106760) of their negative-sense RNA genomes are split over eight distinct RNA molecules (BNID 110337), thereby demonstrating that, even in the case of viruses, genomes are sometimes split up into distinct molecules. This kind of weirdness is even more strikingly demonstrated in the case of the cowpea chlorotic mottle virus (CCMV), whose ≈8000-base genome is separated into three distinct RNA molecules (BNID 106457), each of which is packed into a different capsid—meaning that all three of them need to infect the host in order for the infection to be viable.

Though our favorite bacterium *E. coli* has only a single circular chromosome, many prokaryotes have multiple circular chromosomes. For example, *Vibrio cholerae*, the pathogen that is responsible for cholera, has two circular chromosomes, one with a length of 2.9 Mb and the other with a length of 1.1 Mb. A more bizarre example is found in the bacterium *Borrellia burgdorferi*, which sometimes causes Lyme disease after animals suffer a bite from a Borrellia-infected tick. This bacterium contains 11 plasmids containing 430 genes beyond its long linear chromosome (BNID 111258). The microscopic world of archaea seems to have similar chromosomal distributions to those found in bacteria, though *M. jannaschii* has three different circular DNA molecules with lengths of roughly 1.6 Mbp, 58.5 kbp, and 16.5 kbp—showing again the wide and varied personalities of microbial chromosomes. The picture of microbial genomes is further complicated by the fact that our tidy picture of circular Mb-sized chromosomes is woefully incomplete since it ignores the genes that are shuttled around on small (that is, roughly 5-kbp) plasmids.

Ultimately, for most of us, our mental picture of chromosomes is largely based upon images from eukaryotic organisms like those shown in Figure 5-3. As listed in Table 5-1, there is a great variety in the number of pairs of chromosomes in different organisms. We would think that at least the two model fungi, budding yeast and fission yeast, would have similar numbers of chromosomes. Yet surprisingly, the budding yeast *S. cerevisiae* has 16 chromosomes, whereas the fission yeast *S. pombe* has only three chromosomes. Similarly, other classic model organisms do not show any consistent pattern: *C. elegans* has six chromosomes, fruit fly *Drosophila melanogaster* has four chromosomes, and mouse *Mus musculus* has 20 chromosomes. Comparing budding yeast and the fly shows how a ≈ 10-fold larger genome in the case of *Drosophila* can be accommodated with one-quarter as many chromosomes as the 16 found in budding yeast. Among animals, the red vizcacha rat has the largest number of chromosomes at 102 (BNID 110010). These examples demonstrate that the number of chromosomes is not at all dictated by the physical size of the animal. They also overturn the long-held belief that animals cannot be polyploid, because the red vizcacha rat is tetrapolid—that is, it has four copies of each chromosome rather than the two found in humans and other diploids.

As most of us learn in high school, humans have 23 pairs of chromosomes. Given the 3×10^9 base pairs in the human haploid genome, this means that each chromosome harbors on average roughly 130 Mbp of DNA, with the smallest—Chromosome 21—carrying ≈ 50 Mbp and the largest—Chromosome 1—at ≈ 250 Mbp. Some of the most insidious genetic diseases are the result of extra copies of chromosomes. For example, Down syndrome results from an extra copy of Chromosome 21, and there are more of these so-called "trisomies" associated with other chromosomes and leading to other (mainly lethal) syndromes.

One of the stories that elicits the most fascination in all of biology centers on the question of human evolution and its relation to chromosome number. As shown in **Figure 5-4**, humans have 23 pairs of chromosomes, whereas chimps, gorillas, and orangutans have 24 such pairs. The figure compares the structure of Chromosome 2 in humans and of two related chromosomes

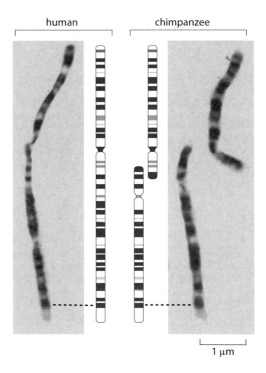

Figure 5-4 Chromosomal banding patterns in late-prophase chromosomes. (Adapted from Yunis JJ & Prakash O [1982] *Science* 215:1525–1530.)

(called 2p and 2q) in one of our closest primate relatives, the chimpanzee. A comparison of the banding patterns in late-prophase chromosomes has been invoked as a key piece of evidence for common chromosomal ancestry (the reader is invited to examine the highly stereotyped chromosomal patterns in the rest of the chromosomes in the original papers). A head-to-head fusion of the 2p and 2q primate chromosomes led to the formation of the human Chromosome 2. This picture was lent much more credence as a result of recent DNA sequencing, which found evidence within human Chromosome 2, such as a defunct centromeric sequence corresponding to the centromere from one of the chimp chromosomes, as well as a vestigial telomere on our Chromosome 2. This story has garnered great interest on the internet, where nonscientists who take issue with both the fact and theory of evolution espouse various refutations and untestable conspiratorial speculations on this fascinating chromosomal history.

Another exciting recent experimental development in the study of genome organization has been the ability to explore the relative spatial organization of different chromosomes. The existence of well-defined chromosome territories has been discovered in both prokaryotes and eukaryotes. **Figure 5-5** shows an example for the nucleus of a human fibroblast cell. Hybridization of fluorescent probes led to the false-color representation of chromosome territories in the mid section of the nucleus. Using more recent tools known as "chromosome capture," even the chromosome territories of the human genome have been mapped out. In these chromosome-capture methods, physical cross-linking of parts of the genome that are near each other are used to build a proximity map. The maps make the chromosomes look like crumpled globules, which would not be the case if they behaved like equilibrated linear polymers, but are rather the result of active structuring taking place inside the nucleus, leading to nuclear and chromosomal territories. Interestingly, disorders in such territories are now suggested to cause diseases, such as the very early aging in progeria due to a mutation in a critical component of the nuclear lamina that leads to displacement of some inactive genes and therefore to their up-regulation.[*] In yeast there is no proof of such structure, and the use of polymer physics ideas on equilibrium polymers appears to be a valid representation. At finer resolution, chromosomes are further subdivided into "domains." That is, parts of one chromosome are to a large extent territorially segregated from each other. This might enable the actual number of chromosomes to change quite a lot without severely affecting genome spatial regulation. Finally, there is heterogeneity in location, where while chromosomes are segregated, the specific "geography" of territories might be different for either different cells, or even for one cell over time.

---

[*]Tai PWL, Zaidi SK, Wu H et al. (2014) The dynamic architectural and epigenetic nuclear landscape: developing the genomic almanac of biology and disease. *J Cell Physiol* 229:711–727.

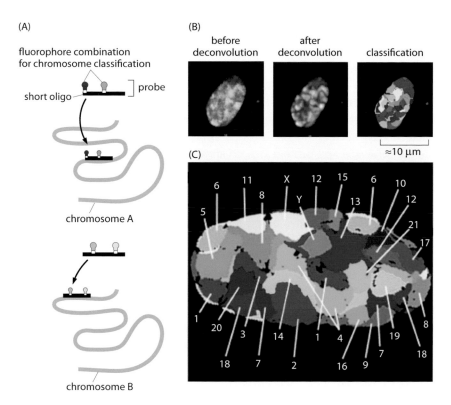

Figure 5-5 Localization of chromosome territories revealed using confocal microscopy. (A) Chromosome probes for all 24 chromosome types (1–22, plus X and Y) were labeled using a combinatorial labeling scheme with seven differentially labeled nucleotides. (B, C) Classification of chromosomes in a human fibroblast nucleus is based on 24-color 3D FISH experiments. (Adapted from Bolzer A, Kreth G, Solovei I et al. [2005] *PLOS Biol* 3:e157.)

Despite the many interesting stories that color this vignette, we are curious to see if new research will associate any deeper functional significance to the chromosome count and spatial organization.

How many genes are in a genome?

We have already examined the great diversity in genome sizes across the living world (see Table 5-1). As a first step in refining our understanding of the information content of these genomes, we need a sense of the number of genes that they harbor. When we refer to genes, we will be thinking of protein-coding genes, not the ever-expanding collection of RNA-coding regions in genomes.

Over the whole tree of life, though genome sizes differ by as much as eight orders of magnitude—from <2 kb for hepatitis D virus (BNID 105570) to >100 Gbp for the marbled lungfish (BNID 100597) and certain *Fritillaria* flowers (BNID 102726)—the range in the number of genes varies by less than five orders of magnitude (from viruses like MS2 and QB bacteriophages, which have only four genes, to about 100,000 in wheat). Many bacteria have several thousand genes. This gene content is proportional to the genome size and protein size, as shown below. Interestingly, eukaryotic genomes, which are often a thousand times or more larger than those in prokaryotes, contain only an order of magnitude more genes than their prokaryotic counterparts. The inability to successfully estimate the number of genes in eukaryotes based on a knowledge of the gene content of prokaryotes was one of the unexpected twists of modern biology.

The simplest estimate of the number of genes in a genome unfolds by assuming that the entirety of the genome codes for proteins. To make further progress with the estimate, we need to have a measure of the number of amino acids in a typical protein, which we will take to be roughly 300, cognizant however of the fact that like genomes, proteins come in a wide variety of sizes themselves, as is revealed in the vignette entitled "How big is the 'average' protein?" (pg. 45). On the basis of this meager assumption, we see that the number of bases needed to code for our typical protein is roughly 1000 (3 bp per amino acid). Hence, within this mindset, the number of genes contained in a genome is estimated to be the genome size/1000. For bacterial genomes, this strategy works surprisingly well, as can be seen in **Table 5-2** and **Figure 5-6**. For example, when applied to the *E. coli* K-12 genome of 4.6×10^6 bp, this rule of thumb leads to an estimate of 4600 genes, which can be compared to the current best knowledge of this quantity, which is 4225. In going through a dozen representative bacteria and archeal genomes in the table, a similarly striking predictive power to within about 10% is observed. On the other hand, this strategy fails spectacularly when we apply it to eukaryotic genomes, resulting for example in the estimate that the number of genes in the human genome should be 3,000,000—a gross overestimate. The unreliability of this estimate helps explain the existence of the Genesweep betting pool, which as recently as the early 2000s had people betting on the number of genes in the human genome, with people's estimates varying by more than a factor of 10.

What explains this spectacular failure of the most naïve estimate, and what does it teach us about the information organized in genomes? Eukaryotic genomes, especially those associated with multicellular organisms, are characterized by a host of intriguing features that disrupt the simple coding picture exploited in the naïve estimate. These differences in genome usage are depicted pictorially in **Figure 5-7**, which shows the percentage of the genome used for purposes other than protein coding. As evident in Figure 5-6,

organism	# of protein-coding genes	# of genes naïve estimate: (genome size /1000)	BNID
viruses			
HIV 1	9	10	105769
Influenza A virus	10–11	14	105767
Bacteriophage λ	66	49	105770
Epstein-Barr virus	80	170	103246
prokaryotes			
Buchnera sp.	610	640	105757
T. maritima	1900	1900	105766
S. aureus	2700	2900	105500
V. cholerae	3900	4000	105760
B. subtilis	4400	4200	111448
E. coli	4300	4600	105443
eukaryotes			
S. cerevisiae	6600	12,000	105444
C. elegans	20,000	100,000	101364
A. thaliana	27,000	140,000	111380
D. melanogaster	14,000	140,000	111379
F. rubripes	19,000	400,000	111375
Z. mays	33,000	2,300,000	110565
M. musculus	20,000	2,800,000	100308
H. sapiens	21,000	3,200,000	100399, 111378
T. aestivum (hexaploid)	95,000	16,800,000	105448, 102713

Table 5-2 A comparison between the number of genes in an organism and a naïve estimate based on the genome size divided by a constant factor of 1000 bp/gene—that is, predicted number of genes = genome size/1000. This crude rule of thumb works surprisingly well for many bacteria and archaea, but fails miserably for multicellular organisms.

prokaryotes can efficiently compact their protein-coding sequences such that they are almost continuous and result in less than 10% of their genomes being assigned to noncoding DNA (12% in *E. coli*; BNID 105750), whereas in humans over 98% (BNID 103748) is non-protein-coding.

The discovery of these other uses of the genome constitute some of the most important insights into DNA, and biology more generally, from the last 60 years. One of these

Figure 5-6 Number of genes as a function of genome size. The figure shows data for a variety of bacteria and archaea, with the slope of the data line confirming the simple rule of thumb relating genome size and gene number. (Adapted from Lynch M [2007] The Origins of Genome Architecture. Sinauer Associates.)

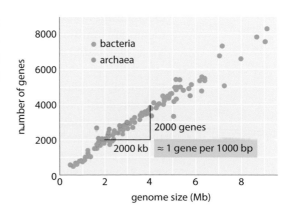

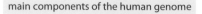

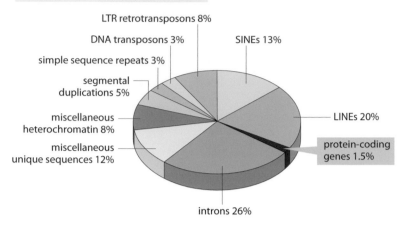

Figure 5-7 The different sequence components making up the human genome. About 1.5% of the genome consists of the ≈ 20,000 protein-coding sequences, which are interspersed by the noncoding introns, making up about 26%. Transposable elements are the largest fraction (40–50%), including, for example, long interspersed nuclear elements (LINEs) and short interspersed nuclear elements (SINEs). Most transposable elements are genomic remnants, which are currently defunct. (BNID 110283) (Adapted from Gregory TR [2005] *Nat Rev Genet* 9:699-708.)

alternative uses for genomic real estate is the regulatory genome—namely, the way in which large chunks of the genome are used as targets for the binding of regulatory proteins that give rise to the combinatorial control so typical of genomes in multicellular organisms. Another of the key features of eukaryotic genomes is the organization of their genes into introns and exons, with the expressed exons being much smaller than the intervening and spliced-out introns. Beyond these features, there are endogenous retro-viruses—fossil relics of former viral infections and, strikingly, over 50% of the genome is taken up by the existence of repeating elements and transposons, various forms of which can perhaps be interpreted as selfish genes that have mechanisms to proliferate in a host genome. Some of these repeating elements and transposons are still active today, whereas others have remained a relic after losing the ability to further proliferate in the genome.

In conclusion, genomes can be partitioned into two main classes: compact and expansive. The compact genomes are gene dense, with only about 10% of noncoding region and strict proportionality between genome size and genome number. This group extends to genomes of sizes up to about 10 Mbp, covering viruses, bacteria, archaea, and some unicellular eukary-otes. The expansive genomes show no clear correlation between genome size and gene number; they are composed mostly of noncoding elements; and they cover all multicellular organisms.

How genetically similar are two random people?

Understanding the similarities and differences among people occupies psychologists, anthropologists, artists, doctors, and many biologists. Even when zooming in on only the genetic differences among people, there is a dazzling range of issues to discuss. The day that DNA extracted at a crime scene can lead to a mugshot portrait seems to have already arrived, at least according to a recent publication on modeling 3D facial shape from DNA.[†] In the spirit of cell biology by the numbers, can we get some basic intuition from logically analyzing the implications of a few key numbers that pertain to the question of genetic diversity in humans?

We begin by focusing on single-base-pair differences, or single-nucleotide polymorphisms (SNPs). Other components of variation, like insertions and deletions, varying numbers of gene repeats (part of what are known as copy number variations, or CNVs), and transposable elements, will be touched upon below. How many single-base-pair variations would you expect between yourself and a randomly selected person from a street corner? Sequencing efforts such as the 1000 Genomes Project give us a rule of thumb. They find about one SNP per 1000 bases. That is, other components set aside, the basis for the claim that people are 99.9% genetically similar. But this genetic similarity begs the question: How come we feel so different from that person we run into on the street? Well, keep on reading to learn of other genetic differences, but we should also appreciate how our brains are tuned to notice and amplify differences and dispense the unifying properties, such as all of us having two hands, one nose, a big brain, and so forth. To an alien we probably would all look identical, just like you may see two mice, and if their fur coats were the same, they would seem like clones, even if one were the Richard Feynman of his clan and the other the Winston Churchill.

Back to the numbers. Let's check on the accuracy and implications of the rule of thumb of one SNP per 1000 bases. The human genome is about 3 Gbp long. This suggests about 3 million SNPs among two random people. This is indeed the reported value to within 10%, which is no surprise because this is the origin of the rule of thumb (BNID 110117). What else can we say about this number? With about 20,000 genes, each having a coding sequence (exons) about 1.5 kb long (that is, an average of about 500 amino acids per protein), the human coding sequence covers 30 Mbp, or about 1% of the genome. If SNPs were randomly distributed along the genome, that would suggest about 30,000 SNPs across the genome coding sequence, or just over one per gene-coding sequence. The measured value is about 20,000 SNPs,

---

[†] Claes P, Liberton DK, Daniels K et al. (2014) Modeling 3D facial shape from DNA. *PLOS Genetics* 10:e1004224.

which gives a sense of how accurate we were in our assumption that the SNPs were distributed randomly. Statistically speaking, we are wrong as any statistical test would give an impressively low probability for this difference to appear by chance. The difference is probably an indication of stronger purifying selection on coding regions. At the same time, from a practical perspective, this less-than-twofold variation suggests that this bias is not very strong and that the one SNP per gene is a reasonable rule of thumb.

How does this distribution of SNPs translate into changes in amino acid in proteins? Let's again assume a homogeneous distribution among amino-acid-changing mutations (non-synonymous) and those that do not affect the amino acid identity (synonymous). From the genetic code, the number of non-synonymous changes, when there is no selection or bias of any sort, should be about four times that of synonymous mutations (that is, synonymous mutations are about 20% of the possible mutations; BNID 111167). That is because there are more base substitutions that change an amino acid than ones that keep the amino acid identity the same. What do we find in reality? About 10,000 mutations of each type are actually found (BNID 110117), showing that indeed there is a bias towards under-representation of non-synonymous mutations, but in our order-of-magnitude worldview, it is not a major one.

One type of mutation that can be especially important, though, is the nonsense mutation that creates a stop codon that will terminate translation early. How often might we naïvely expect to find such mutations given the overall load of SNPs? Three of the 64 codons are stop codons, so we would crudely expect

$$20{,}000 \times (3/64) \approx 1000 \text{ early stop mutations.} \qquad (5.1)$$

Observations show about 100 such nonsense mutations, indicating a strong selective bias against such mutations. Still, we find it interesting to look at the person next to us and wonder what 100 proteins in our genomes are differentially truncated. Thanks to the diploid nature of our genomes, there is usually another fully intact copy of the gene (the situation is known as heterozygosity) that can serve as backup.

How different is your genotype from that of each of your parents? Assuming they have unrelated genotypes, the values above should be cut in half, because you share half of your father's and mother's genomes. So, you still have quite a few truncated genes and substituted amino acids. The situation with your brother or sister is quantitatively similar because you again share, on average, half of your genomes (assuming you are not identical twins). Actually, for about one-fourth of your genome, you and your sibling are like identical twins—that is, you have the same two parental copies of the DNA.

Insertions and deletions (nicknamed indels) of up to about 100 bases are harder to enumerate, but an order of magnitude of 1 million per genome is observed, about 3000 of them in coding regions (so an underrepresentation of about half an order of magnitude). Larger variations of longer stretches, including copy-number variations, are in the tens of thousands per genome, but because they are such long stretches, their summed length might be longer than the number of bases in SNPs.

The ability to comprehensively characterize these variations is a very recent scientific achievement, starting only in the third millennium with the memorable race between the Human Genome Project consortium and the group led by Craig Venter. In comparing the results between these two teams, we find that in comparing the genome of Craig Venter to that of the consensus human genome reference sequence, there is about 1.2% difference when indels and CNVs are considered, 0.1% when SNPs are considered, and ≈0.3% when inversions are considered—a grand total of 1.6% (BNID 110248). In the decade that followed the sequencing of the human genome, technologies were moving forward extremely rapidly, leading to the 1000 Genomes Project, which might seem like a rotation project to some of our readers by the time they read these words. Who knows how soon the reader will be able to actually check on our quoted numbers by loading his or her genome from his or her medical report and compare it to that of some random friend?

MUTATIONS AND ERRORS

What is the mutation rate during genome replication?

Mutation is a highly acclaimed chisel with which evolution sculpts organisms. Together with recombination, duplication events (of genes, chromosomes, and whole genomes), and lateral gene transfer, mutations are a source of the generation-to-generation variability that is one of the central ingredients of the evolutionary process as articulated by Charles Darwin and Alfred R. Wallace. As is often the case in biology, the qualitative discovery of the existence of a process, such as mutations during DNA replication, and even the exploration of its implications is quite different from the ability to precisely quantify that process. To quantify the average rates of mutation, what we want is measurements of the number of mutations per base pair for each replication event. What are typical rates for such genomic alterations, and how are they measured?

The genomic era has ushered in the ability to read out mutation rates directly. It replaced older methods of inference that were based on indirect evolutionary comparisons or studies of mutations that were visually remarkable, such as those resulting in color changes of an organism or changes in pathogenic outcomes. A landmark effort at chasing down mutations in bacteria is a long-term experiment in evolution that has been running for more than two decades in the group of Richard Lenski. In this case, it is possible to query the genome directly through sequencing at different time points in the evolutionary process and to examine both where these mutations occur, as shown in **Figure 5-8**, as well as how they

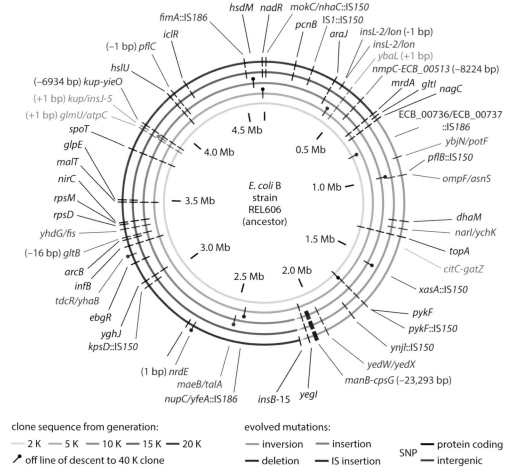

Figure 5-8 Sequencing measurements of fixed mutations over 20,000 generations in E. coli. Because of this long-term experiment, it is possible to compare the full genome sequence at different times to the reference sequence for the genome at the time the experiment started. The labels in the outer ring show the specific mutations that were present after 20,000 generations. (Adapted from Barrick JE, Yu DS, Yoon SH et al. [2009] *Nature* 461:1243–1247.)

accumulate with time, as shown in **Figure 5-9**. The sequencing of 19 whole genomes detected 25 synonymous mutations (indicating neutral rather than selective changes) that got fixed in the 40,000 generations of the experiment. This measurement enabled the inference that the mutation rate is about 10^{-10} mutations per bp per replication in the measured conditions (BNID 111229).

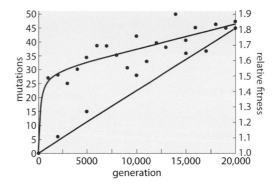

Figure 5-9 Mutation accumulation and fitness over time. Sequencing measurements make it possible to examine the rate of mutation accumulation and the corresponding fitness over time. (Adapted from Barrick JE, Yu DS, Yoon SH et al. [2009] *Nature* 461:1243–1247.)

What are the implications of an *E. coli* mutation rate on the order of 10^{-10} mutations/bp/replication? Given a genome size of 5×10^6, this mutation rate leads to about one mutation per 1000 generations anywhere throughout the genome. At the same time, because an overnight culture test tube often contains over 10^9 bacterial cells per mL, we find that every possible single-base-pair mutation is present, as worked out in **Estimate 5-1**. Mutation rates vary with the environmental conditions and become higher under stressful conditions, such as those prevailing in stationary phase. A collection of mutation rates in a range of organisms is provided in **Table 5-3**.

In humans, a mutation rate of about 10^{-8} mutations/bp/generation (BNID 105813) was inferred from projects where both parents and their children were sequenced at high coverage. Note that the value of the mutation rate is on a per-generation basis and is thus the accumulation in the gametes of mutations occurring over several tens of genome replications between the fertilization of the egg all the way until the formation

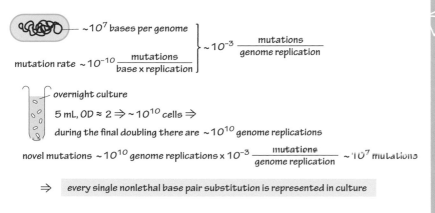

The ease of achieving any specific simple base-pair mutation

$\sim 10^7$ bases per genome

mutation rate $\sim 10^{-10} \dfrac{\text{mutations}}{\text{base} \times \text{replication}}$

$\Big\} \sim 10^{-3} \dfrac{\text{mutations}}{\text{genome replication}}$

overnight culture

5 mL, OD $\approx 2 \Rightarrow \sim 10^{10}$ cells $\Rightarrow$

during the final doubling there are $\sim 10^{10}$ genome replications

novel mutations $\sim 10^{10}$ genome replications $\times 10^{-3} \dfrac{\text{mutations}}{\text{genome replication}} \sim 10^7$ mutations

$\Rightarrow$ every single nonlethal base pair substitution is represented in culture

Estimate 5-1

organism	mutations/ base pair/ replication	mutations/ base pair/ generation	mutations/ genome/ replication	BNID
multicellular				
human *H. sapiens*	10^{-10}	1–4×10^{-8} (mitochondria: 3×10^{-5})	0.2–1	105813, 100417, 105095, 108040, 109959, 105813, 110292, 111227, 111228
mouse *M. musculus*	2×10^{-10}	10^{-8}	0.5	100315, 106792, 100320
D. melanogaster	3×10^{-10}	10^{-8}	0.06	100365, 106793, 100370
C. elegans	10^{-10}–10^{-9}	10^{-8}	0.02–0.2	100290, 100287, 109959, 103520, 107886
unicellular				
bread mold *N. crassa*		10^{-10}	0.003	100355, 100359, 106747
budding yeast		10^{-10}–10^{-9}	0.003	100458, 100457, 109959, 110018
E. coli		10^{-10}–10^{-9}	0.0005–0.005	106748, 100269, 100263
DNA viruses				
bacteriophage T2 & T4		2×10^{-8}	0.004	103918, 103918
bacteriophage lambda		10^{-7}	0.004	100222, 105770, 100220
bacteriophage M13		10^{-6}	0.005	106788
RNA viruses				
bacteriophage Qβ		10^{-3}	7	106762
poliovirus		10^{-4}	1	106760
vesicular stomatitis virus		3×10^{-4}	4	106760
influenza A		10^{-5}	1	106760
RNA retroviruses				
spleen necrosis virus		2×10^{-5}	0.2	106762
moloney murine leukemia virus		4×10^{-6}	0.03	106760
rous sarcoma virus		5×10^{-5}	0.4	106762

Table 5-3 Mutation rates of different organisms from different domains of life. RNA virus mutation rates are especially high, partially due to not having a proofreading mechanism. For multicellular organisms, a distinction is made between mutations per replication versus mutations per generation, which includes many replications from gamete to gamete—see vignette entitled "How quickly do different cells in the body replace themselves?" (pg. 278). To arrive at the mutation rate per genome, the rates per base pair are multiplied by the genome length. Mutation rates in the mitochondrial genome are usually an order of magnitude higher (BNID 109959).

of the next generation of gametes. The characteristic number of such replications is discussed in the vignette entitled "How many chromosome replications occur per generation?" (pg. 319). In humans it is estimated that there are about 20–30 genome replications between the fertilized egg and the female gametes (BNID 105585) and about 10 times that for males, with large variation depending on age (BNID 105574). With $\approx 3 \times 10^9$ bp in the human genome, the mutation rate leads to about

$$10^{-8} \text{ mutations/bp/generation} \times 3 \times 10^9 \text{ bp/genome}$$
$$\approx 10\text{–}100 \text{ mutations per genome per generation} \quad (5.2)$$

(BNID 110293). Using an order of magnitude of 100 replications per generation, we arrive at 0.1–1 mutations per genome per replication. Though we discuss mutations on a per-replication or per-generation rate, nondividing cells will also have damage caused to their genomes through mechanisms such as radiation and reactive oxygen species. When the damage is corrected, mutations accumulate with time at rates that are still not well constrained experimentally. Yet it is clear that with the aid of the sequencing revolution, we will soon know much more.

The numbers for humans can be compared to the mutation rates in the model plant *Arabidopsis thaliana*, where a similar study was undertaken. Five plants derived from 30 generations of single-seed descent were sequenced and compared. The full complement of observed mutations is shown in **Figure 5-10A**. The spontaneous single-base-pair mutation rate was found to be roughly 7×10^{-9} per bp per generation. Given that there are an estimated 30 replications per generation [see the vignette entitled "How many chromosome replications occur per generation?" (pg. 319)], this leads to about 2×10^{-10} mutations per bp per replication. Note that many different classes of point mutations can be categorized as a result of such sequencing experiments, giving a picture of whether the mutations are synonymous or non-synonymous, and whether the mutation event is a transition or a transversion. Different mutations are not evenly distributed, as we show in **Figure 5-10B**. They are dominated by a G-C base pair being transformed into an A-T base pair. This arises due to the biochemical susceptibility of the nucleotides to being mutated. Other common types of mutation in the genome are insertion and deletion events, defined two pages earlier. With the same approach as that outlined above, the rates of 1–3 bp insertions and deletions were estimated to be an order of magnitude less abundant than single-base-pair substitutions at 0.6×10^{-9} and 0.3×10^{-9} per bp per generation, respectively. Deletions larger than 3 bp occur at a frequency of $0.5 \pm 0.2 \times 10^{-9}$ per site per generation, and remove on average 800 ± 1900 bp per event (BNID 110372). Note that the distribution is so wide that the standard deviation is larger than the mean (this can occur due to many small deletions and some very large deletions). Beyond these often-discussed forms of genome alteration through mutation, genomes show surprising dynamism, as revealed by other forms of genome rearrangement, such as the "jumping genes" discovered by Barbara McClintock, many of which still defy even rudimentary quantification.

Given the existence of these various mechanisms of genome rearrangement, it is interesting to consider the extent to which the space of possible genomic mutations is explored. A recurring class of estimates in various contexts, such as the famed Levinthal paradox, center on how well biological systems "explore" the space of all possible outcomes. In many of these examples (protein folding, space of possible genomes, etc.), the

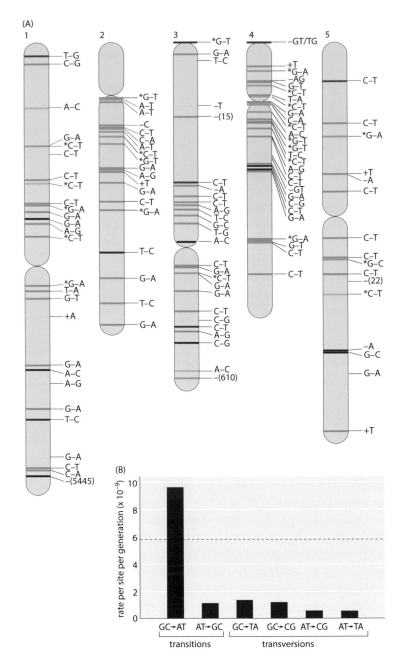

Figure 5-10 (A) Spontaneous mutations across the *A. thaliana* chromosomes after 30 generations (single seed dependents). Color definitions: pink, intergenic region; yellow, intron; dark blue, non-synonymous substitution, shift of reading frame for short indels, or gene deletion for large deletions; green, synonymous substitution; purple, UTR; and light blue, transposable element. + and – refer to insertions and deletions, respectively. Asterisk denotes methylated cytosine. (B) The rate of mutations varies across different base pairs. Mutation rates are shown per site per generation. The overall mutation rate, which is the average of the total mutation rates at A:T and G:C sites, and its standard error in gray are shown in the background. The total mutation rate sums, for example, for the base pair A:T the rates of change to C:G, G:C, and T:A. (Adapted from Ossowski S, Schneeberger K, Lucas-Lledó JI et al. [2010] *Science* 327:92–94.)

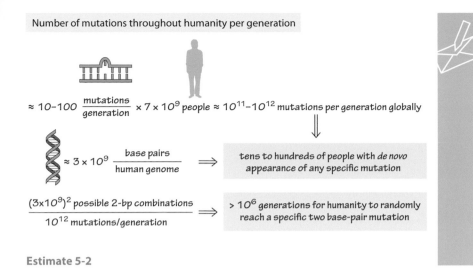

Number of mutations throughout humanity per generation

$\approx 10\text{-}100 \ \dfrac{\text{mutations}}{\text{generation}} \times 7 \times 10^9 \ \text{people} \approx 10^{11}\text{-}10^{12}$ mutations per generation globally

$\approx 3 \times 10^9 \ \dfrac{\text{base pairs}}{\text{human genome}} \implies$ tens to hundreds of people with *de novo* appearance of any specific mutation

$\dfrac{(3\times 10^9)^2 \ \text{possible 2-bp combinations}}{10^{12} \ \text{mutations/generation}} \implies > 10^6$ generations for humanity to randomly reach a specific two base-pair mutation

Estimate 5-2

astronomical numbers of possible outcomes are simply staggering. As a result, it is easy to wonder how thoroughly the space of possible mutations is "searched" within the human population. We explore how such an estimate might go in **Estimate 5-2**. Given that there are about 7 billion people on Earth, with on the order of ≈ 10 mutations per generation, we estimate that the current human occupants of the planet explore roughly $7 \times 10^9 \times 10 \approx 10^{11}$ new mutations during the turnover from one generation to the next. This means that if we focus our attention on any single site within the 3 billion base-pair human genome, dozens of humans harbor a mutation in that particular site. As a result, the space of single-base-pair mutations is fully explored amongst the entire population of humans on Earth. On the other hand, if we consider a specific two-base-pair mutation, we find that by random mutation it would require on the order of 10^7 generations of the human population to achieve it by chance!

What is the error rate in transcription and translation?

The central dogma recognizes the flow of genomic information from the DNA into functional proteins via the act of transcription, which results in the synthesis of messenger RNA, and the subsequent process of translation of that RNA into the string of amino acids that make up a protein. This chain of events is presented in textbooks as a steady and deterministic process, but is, in fact, full of glitches in the form of errors in both

organism	errors per base or codon	BNID and measurement methods
transcription		
E. coli	10^{-4}	111146, transition mutations based on sequencing at very high (10^6) coverage (2013)
E. coli	10^{-5}	105212, *in vitro* selection for rifampicin resistance and increased leakiness of an early, strongly polar nonsense mutation of lacZ (1983, 1986)
E. coli	10^{-4}	103453, activity in strains carrying lacZ mutations (1981)
S. cerevisiae	2×10^{-6}	110019, RNA pol II, determined *in vitro* (2008)
S. cerevisiae	2×10^{-4}	105213, RNA pol III, determined based on selectivity (2007)
C. elegans	4×10^{-6}	111144, determined using bar coded sequencing (2013)
translation		
E. coli	3×10^{-4}	105069, Lys-tRNA, reporter system for frequency of each type of misreading error (2007)
E. coli	$1–4 \times 10^{-3}$	105215, identify cases that do not contain the amino acid cysteine responsible for the missense substitution (1983)
E. coli	$10^{-4}–10^{-3}$	103454, identify cases that do not contain the amino acid cysteine responsible for the missense substitution (1977, 1983)
B. subtilis	4×10^{-3}	105466, GFP with nonsense mutation, also find 2.4% for frameshift (!) (2010)
S. cerevisiae	$0.5–2 \times 10^{-5}$	105216, measurement of rescue rate of inactivating mutations of type III chloramphenicol acetyl transferase (1998)

Table 5-4 Error rates in transcription and translation. For transcription, the error rates are given per base, whereas for translation, the error rates are per codon (that is, amino acid).

the incorporation of nucleotides into RNA and the incorporation of amino acids into proteins. In this vignette we ask, "How common are these mistakes?"

One approach to measuring the error rate in transcription is to use an *E. coli* mutant carrying a nonsense mutation in lacZ (that is, one that puts a premature stop codon conferring a loss of function) and then assay for the activity of this protein, which enables utilization of the sugar lactose. The idea of the experiment is that *functional* LacZ will be produced through rare cases of erroneous transcription resulting from a misincorporation event that bypasses the mutation. The sensitivity of the assay makes it possible to measure this residual activity due to "incorrect" transcripts, giving an indication of an error rate in transcription of $\approx 10^{-4}$ per base (BNID 103453; **Table 5-4**), which in this well-orchestrated experiment changed the spurious stop codon to a codon responsible instead for some other amino acid, thus resurrecting the functional protein. Later measurements suggested a value an order of magnitude better of 10^{-5} (BNID 105212[†]). Jacques Ninio's analysis of these error rates led to the hypothesis of an error correction mechanism called kinetic proofreading, paralleling a similar analysis performed by John Hopfield for protein synthesis. Recently, GFP was incorporated into the genome in the wrong reading frame, thus enabling the study of error rates for those processes resulting in frameshifts in the bacterium

[†] Ninio J (1991) Connections between translation, transcription and replication error-rates. *Biochimie* 73:1517–1523.

B. subtilis. A high error rate of about 2% was observed (BNID 105465), which could arise at either the transcriptional or translational levels, because both could bypass the inserted mutation. The combined error rate for the frameshift is much higher than the estimated values for substitution mutations, thus indicating that the prevalence and implications of errors are still far from completely understood. Like with many of the measurements described throughout our book, often the extremely clever initial measurements of key parameters have been superseded in recent years by the advent of sequencing-based methods. The study of transcription error rates is no exception, with recent RNA-Seq experiments making it possible to simply read out the transcriptional errors directly, though these measurements are fraught with challenges since sequencing error rates are comparable to the transcriptional error rates (10^{-4}–10^{-5}) that are being measured.

The error rate of RNA polymerase III, the enzyme that carries out transcription of tRNA in yeast, has also been measured. The authors were able to tease apart the contribution to transcriptional fidelity arising from several different steps in the process. First, there is the initial selectivity itself. This is followed by a second error-correcting step that involves proofreading. The total error rate was estimated to be 10^{-7}, which should be viewed as a product of two error rates, $\approx 10^{-4}$ arising from initial selectivity and an extra factor of $\approx 10^{-3}$ arising from proofreading (BNID 105213, 105214). Perhaps the best way to develop intuition for these error rates is through an analogy. An error rate of 10^{-4} corresponds to the authors of this book making one typo every several vignettes. An error rate of 10^{-7} corresponds more impressively to one error in a 1000-page textbook (an impossibility for most book authors).

Error rates in translation (10^{-4}–10^{-3}) are generally thought to be about an order of magnitude higher than those in transcription (10^{-5}–10^{-4}), as roughly observed in Table 5-4. For a characteristic 1000 bp/300 aa gene, this suggests on the order of one error per 30 transcripts synthesized and one error per 10 proteins formed. Like with measurements of errors in transcription, one of the ways that researchers have gone about determining translational error rates is by looking for the incorporation of amino acids that are known to not be present in the wild-type protein. For example, a number of proteins are known to have no cysteine residues. The experiment then consists of using radioactive isotopes of sulfur present in cysteine and measuring the resulting radioactivity of the newly synthesized proteins. Rates in *E. coli* using this methodology yield mistranslation rates of 1–4×10^{-3} per residue (BNID 105215). One interpretation of the evolutionary underpinnings of the lower error rate in transcription relative to translation is that an error in transcription would lead to many erroneous protein copies, whereas an error in translation affects only one protein copy. Moreover, the correspondence of three nucleotides to one amino acid means that mRNA messages require higher fidelity per "letter" to achieve the same overall error rate. Note also that in addition to the mistranslation of mRNAs, the protein synthesis process can also be contaminated by the incorrect charging of the tRNAs themselves, though the incorporation of the wrong amino acid on a given tRNA has been measured to occur with error rates of 10^{-6}.

A standing challenge is to elucidate what limits the possibility to decrease the error rates in these crucial processes in the central dogma even further, say, to values similar to those achieved by DNA polymerase. Is there a biophysical tradeoff in play, or maybe the observed error rates have some selective advantages?

What is the rate of recombination?

In his autobiography, Darwin mused with regret at his failure to learn more mathematics, observing that those with an understanding of the "great leading principles of mathematics . . . seem to have an extra sense." This extra sense is beautifully exemplified in a subject that was near to Darwin's heart—namely, the origins of heredity, the study of which gave rise to modern genetics. Gregor Mendel was intrigued by the same question that has perplexed naturalists as well as parents for countless generations— namely, what are the rules governing the similarities and differences of parents and their offspring? His approach required the painstaking and meticulous act of counting frequencies of various traits such as pea shape from carefully constructed plant crosses, where he found that out of a total of 7324 garden peas, 5474 of them were round and 1850 were wrinkled. The subsequent analysis of the data showed for this case a ratio of these traits in the second generation of crosses of 2.96 to 1, thereby providing a critical clue that permitted Mendel to posit the existence of the abstract particles of inheritance we now call genes.

To cause a sea change in biological research required going beyond phenomenological observations to a situation where genetic manipulations could be more easily performed and more detailed predictions made. This came about when Thomas Hunt Morgan, head of a lab already overflowing with studies of pigeons and starfish, undertook with his students an object of study with minimal space requirements and faster generation times. So came to the scene one of the great protagonists of modern genetics, the fruit fly *Drosophila melanogaster*. As Morgan's lab transformed into what became known as the "fly room" (first at Columbia University, then at

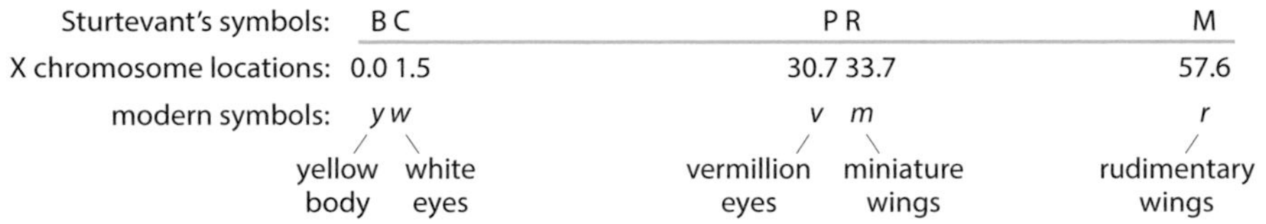

Sturtevant's symbols:	B C		P R		M
X chromosome locations:	0.0 1.5		30.7 33.7		57.6
modern symbols:	y w		v m		r

	yellow	white	vermillion	miniature	rudimentary
	body	eyes	eyes	wings	wings

Figure 5-11 Schematic of the first genetic map of the X chromosome of *Drosophila* redrawn with modern symbols. Sturtevant's map included five genes on the X chromosome of *Drosophila*. (Adapted from Pierce BA [2006] Genetics: A Conceptual Approach, 2nd ed. W. H. Freeman.)

Caltech), it harbored flies with several distinct morphological properties akin to Mendel's mottled and different-colored peas. Systematic crosses of these mutant flies showed deviations from the predictions of Mendelian genetics on the relative fractions of different progeny. An inquisitive Columbia University undergraduate student in Morgan's lab decided to analyze the frequencies of linkage—that is, of pairs of co-inherited traits. During a long night that was supposed to be devoted to homework for his undergraduate studies, the young Alfred Sturtevant instead made a conceptual leap that was to become textbook material and a cherished story from the history of science. He found that the tendency of the traits they studied to be inherited together, such as white eyes instead of red eyes or a more yellow body color could be quantitatively explained if we assume that the genes for these traits are ordered along a line (chromosome), and the tendency not to be inherited together is then reasonably predicted as increasing linearly with their distance. Using this logic, that night Sturtevant created the first genetic map, which we have reproduced in **Figure 5-11**.

The mechanism explaining the frequency with which characteristics are inherited together is that of recombination. This is an act of two chromosomes of similar composition coming together and performing a molecular crossover, thereby exchanging genetic content. Two genes on the chromosome that have a 1% chance of crossover per generation are defined to be at a distance of one centimorgan, or cM. In humans, the average rate of recombination is about 1cM per 1Mbp (BNID 107023)—that is, for every million base pairs, there is a 1 in 100 chance of crossover on average per generation. The variation in the rate of recombination is shown in **Table 5-5**. It tends to scale inversely with genomic length. This interesting scaling property can be simply understood by noting that in most species there are one to two crossover events per chromosome per replication. This results in an organism-wide rule of thumb of one recombination event per chromosome, as demonstrated in the right-most column of Table 5-5, or equivalently as 100 cM (that is, one morgan or one crossover) per chromosome per replication. Beyond general rules of thumb, we now also know that some locations along chromosomes are hotspots that are more labile

species	genome size (Mb)	chromosome number (*n*)	genetic map length (cM)	recombination rate (cM/Mb)	recombination events per chromosome
dog	2500	39	3900	1.6	1.0
human	3000	23	3600	1.2	1.6
sheep	3000	27	3600	1.2	1.3
cat	3000	19	3300	1.1	1.7
cow	3000	30	3200	1.1	1.1
horse	2700	32	2800	1.0	0.9
pig	3000	19	2300	0.8	1.2
macaque	3100	21	2300	0.7	1.1
baboon	3100	21	2000	0.6	1.0
rat	2800	21	1500	0.6	0.7
mouse	2600	20	1400	0.5	0.7
wallaby	3700	8	830	0.2	1.0
opossum	3500	11	640	0.2	0.6

Table 5-5 Recombination rates in various mammals and marsupials of similar genome sizes. Genetic map length is the sum of genetic map lengths summing in units of cM over all chromosomes in each genome. The right-most column, recombination events per chromosome, is calculated by dividing the genetic map length (cM/100) by the number of chromosomes. Note how this genetic map length per chromosome is close to one over the range of organisms. (BNID 107023) (Adapted from Dumont BL & Payseur BA [2008] *Evolution* 62:276–294; chromosome numbers are from http://www.genomesize.com.)

for crossovers. Finally, human females have ≈50% higher recombination rates than males (42 vs. 28 on average in one recent study; BNID 109268). So, even though you tend to get more of your single-base mutations from your father, as discussed in the vignette entitled "How many chromosome replications occur per generation?" (pg. 319), your crossovers are mostly thanks to your mother.

Recent breakthroughs in genotyping have made it possible to perform a single-cell analysis of recombination activity. SNPs are locations in the human genome where there is variation between people such that, say, more than 1% of the population has a nucleotide different than the majority of the population. For the human population, there are on the order of 10^6 such locations on the genome. Here is how this can be used to infer the number and location of recombination events. The chromosomes of a male are separated in a microfluidic device (arbitrarily marked as left and right for each of the 22 pairs), and then each chromosome is separately analyzed for the variant of nucleotide it carries by a microarray technology. The same process is repeated for a sperm cell, leading to maps such as that shown in **Figure 5-12**. At the locations where it is known that there is polymorphism in the genome, it is checked if the variant in the sperm cells is the one that appears in one chromosome but not the other, and if so, its location is marked as a blue stripe on the relevant chromosome. The events

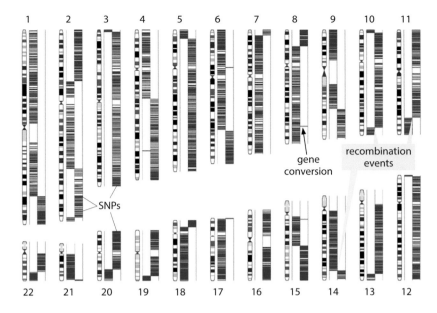

Figure 5-12 Detection of recombination events based on mapping of nucleotide polymorphisms of a single sperm cell. The two columns in each chromosome represent the two homologous chromosomes carried by the subject, which were analyzed separately for their sequence at locations of known polymorphism in the human genome. The source of the sperm single-chromosome copy can be traced to one or the other of the homologous chromosomes based on the single-nucleotide polymorphisms that it carries. In all cases it appears in one chromosome but not the other. Blue lines show the association of the sperm sequence to the two chromosome sets based on those single-nucleotide polymorphisms. Each switch (haplotype block) indicates a recombination event. Not all detected single-nucleotide polymorphisms are shown in the figure. (Adapted from Wang J, Fan HC, Behr B & Quake SR [2012] *Cell* 150:402–412.)

of recombination are clearly seen as switches of those polymorphism locations from one arm to the other. On average, 23 recombination events were found for a human sperm cell (BNID 108035). Short stretches consisting of a single SNP switching chromosome, as highlighted in Chromosome 8, are cases of what has been called gene conversion, where one allele (gene copy) has performed homologous recombination that makes it replace the other copy (its heterozygous allele). Such analysis at the single-cell level, in contrast to inference at the population level or from studying progeny in a family, makes it possible to see the rates of events, such as recombination and mutation in the gametes, including for those gametes that will not lead to viable progeny. This is relevant because the human monthly fecundity rate—that is, the chance of a menstrual cycle leading to pregnancy, is only about 25% (BNID 108080) even at the peak ages of 20–30. Aberrations in the genome content are often detected naturally early in development, within the first few weeks following conception, and lead to natural termination of the pregnancy even before the woman is aware she is pregnant.

Today recombination also serves researchers as a key tool in genetic engineering for creating designer genomes. Homologous recombination enables the incorporation of a DNA sequence at a prescribed location within the genome. Its use has transformed our ability to tag genes of interest and has resulted in genome-wide libraries that enable the high-throughput analysis of key cellular properties ranging from the localization of proteins to different cellular locations to genome-wide assessments of protein levels and variability. Though recombineering, as it is called, is incredibly powerful, unfortunately it can only be used in some organisms and not others, giving those lucky organisms a strong selective pressure in labs around the world as attractive model systems. Outstanding examples are the budding yeast and the moss *physcomitrella patens*. The method of homologous recombination requires a sequence of homology flanking the integrated sequence. The length of this sequence varies depending on the organism, the gene of interest, and the specific technique and protocol employed. Some characteristic values are ≈30–50 bp in budding yeast (BNID 101986), whereas in mouse it is ≈3–5 kbp (BNID 101987). With the longer stretches also comes a much lower efficiency of performing the act of homologous recombination, thus complicating the lives of molecular biologists, though modern CRISPR (clustered, regularly interspaced, short palindromic repeats) techniques have effected a new revolution in genome editing that may largely supersede recombineering methods.

DIVERSITY

Chapter 6: A Quantitative Miscellany

In our introductory chapter, we spoke of giving a friendly alien a single pub-lication to learn about what our society and daily lives look like. There we noted that our favorite suggestion would be some report from the Bureau of Statistics that details everything from income to age at marriage. Over the last five chapters, we have performed a systematic analysis of many of the key questions that we can address as part of the main substance of the bureau of statistics of the cell—namely, how big, how many, how forceful, how fast? Many statistics about our society are obscure, but still interest-ing, such as the number of deaths from falling, an always surprising statistic given the frighteningly high numbers that surpass the number from food poisoning, snake bites, and airplane crashes combined. Similarly, many interesting biological quantities defy simple categorization, and yet they deserve mention in our pantheon of bionumbers. That is the purpose of this final chapter, where we bring together some important numbers that help us understand the world of the cell but that did not fit into the categories heading the other chapters.

We now turn to a quantitative miscellany of topics that runs the gamut from exploring the permeability of membranes to the "burst size" of viruses that tell us how many new viruses will erupt from an infected cell. In each of these cases, we invite the reader to continue with the style of arguments that have been made in vignettes throughout the book, and more importantly, to imagine what other interesting bionumbers would end up on their own personal quantitative miscellany.

To whet the appetite for the current chapter, we thought it would be of interest to our readers to hear something more about the statistics of the searches that are made on the BioNumbers website itself. About 200 researchers every day, from across the globe, find themselves curious about a very wide spectrum of different quantities that characterize the liv-ing world. The most popular queries are independently searched for many hundreds of times each year. Some of those queries fall right within the framework of our main chapters throughout the book, such as how heavy is the tobacco mosaic virus, how rapid is DNA replication in humans, or the microbiologist's favorite, what is the conversion from optical density units to number of cells? But many other search queries do not fit at all into the framework laid out in our various chapter headings. For example, one favorite is what is the average spacing between the origins of replication on human chromosomes? Or, how many cells are in a colony? Finally, the number of hairs on a human head and the duration of the blink of an eye command great interest among internet searchers. For us, the database searches show that the need for knowing the numbers that govern life is widespread and takes many forms. We hope to have given the reader a bit more of an overview of what these numbers are and how knowing them can lead to unexpected insights.

HOW MANY CELLS ARE THERE IN AN ORGANISM?

The fact that all organisms are built of basic units—namely, cells—is one of the great revelations of biology. Even though this knowledge is often now taken as a triviality, it is one of the deepest insights in the history of biology and serves as a unifying principle in a field where diversity is the rule rather than the exception. But how many cells are there in a given organism, and what controls this number and their size? The answers to these questions can vary for different individuals within a species and depend critically on the stage in life. **Table 6-1** attempts to provide a feel for the range of different cell counts, based upon both measurements and simple estimates. This will lead us to approach the classic conundrum: Does a whale vary from a mouse mostly in the number of cells, or is it the sizes of the cells that confer these differences in overall body size?

Perhaps the most intriguing answer to the question of cell counts is given by the case of *C. elegans*, which is remarkable for the fact that every individual has the same cell lineage, resulting in precisely 1031 cells (BNID 100582) from one individual to the next for males and 959 cells (BNID 100581)

organism	stage in life cycle or organ	estimated cell count	BNID
human	adult	$3.7\pm0.8\times10^{13}$	109716
D. melanogaster	embryo cycle 14	6000 (nuclei)	106463
C. elegans	adult male (somatic)	1031	100582
C. elegans	adult hermaphrodite (somatic)	959	100581
C. elegans	hatched larvae	558	101366

Table 6-1 Number of cells in selected organisms. All values, save human, are based on counting using light or electron microscopy.

for hermaphrodites (females also capable of self-fertilization). Specific knowledge of the cell inventory in *C. elegans* makes it possible to count the number of cellular participants in every tissue type and reaches its pinnacle in the mapping of most synaptic connections among cells of the nervous system (including the worm "brain"), where every worm contains exactly 302 neurons. These surprising regularities have made the worm an unexpected leading figure in developmental biology and neuroscience. It is also possible to track down the 131 cells (BNID 101367) that are subject to programmed cell death (apoptosis) during embryonic development. Though not examined to the same level of detail, other organisms besides *C. elegans* have a constant number of cells, and some reveal the same sort of stereotyped development with specific, deterministic lineages of all cells in the organism. Organisms that contain a fixed cell number are said to be eutelic. Examples include many but not all nematodes, as well as tardigrades (that is, water bears) and rotifers. Some of our closest invertebrate relatives, ascidians such as *Ciona*, have an apparently fixed lineage as embryos, but they do not have a fixed number of cells as adults, which arise from metamorphosis of their nearly eutelic larvae. Having a constant number of cells therefore does not seem to have any particular evolutionary origin, but rather seems to be a common characteristic of rapidly developing animals with relatively small cell numbers (on the order of 1000 somatic cells).

In larger organisms, the cellular census is considerably more challenging. One route for making an estimate of the cellular census is to resort to estimates based upon volume, as shown in **Estimate 6-1**. For example, a human with a mass of ≈ 100 kg will have a volume of $\approx 10^{-1}$ m^3. Mammalian cells are usually in the volume range 10^3–10^4 μm^3 = 10^{-15}–10^{-14} m^3, implying that the number of cells is $\approx 10^{13}$–10^{14}, which is the range quoted in the literature (BNID 102390). Though the sizes (linear dimension) of eukaryotic organisms can vary by more than 10 orders of magnitude, the size of their cells measured by the "radius," for example, usually varies

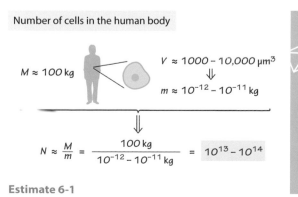

Number of cells in the human body

$M \approx 100$ kg

$V \approx 1000 - 10,000$ μm^3

$\Downarrow$

$m \approx 10^{-12} - 10^{-11}$ kg

$\Downarrow$

$N \approx \dfrac{M}{m} = \dfrac{100\ \text{kg}}{10^{-12} - 10^{-11}\ \text{kg}} = 10^{13} - 10^{14}$

Estimate 6-1

by only a factor of 10 at most, except for intriguing exceptions such as the cells of the nervous system and oocytes. However, the level of accuracy of estimates like those given above on the basis of volume should be viewed with a measure of skepticism, as can be seen by considering your recent blood test results. The normal red blood cell count is 4–6 million such cells per microliter. With about 5 L of blood in an adult, this results in an estimate of 3×10^{13} such cells rushing about in your bloodstream, which is already for this cell type alone as many as the total number of cells in a human body that we estimated using volume arguments. The disagreement with the estimate above results from the fact that red blood cells are much smaller than the characteristic mammalian cell at about $10^2 \ \mu m^3$ in volume. This shows how the above estimate should in fact be increased (and several textbooks revised). A census of the cells in the body was achieved by methodically analyzing different cell types and tissues, thus arriving at a value of $3.7 \pm 0.8 \times 10^{13}$ cells in a human adult (BNID 109716). The breakdown by cell type for the major contributors is shown in **Figure 6-1**. The numerical dominance of red blood cells is visually clear. We do not account for bacterial cells or other residents in our body, and the number of cells composing this so-called microbiota outnumber our human cells by a factor still unknown, but probably closer to 100 than to the often-quoted value of 10.

What is the connection between organism size, cell size, and cell number? Or to add some melodrama, does a whale mostly have larger cells or more cells than a mouse? In studying the large variation in fruit organ

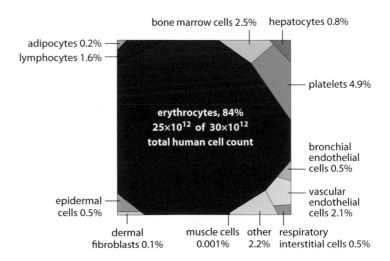

Figure 6-1 Estimate of the number of cells in an adult human partitioned according to cell type. Each cell type in the human body is represented as a polygon with an area proportional to the number of cells. The dominant component is red blood cells. (Based on data from Sender R et al. [2015] in preparation.)

(A)

(B)

Figure 6-2 Plant and organ size changes from domestication, breeding hybridization, and transgenic modification. These variations are found to be mostly driven by change in cell number. Fruit size of wild and domesticated species: (A) wild relative species of pepper, *Capsicum annuum* cv. *Chiltepin* (left) and bell pepper (right); (B) wild relative species of tomato, *Solanum* (left), *Solanum esculentum* cv Giant Red (right). (B, adapted from Guo M & Simmons CR [2011] *Plant Sci* 181:1–7.)

size, as shown in **Figure 6-2**, it was found that the change in the number of cells is the predominant factor driving size variability. In the model plant *Arabidopsis thaliana*, early versus later leaves vary in total leaf area from 30 to 200 mm^2 (BNID 107043). This variation comes about as a result of a concomitant change in cell number from 20,000 to 130,000, with cell area remaining almost constant at 1600 μm^2 (BNID 107044). In contrast, in the green revolution that tripled yields of rice and wheat in the 1970s, a major factor was the introduction of miniature strains, where the smaller size makes it possible for the plant to support bigger grains without falling over. The smaller cultivars were achieved through breeding for less response to the plant hormones gibberellins, which affects stem cell elongation. In this case, a decrease in cell size, not cell number, is the dominant factor—a change in the underlying biology of these plants that helps feed over a billion people.

When the ploidy of the genome is changed, the cells tend to change size accordingly. For example, cells in a tetraploid salamander are twice the size of those in a diploid salamander, although the corresponding organs in the two animals have the same size. Everything fits well because the

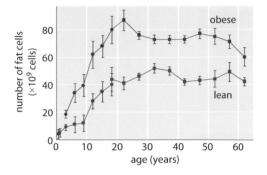

Figure 6-3 Adipocyte number remains stable in adulthood, although significant weight loss can result in a decrease in adipocyte volume. Total adipocyte number from adult individuals (squares) was combined with previous results for children and adolescents (circles). The adipocyte number increases in childhood and adolescence. Lean is defined as having a body mass index <25 and obese is >30. (Adapted from Spalding KL, Arner E, Westermark PO et al. [2008] *Nature* 453:783–787.)

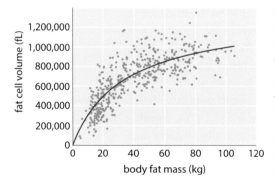

Figure 6-4 Average fat cell size as a function of body fat mass. As the fat content of a person increases, the average adipocyte volume initially increases almost linearly and then saturates. Thus, the change in total fat among humans can be attributed mostly to larger cells of a similar number, and at more extreme disparities, to change in the number of fat cells, too. (Adapted from Spalding KL, Arner E, Westermark PO et al. [2008] *Nature* 453:783–787.)

tetraploid salamander contains half as many cells as the diploid (BNID 111481).

When two people differ in size, is it due to a difference in the number of cells or in the average size of cells? We can begin to answer such questions by appealing to data for lean versus obese humans. Obese adults have on the average almost twice as many fat cells. This difference between lean and obese human adults is usually established at an early age, as shown in **Figure 6-3**. What about the average cell size? **Figure 6-4** shows the variation in the average volume of a fat cell as a function of the body fat mass. At low body fat masses, the close-to-linear increase, passing through the origin, indicates that, in this regime, differences are mostly driven by a change in the volume of the cells—that is, the total number of cells remains relatively constant. At the high body fat range, the cell volume increase is sublinear, indicating that an increase in the number of cells is becoming important. The extra fat weight, which in obese individuals can reach 100 kg, is accompanied by a change in the number of fat cells (as shown in Figure 6-3) of less than 10^{11}, which is much less than 1% of the total number of cells in the body estimated above. Thus, we conclude that between lean and obese people, the main change is a change in fat cell volume rather than total number of cells in the body. This is contrasted by what happens when comparing across organisms of very different sizes, say, between a human and a mouse. Both organisms have cells that are usually of similar size, though a person weighs more than 1000 times more. Thus, in this case, and we claim this is often the case across multicellular organisms that are orders of magnitude apart, the number of cells is the main driver of size differences. With elephants having red blood cells (BNID 109091), as well as other cells, of sizes not unlike ours (BNID 109094), we hypothesize that this is also true for them.

HOW MANY CHROMOSOME REPLICATIONS OCCUR PER GENERATION?

A look in the mirror shows us that we are made of many different types of cells. In terms of genetics and evolution, the most important distinction is between the cells of the germ line, those cells that have the potential to culminate in new offspring, and somatic cells, those cells that build the rest of the body but do not propagate to the next generation. This dichotomy is more relaxed in plants, but in principle is similar. An important factor contributing to the fidelity with which the genetic information is transferred from one generation to the next is the number of cell divisions each germ cell will make on average before the actual fertilization event.

In a previous vignette on "How many cells are there in an organism?" (pg. 314), we made the estimate that a human is made up of 3×10^{13} cells (BNID 109716). But cells are constantly being born and dying. Given this turnover, how many cells does a person make in a lifetime? Though the question touches our very own composition, we could only find one passing mention of about 10^{16} cell divisions in total during a human lifespan (BNID 100379). Our sanity check on this value relies on knowing that red blood cells are the dominant cell type by sheer number in the body (bacteria aside) and that their lifetime is on the order of 100 days. So, in 100 years of life, there will be about 300 cycles of replacement for these red blood cells, and the inferred total number of cell divisions is indeed of order 10^{16} ($\approx 2 \times 10^{13}$ rbc cells/person $\times$ 300 cycles in lifetime). We proceed to analyze two very naïve and extreme models regarding how many replications of the chromosomes are required to obtain the somatic cells that lead to the next generation. In the first simplified model we assume, as depicted in **Estimate 6-2**, that all cells divide in a symmetric manner like a binary bifurcating tree. The number of cells progresses in a geometrical series starting from 1 at the first generation to 2, 4, 8, 16, etc. We will thus have 2^n cells after n replication rounds, and 10^{13} cells will be reached after $\log_2(10^{13}) \approx 40$ replication rounds and 10^{16} cells after ≈ 50 replication rounds.

In our second toy model, we imagine an idealized process, schematically drawn in Estimate 6-2, in which every cell in the body is a direct descendant of some single "stem cell." In this case, the generation of the above-mentioned full complement of a lifetime of cells in the human body would require 10^{16} replication rounds (maybe minus one, to be accurate) for the lifetime repository of cells. Such a cell lineage model would place enormous demands on the fidelity of the replication process, because mutations would accumulate, as discussed in the vignette entitled "What is the mutation rate during genome replication?" (pg. 297).

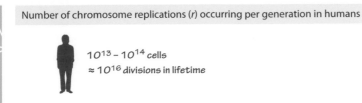

Number of chromosome replications (*r*) occurring per generation in humans

$10^{13} - 10^{14}$ cells

$\approx 10^{16}$ divisions in lifetime

three idealized scenarios:

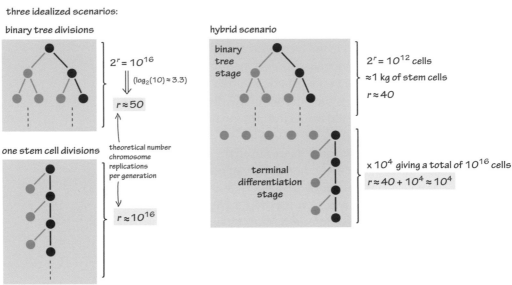

binary tree divisions

$2^r = 10^{16}$

$(\log_2(10) \approx 3.3)$

$r \approx 50$

one stem cell divisions

theoretical number chromosome replications per generation

$r \approx 10^{16}$

hybrid scenario

binary tree stage

$2^r = 10^{12}$ cells

≈ 1 kg of stem cells

$r \approx 40$

terminal differentiation stage

x 10^4 giving a total of 10^{16} cells

$r \approx 40 + 10^4 \approx 10^4$

Estimate 6-2

The cell lineage from egg to adult is closer to the tree of binary divisions described in the first model. Nowhere has this been illustrated more dramatically than in the stunning experiments carried out to map the fates and histories of every cell in the nematode *C. elegans*. In a Herculean effort in the 1970s, Sir John Sulston and coworkers delineated the full tree for *C. elegans*, which is redrawn in **Figure 6-5**, by careful microscope observations of the development of this transparent nematode. We can see that for this remarkable organism with its extremely conserved developmental strategy, the depth of this tree from egg to egg is ≈ 9 replications (BNID 105572).

In larger multicellular organisms, the picture is very complex and can be thought of as a hybrid between the two simplified models, as also noted in Estimate 6-2, starting from a binary tree expansion stage that then turns into a terminal differentiation stage. Such complex structured models were observed in the crypts of the colon and in the apical meristem of plants. In plants, for example, a small set of stem cells, say, about 10 in a plant apical meristem, divide slowly in what is called the quiescent center. These stem cells lead to a larger population of, say,

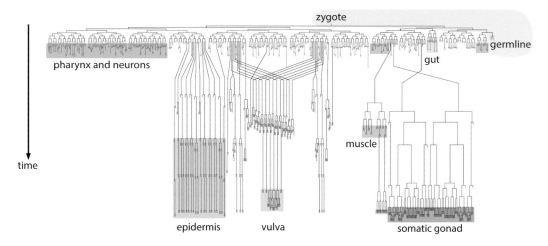

Figure 6-5 *C. elegans* lineage tree as deciphered through light microscopy. The path to the germline, not showing all divisions and germ cells, is highlighted in yellow. (Adapted from the WormAtlas Database.)

100 cells, which divide rapidly to give the majority of cells. The rapidly dividing cells, which accumulate mutations, are slowly replaced before they accumulate too many mutations by the progeny of the slowly dividing stem cells.

Estimates of the average number of replication rounds leading to adult cells in a range of organisms are given in **Table 6-2**. Normal mammalian cells that are not stem cells or cancerous cell lines usually stop dividing after about 40–60 cell cycles (BNID 105586). In humans, eggs are produced much earlier and in more restricted numbers than sperm cells. Indeed, there are much fewer chromosome replications from egg to egg (23; BNID 105585) versus sperm to sperm (from 35 at age 15, to >800 at age 50; BNID 105574). As mutations accumulate with replication rounds, most of the mutations arise in the male lineage. Indeed, the ratio of mutations passed on by older fathers compared to mothers is about 4 to 1 (BNID 110290).

organism	number of chromosome replications leading to:		BNID
	male sperm	female ovum/ovule	
nematode *C. elegans*	8	10	105572
D. melanogaster	34–39	36	103523
Arabidopsis thaliana			106749
mouse	62	25	105576
human	40–1000 [age range 15–50]	23	105574, 105585

Table 6-2 Number of chromosome replications leading to male sperm and to female ovule in different organisms.

Mothers are reported to pass about 15 mutations on average irrespective of age (BNID 110295), while 20-year-old fathers transmit about 25 mutations and 40-year-old fathers transmit about 65 mutations (BNID 11294). That is about two extra mutations per extra year of age of the father (BNID 110291). Some have gone so far as to suggest that the fact that older fathers are chiefly responsible for introducing mutations into the population can provide a potential explanation for hemophilia in the British royal family, whose kings kept on having children at advanced ages.

Should the number of divisions have implications for the occurrence of cancer, which has mutation and replication at its essence? Different types of cancers are known to have very different lifetime risks that span several orders of magnitude. Recently, the number of stem cells and their division rates are becoming available. In a recent study,[*] researchers collected the number of total stem cell divisions in a lifetime for 31 tissue types and correlated it to the lifetime risk of cancer occurring in that tissue. The correlation was found to be striking at about 0.8. This high correlation leaves only a much smaller fraction to be explained by environmental factors or genetic predispositions, though these have been at the center of research for decades. In our perspective, this is a striking example of how paying careful attention to the numbers can still today bring simple insights into view.

HOW MANY RIBOSOMAL RNA GENE COPIES ARE IN THE GENOME?

rRNA is the ribosomal RNA, a major constituent of the ribosome, accounting for about two-thirds of its mass (BNID 100119). In an earlier vignette entitled "How many ribosomes are in a cell?" (pg. 147), we discussed the large number of ribosomes required just to keep the steady pace of protein production moving. As a result of this high demand for protein production, under many growth conditions, one copy of the rRNA gene will not be enough to supply the ribosomal needs for cell growth, even if that locus is being transcribed as fast as possible. Consider the budding yeast, as depicted in **Estimate 6-3**. Under fast exponential growth, it is estimated to harbor about 200,000 ribosomes (BNID 100267). For a cell cycle time of 100 min, these cells need to produce ≈ 30 rRNA per second just to keep up with the demand for new ribosomes. Can the cell

---

[*] Tomasetti C & Vogelstein B (2015) Variation in cancer risk among tissues can be explained by the number of stem cell divisions. *Science* 347:78,–81.

How many rRNA genes are needed?

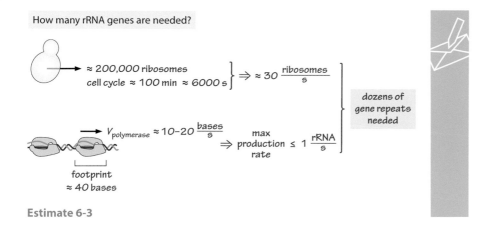

Estimate 6-3

achieve this production rate with one gene copy? In yeast, transcription is performed by the RNA polymerase at an average speed of 10–20 bp/s (BNID 103012, 103657). The polymerase size, or footprint over the DNA, is about 40 bp (BNID 107873). Even if these genes were packed with RNA polymerase in a sequential array like cars in a traffic jam and all were moving at maximal speed, the net transcription rate would still be less than 1 rRNA/s. We thus need tens of copies of the ribosomal rRNA gene (rrna) to enable enough transcription to supply new ribosomes at the necessary rate. Indeed, budding yeast is known to have about 150 rrn (encoding the rRNA) gene copies (BNID 101733, 100243). As an aside, we note how these rrn genes are suggested to be important players in causing aging in budding yeast.[†]

Even in *E. coli*, where very few genes appear with more than one copy per chromosome, there are seven copies of the rRNA genes, as shown in **Figure 6-6** (BNID 102219). A qualitative way to think about why there must be several copies of the rRNA gene is that rRNA is not translated, so there is no second amplification step in translation as there would be for many proteins. Therefore, the only way to achieve the necessary concentration of ribosomes is to have many gene copies. The number of ribosomes in cells has been suggested to be limited by the number of operon copies, with ribosomal proteins being regulated to match the synthesis rate of rRNA. The importance of the number of copies of rrn genes has been tested in a study in *E. coli*, where rrn operons were deleted, and the resulting growth rate was measured as a function of the rrn copy number for a range of 1–7 copies. With less than six copies, there was a significant decrease in growth rate. In other experiments, extra copies beyond seven have also been shown to be detrimental to growth, possibly because of an increase

[†] Sinclair DA & Guarente L (1997) Extrachromosomal rDNA circles—a cause of aging in yeast. *Cell* 91:1033,–1042.

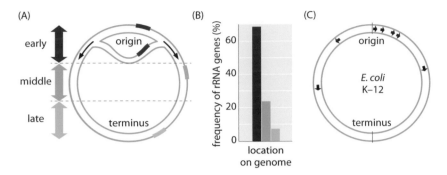

Figure 6-6 Ribosomal RNA genes across microbial genomes. (A, B) Frequency in each third of the chromosome of rDNA operons in 68 bacterial genomes. (C) Locations of the rRNA genes on the circular genome of *E. coli*. Note that all copies of the RNA genes are on the leading strand and none are on the lagging strand. (A and B, adapted from Rocha EPC [2004] *Microbiology* 150:1609–1627. With permission from the Society for General Microbiology. C, adapted from Nomura M [1999] *Proc Natl Acad Sci USA* 96:1820–1822.)

in the diffusion times as the cytoplasm becomes ever more packed with ribosomes.[‡]

The number of rRNA copies in *E. coli* has apparently remained constant at seven copies per genome over the ≈ 150 million years since it diverged from *S. typhimurium* (BNID 107087, 107867). Because *E. coli* can have nested replication forks, resulting in multiple DNA replication operons close to the origin, under fast growth rates there will be more copies of the origin. Indeed, the rrn operons tend to be close to the origin, and the number of copies per cell can be much higher than seven—at high growth rate this number can be as high as 36 (BNID 102359). Advances in genomics allow us to retrieve the number of rRNA genes of hundreds of organisms, and these numbers are now summarized in databases (BNID 104390). The database tables show how the number of copies tends to be higher in faster-growing cells. Information on their location along the genome can show, for example, how common is the tendency to cluster near the origin, as shown in Figure 6-6B. We can also observe the tendency of rRNA to be coded on the leading strand (arrow direction facing away from the origin in Figure 6-6A) rather than on the lagging strand in replication, as discussed further in the caption. This

[‡] Asai T, Condon C, Voulgaris J et al. (1999) Construction and initial characterization of *Escherichia coli* strains with few or no intact chromosomal rRNA operons. *J Bacteriol* 181:3803–3809; Schneider DA & Gourse RL (2003) Changes in *Escherichia coli* rRNA promoter activity correlate with changes in initiating nucleoside triphosphate and guanosine 5' diphosphate 3'-diphosphate concentrations after induction of feedback control of ribosome synthesis. *J Bacteriol* 185:6185–6191; Tadmor AD & Tlusty T (2008) A coarse-grained biophysical model of *E. coli* and its application to perturbation of the rRNA operon copy number. *PLOS Comp Biol* 4:e1000038.

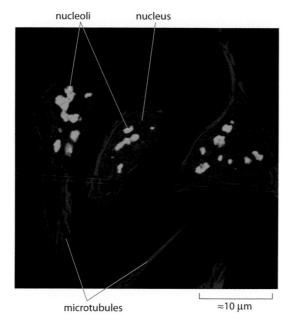

Figure 6-7 Visualizing the nucleolus, the location of ribosomal RNA transcription and assembly in the human cell line U-2 OS. Immunofluorescent staining for the gene RRP15 that is located in the nucleolus and functions in the maturation of the ribosomal subunit. Staining of gene (nucleoli) in green, with additional staining of nucleus in blue and microtubules in red. (Courtesy of the Human Protein Atlas Project.)

can be understood through the "collision avoidance" model. When the replication fork overtakes a transcribing RNA polymerase moving in the same direction, there is replication slowdown until transcription ends. When the RNA polymerase is moving in the direction opposite to the fork advance, there is a head-on collision that leads to a much more problematic replication arrest and transcription abortion. Avoidance of these latter cases selects for locating RNA genes on the leading strand.[§]

How does this question of number of ribosomal RNA genes play out in the case of *Homo sapiens*? Humans carry about 200 copies of the rrn genes in their genome (BNID 107865), and these are organized in five clusters known as the nucleolus organizers, each of which contains multiple copies of the rRNA operon. In phase microscopy of human cells, these areas of extensive transcriptional activity known as nucleoli are vividly seen as black dots inside the nucleus and even more strikingly in fluorescent microscopy, as seen in **Figure 6-7**. In organisms ranging from bacteria to humans, the number of ribosomes is a critical cellular parameter, and

---

[§] Rocha EPC (2008) The organization of the bacterial genome. *Annu Rev Genet* 42:211–233.

one of the main ways that it is regulated is through the gene copy number itself.

WHAT IS THE PERMEABILITY OF THE CELL MEMBRANE?

One of the signature characteristics of all living organisms is that they contain a distinctive mixture of ions and small molecules. The composition not only differs from the environment but can also vary within the cell. For example, the concentration of hydrogen ions in some cellular compartments can be 10^4 times greater than in others (the mitochondria can reach a pH as high as 8, whereas the lysosomes can have a pH as low as 4; BNID 107521, 106074). The ratio of the concentrations of Ca^{2+} ions in the extracellular and intracellular fluid compartments can once again be 10^4-fold (BNID 104083). This concentration difference is so large that transporting a Ca^{2+} ion across the membrane, from the intracelluar to the extracellular compartment, requires the energy of more than one proton or sodium ion flowing down the proton-motive force gradient. To see this, the reader should remember the rule of thumb from our tricks of the trade list—namely, that to establish an order of magnitude potential difference requires 6 kJ/mol (≈ 2 kBT) of energy. This energy can be attained, for example, by the transport of one electric charge through a 60-mV potential difference. To achieve a four orders of magnitude concentration ratio would then require a charge to travel down about 240 mV of electron motive force (actually even more due to the double charge of the calcium ion). This is very close to the breakdown voltage of the membrane, as discussed in the vignette entitled "What is the electric potential difference across membranes?" (pg. 196). Indeed, the high concentration ratio of Ca^{2+} is usually achieved by coupling to the transport of three sodium ions or the hydrolysis of ATP, which helps achieve the required density difference without dangerously energizing the membrane.

The second law of thermodynamics teaches us that, in general, the presence of concentration gradients will eventually be bled off by mass transport processes, which steadily drive systems to a state of equilibrium. However, although the second law of thermodynamics tells us the nature of the ultimate state of a system (for example, uniform concentrations), it doesn't tell us how long it will take to achieve that state. Membranes have evolved to form a very effective barrier to the spontaneous transfer of many ionic and molecular species. To estimate the time scale for equalizing concentrations, we need to know the rates of mass transport, which depend upon key material properties such as diffusion constants and permeabilities.

A hugely successful class of "laws," which describe the behavior of systems that have suffered some small departure from equilibrium, are the linear transport laws. These laws posit a simple linear relation between the rate of transport of some quantity of interest and the associated driving force. For mass transport, there is a linear relation between the flux (that is, the number of molecules crossing unit area per unit time) and the concentration difference (which serves as the relevant driving force). For transport across membranes, these ideas have been codified in the simple equation (for neutral solute),

$$j = -p \cdot (c_{in} - c_{out}),$$ (6.1)

where j is the net flux into the cell, c_{in} and c_{out} are the concentrations on the inside and outside of the membrane-bound region, respectively, and p is a material parameter known as the permeability. The units of p can be deduced by noting that flux has units of number/(area × time) and the concentration has units of number/volume, implying that the units of p itself are length/time. Like many transport quantities (for example, electrical conductivities of materials, which span over 30 orders of magnitude), the permeability has a very large dynamic range, as illustrated in **Figure 6-8**. As seen in the figure, lipid bilayers have a nearly 10^{10}-fold range of permeabilities.

What physicochemical parameters guide the location of a compound on this scale of permeabilities? One rule of thumb is that small molecules have higher permeabilities than larger molecules. Another rule of thumb is that

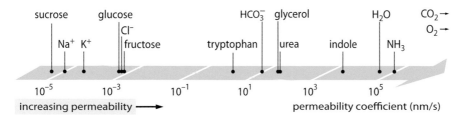

Figure 6-8 The wide range of membrane permeabilities of different compounds in the cell. Membranes are more permeable to uncharged compounds and least permeable to charged ions. Note that the existence of ion channels will make the apparent permeability several orders of magnitude higher when these channels are open. The units are chosen as nm/s, and several nm is the characteristic membrane width. (Adapted from Robertson RN [1983] The Lively Membranes. Cambridge University Press. The value for glucose, which is smaller than in Robertson, is based on several sources, such as BNID 110830, 110807. Other sources of data: BNID 110729, 110731, 110816, 110824, 110806.)

neutral compounds can cross the membrane many orders of magnitude faster than similar charged compounds. Among the charged compounds, negative (anionic) compounds tend to have much higher permeabilities than positive (cationic) compounds. The so-called Overton rule states that membrane permeability increases with hydrophobicity, where hydrophobicity is the tendency of a compound to prefer a nonpolar solvent to a polar (aqueous) solvent. The Overton rule predicts that charged species (nonhydrophobic), such as ions, will tend to have low permeabilities because they incur an energetic penalty associated with penetrating the membrane, whereas dissolved gases such as O_2 and CO_2, which are hydrophobic (because they are uncharged and symmetric), will have high permeabilities. Indeed, the permeabilities of lipid bilayer membranes to CO_2 give values that are 0.01–1 cm/s (yes, permeability measurements have very high uncertainties among different labs; BNID 110004, 110617, 102624), higher than all other values shown in Figure 6-8. This value shows that the barrier created by the cell membrane is actually less of an obstacle than the barrier caused by the unstirred layer of water engulfing the cell membrane from the outside. Such an inference can be derived by the equation for the permeability coefficient of an obstacle, given by

$$p = K \cdot D / l, \tag{6.2}$$

where l is the width, D is the diffusion coefficient, and K is the partition coefficient between the media and the obstacle material. This is also known as the solubility–diffusion model for permeability, where these denote the K and the D effects, which are two steps affecting the permeability. For an unstirred layer of water, $K = 1$, because it is very similar to the media, but for a membrane, the value for all but the most hydrophobic material is usually several orders of magnitude smaller than 1. This dependence on K is at the heart of the Overton rule mentioned above. The high permeability for CO_2 also suggests that channels such as aquaporins, which were suggested to serve for gas transport into the cell, are not required because the membrane is permeable enough. To see how the membrane properties affect the chemical makeup of metabolites, we turn to calculating the time of leakage for different compounds.

We consider glycerol, for example. The analysis shown in **Estimate 6-4** gives an estimate for the time of its leakage out of the cell if the molecule is not phosphorylated or otherwise converted into a more hydrophilic form. The permeability of the cell membrane to glycerol is $p \approx 10$–100 nm/s (BNID 110824), as can be read from Figure 6-8. The time scale for a glycerol molecule inside the cell to escape back to the surrounding medium, assuming no return flow into the cell ($c_{out} = 0$), can be crudely estimated by noting that the efflux from the cell is $p \cdot A \cdot c_{in}$, where A is the cell surface

Leakage time scale through membrane (rapid if small molecule is uncharged—e.g., glycerol)

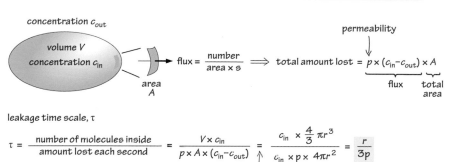

concentration c_{out}

volume V
concentration c_{in}

area
A

$$flux = \frac{number}{area \times s} \implies total\ amount\ lost = \underbrace{\underbrace{p \times (c_{in} - c_{out})}_{flux} \times A}_{total\ area}$$

permeability

leakage time scale, τ

$$\tau = \frac{number\ of\ molecules\ inside}{amount\ lost\ each\ second} = \frac{V \times c_{in}}{p \times A \times (c_{in} - c_{out})} = \frac{c_{in} \times \frac{4}{3}\pi r^3}{c_{in} \times p \times 4\pi r^2} = \frac{r}{3p}$$

assume
$c_{out} = 0$

for glycerol in E. coli

bacterial cell size

$$\tau = \frac{r}{3p} \approx \frac{1}{3} \times \frac{10^{-6}\,m}{4 \times 10^{-8}\,\frac{m}{s}} \approx 10\,s$$

glycerol permeability

so if the similar glyceraldehyde used in glycolysis was not phosphorylated it would rapidly leak from cell

Estimate 6-4

area. The time scale is found by taking the total amount in the cell, $V \cdot c_{in}$ (where V is cell volume, or more accurately the cell water volume), and dividing by this flux, resulting for a bacterial cell ($r \approx 1\,\mu m$) in the following time scale:

$$\tau = V \cdot c_{in}/p \cdot A \cdot c_{in} = (4\pi r^3/3)/(4\pi r^2 \cdot 30\ nm/s) \approx 10\ s. \tag{6.3}$$

This is a crude estimate because we did not account for the decreasing concentration of c_{in} with time, which will give a correction factor of $1/\ln(2)$—that is, less than a twofold increase. What we learn from these estimates is that if the glycolytic intermediates glyceraldehyde or dihydroxyacetone, which are very similar to glycerol, were not phosphorylated, resulting in the addition of a charge, they would be lost to the medium by diffusion through the cell membrane. In lab media, where a carbon source is supplied in abundance, this is not a major issue, but in a natural environment, where cells are often waiting in stationary phase for a lucky pulse of nutrients (E. coli is believed to go through months of no growth after its excretion from the body before it finds a new host), the cell can curb its losses by making sure metabolic intermediates are tagged with a charge that will keep them from recrossing the barrier presented by the lipid bilayer.

HOW MANY PHOTONS DOES IT TAKE TO MAKE A CYANOBACTERIUM?

Autotrophs are those organisms that are able to make a living without resorting to preexisting organic compounds and, as such, are the primary producers of organic matter on planet Earth. One of the most amazing autotrophic lifestyles involves the use of inorganic carbon in the form of CO_2 and the synthesis of organic carbons using light as the energy input—the phenomenon known as photosynthesis. Chemoautotrophs carry out a similar performance, though in their case, the energy source is not light from the sun, but some other terrestrial energy source, such as a thermal vent in the ocean or a reduced inorganic compound such as molecular hydrogen or ferrous iron.

Photoautotrophs are those organisms that take energy from sunlight and convert it into organic compounds that can be oxidized. The most familiar examples are the plants that surround us in our forests and gardens. However, the overall synthetic budget goes well beyond that coming from plants and includes algae and a variety of microscopic organisms, including single-celled eukaryotes (protists) and a whole range of prokaryotes such as cyanobacteria (formerly known as blue-green algae).

The majority of the Earth's surface is covered by water, and photosynthesis in these great aqueous reservoirs is a significant fraction of the total photosynthetic output across the planet as a whole. Aquatic photosynthesis is largely performed by organisms so small that they are invisible to the naked eye. Despite their macroscopic invisibility, these organisms are responsible for fixing ≈ 50 gigatons (BNID 102936; 10^{39} CO_2 molecules) of carbon every year. This accounts for about one-half of the total primary productivity on Earth (BNID 102937), but the vast majority of this fixed carbon is soon returned to the atmosphere following rapid viral attacks, planktonic grazing, and respiration (BNID 102947). The process of transforming inorganic carbon into the building blocks of the organic world occurs through the process of carbon fixation, where the energy from about 10 photons is used to convert CO_2 into a carbohydrate, $(CH_2O)_n$. The H is donated by water as an electron donor, which is thus transformed into oxygen. By comparing the combustion energy stored per carbon in carbohydrates versus the energy in solar flux leading to 10 photosynthetically active photons, the overall theoretical conversion efficiency can be calculated to be about 10%. This is similar to current off-the-shelf photovoltaic cells that result in electricity rather than carbohydrates as output. The biological efficiency that is actually realized is usually lower by another 1–2 orders of magnitude due to respiration, light saturation, and other processes, but on the other hand, these photo-synthetic machines can reproduce and heal themselves, which cannot be said of silicon cells.

The process of oxygenic photosynthesis was invented about 3 billion years ago and has transformed our atmosphere from one with practically no oxygen to one where abundant oxygen allows the existence of animals like us. Much of the carbon fixation happens in small organelles, known as car-boxysomes, like those shown in **Figure 6-9**. Carboxysomes exist in some photosynthetic prokaryotes and are home to an army of molecules that perform the carbon fixation process through a key carboxylating enzyme, Rubisco, considered by many to be the most abundant protein in the bio-sphere. To make its ubiquity more tangible, if we were to distribute it from autotrophs to humans, there would be about 5 kg of this protein per per-son on Earth (BNID 103827).

From an order-of-magnitude perspective, it suffices to think of a cyano-bacterium as similar in chemical composition to a conventional bac-terium, which means that it takes roughly 10^{10} carbons—see vignette entitled "What is the elemental composition of a cell?" (pg. 68)—to supply the building materials for a new cyanobacterium with a volume of 1 μm^3, as depicted in **Estimate 6-5**. Given that it requires roughly

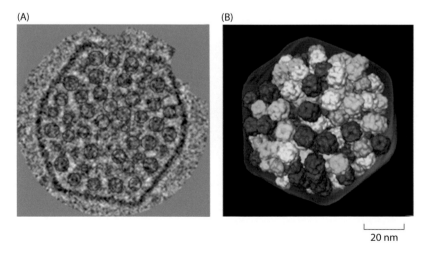

(A)   (B)

20 nm

Figure 6-9 The structure of carboxysomes and the Rubisco octamers occupying them as determined using cryo-electron microscopy. The sizes of individual carboxysomes in this organism (*Synechococcus* strain WH8102) varied from 114 nm to 137 nm, and were approximately icosahedral. There are on average ≈250 Rubisco octamers per carboxysome, organized into three to four concentric layers. *Synechococcus* cells usually contain about 5–10 carboxysomes. (Adapted from Iancu CV, Ding HJ, Morris DM et al. (2007) *J Mol Biol* 372:764–773.)

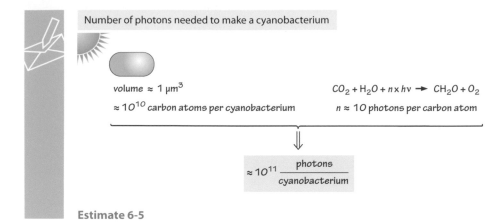

Number of photons needed to make a cyanobacterium

volume $\approx 1\ \mu m^3$

$\approx 10^{10}$ carbon atoms per cyanobacterium

$CO_2 + H_2O + n \times h\nu \longrightarrow CH_2O + O_2$

$n \approx 10$ photons per carbon atom

$\approx 10^{11}\ \dfrac{photons}{cyanobacterium}$

Estimate 6-5

10 photons to fix a carbon atom, this implies roughly 10^{11} photons are absorbed to fix those 10^{10} carbons. This carbon fixation is carried out by roughly 10^4 Rubisco monomers within a given cyanobacterium. What about the energy required for other cellular processes, such as amino acid polymerization into proteins and keeping the membrane potential maintained to drive a multitude of coupled reactions? For bacteria, these energetic requirements were estimated to be on the order of 10^{10} ATP, as discussed in the vignette entitled "What is the power consumption of a cell?" (pg. 199). Given that a photon can be used to produce more than one ATP equivalent (through extruding protons or electron storage in NADPH), we find that the burden of carbon fixation is dominant over these other biosynthetic and maintenance tasks.

HOW MANY VIRIONS RESULT FROM A SINGLE VIRAL INFECTION?

Viruses proliferate in natural environments by infecting cells and hijacking their replication and protein synthesis machinery. After new viral proteins are synthesized and assembled, bursts of viruses are released from the infected (and usually soon-to-be-dead) cells to repeat the process all over again. How many viruses are released from each infected cell? This parameter is referred to as the viral burst size (**Table 6-3**), alluding to the fact that virus emission often leads to either cell lysis (bacteriophage) or cell death (HIV infection of T cells). The emission of new viruses from an infected cell hence occurs as a burst with characteristic numbers of viruses and with time scales lasting from minutes to days, depending upon the kind of virus and host. Burst sizes for different

virus	host cell used	burst size	BNID
multicellular host			
influenza A & B	chicken egg cells	500–1000	101590
influenza A	MDCK cell line	1000–10,000	101605
HIV	*H. sapiens* memory T cells	1000–3000	105872
SIV (model for HIV)	*R. macaque* T cells	40,000–60,000	102377
prokaryotic host			
cyanomyovirus S-PM2	*Synechococcus WH7803*	40	104841
podovirus P60	*Synechococcus WH7803*	80	104842
cyanomyovirus MA-LMM01	*M. aeruginosa*	50–120	103247
bacteriophage S1	*Stenotrophomonas sp.*	80	104855
bacteriophage S3	*Stenotrophomonas sp.*	100	104852
phi EF24C	*E. faecalis*	100	104857
bacteriophage Lambda	*E. coli*	150	105025
bacteriophage T1 to T7	*E. coli*	100–300	105870
bacteriophage MS2	*E. coli*	5000–10,000	109050

Table 6-3 Virions' burst sizes from various host organisms. Table focuses on contrasting prokaryotes versus mammalian cells.

viruses have a large range corresponding, in turn, with the range of different sizes of the host cells. For example, SIV, a cousin and model for the HIV virus, is released from infected T cells with a burst size of ≈ 50,000 (BNID 102377), whereas cyanobacterial viruses have characteristic burst sizes of ≈ 40–80 (BNIDs 103247, 104841, 104842), and phage lambda and other phages (such as T4, T5, and T7) that attack bacteria have burst sizes of ≈ 100–300 (BNID 105025, 105870). An example of a host bacterium prior to the burst process itself is shown in **Figure 6-10**.

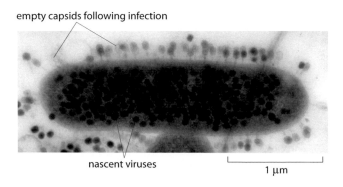

empty capsids following infection

nascent viruses

1 μm

Figure 6-10 A transmission electron micrograph of a thin section of *Escherichia coli* K-12 infected with bacteriophage T4. Dark viruses on the outside are ones that did not eject their DNA into the bacterial host. (Courtesy of John Wertz.)

Fraction of cell volume occupied by bursting virions

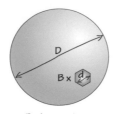

total virions fraction of host cell volume $= \dfrac{d^3 \times B}{D^3}$

e.g.:	$\approx D$ (μm)	$\approx d$ (μm)	$\approx B$	fraction
SIV in T cell	10	0.1	50,000	5%
T phage in *E. coli*	1	0.05	200	2%

B = burst size

Estimate 6-6

One interesting way to garner an impression of the impact of a viral infection on the host metabolism is by thinking about the volume taken up by the newly synthesized viruses in comparison with the size of the host cell. In particular, we ask what fraction of the host cell volume is occupied by all of the viruses making up a viral burst? These volumes can be thought of as a proxy for biomass and thus reflect on the cell's resources that were seized. An SIV virion is roughly 100 nm in diameter, and the host cell has a corresponding diameter of about 10 μm. Given these numbers, 50,000 virions thus represent about 5% of the cell's volume, as shown schematically in **Estimate 6-6**. In the case of bacteria and the viruses that infect them, a T phage with ≈ 50-nm diameter (BNID 105870) shows burst sizes of ≈ 200 in an *E. coli* cell, representing ≈ 2% of the volume. Therefore, the characteristic volume fraction taken up by the viruses in these two very distinct cell types shows a much smaller range (< threefold) than the absolute burst sizes range (>100-fold). This may reflect limits to how much biomass viruses can extract from infected cells. In the marine environment, which is often depleted in phosphorus (which is mostly required for nucleotides), it was suggested that the DNA sizes of the virus and host govern the virion burst size because the virus utilizes the host DNA building blocks.¶ The measurements and estimates throughout this vignette raise the very interesting question of what governs the overall burst size, as well as what fraction of the synthesized viral DNA and proteins actually make it into infectious viruses.

¶ Brown CM, Lawrence JE & Campbell DA (2006) Are phytoplankton population density maxima predictable through analysis of host and viral genomic DNA content? *J Mar Biol Assoc UK* 86:491–498.

Epilogue

In the course of the nearly 100 separate vignettes that make up this book, our work has been animated by several key ideas. First, the overarching theme of the book is that biological numeracy expands our view of the living world in ways that can reveal new insights into organisms and how they work that would otherwise be hidden. It can be thought of as a sixth sense that complements the already powerful arsenal of modern biology. In order to make biological numeracy useful, the values reported for key biological parameters need to be characteristic and actually mean something. To that end, each of our vignettes has tried to report on carefully vetted, state-of-the-art data for a variety of key numbers that dictate the behavior of living matter. But it is not enough to merely quote the numerical values of these quantities. They must also be provided in some context, such that they are actually consonant with what we understand about biological systems. Hence, a second key thrust of our vignettes has been to adopt an attitude of order-of-magnitude thinking to try and use simple estimates to illuminate biological problems in a way that leaves us with an intuition for the meaning of these numbers.

Some challenges make the task of those seeking biological numeracy from reading the literature harder than we might have imagined. One challenge relates to the limited availability of numbers in textbooks and online resources and their often unclear connection to the primary literature. We hope that through the availability of resources such as the BioNumbers database and this book that we have helped remedy some of that challenge. Other challenges we have mentioned several times throughout the book are the misunderstandings that can exist when discussing absolute numbers of some cell component or other property of "the cell" without knowing the cell growth conditions. Differences in cell size can be as much as several fold, and growth rate or different physiological conditions can create even further uncertainty by changing also the per-volume concentrations of numbers of interest. As a result, we strongly believe it is important that every paper that reports a quantitative characterization of cellular properties should at least mention the growth rate, and if referring to copy numbers in cells, aim to measure the cell size, which today can be done with a Coulter counter or FACS machine rather routinely. We hope that referees and editors will make this a "law," though even better yet is that researchers will make it an intrinsic norm of our trade.

There were many more questions that intrigued us than we actually included in our text. In some cases, this was because we did not know how to answer them. In others, we did not sense that the numbers told any compelling story just yet. In the hope that our readers might have insights into answering these questions or some inspiration about how to attack them, we decided to make those questions available here. We are anxious to hear ideas, concrete data, or insights on any of them (just as on any of the vignettes that form the core of the book).

- How many different genes are in a gram of soil, ocean water, and dung?
- How big are vacuoles?
- How long are axons (for example, what happens in a whale)?
- What is the diversity of antibodies in a human?
- How large are the openings in cell membranes?
- What are the concentrations of noncoding RNAs?
- How many of each type of organelles are found in the cell?
- What is the energy cost associated with membrane rearrangements?
- How much sugar is needed to make and power a cell?
- What is the energy invested in carbon and nitrogen assimilation?
- How much force can be exerted by molecular motors?
- How big are osmotic and turgor pressures in cells?
- What is the rate of protein folding?
- What are the mass-specific polymerization rates of the machines of the central dogma?
- What is the rate of post-translational modifications of proteins (for example, glycosylation)?
- How fast does a signal propagate from a receptor to the nucleus?
- What are the maximal growth rates of different organisms?
- How long does apoptosis take?
- How fast is signal transduction in the cell?
- How fast does the molecular clock tick?
- How many carbon fixation pathways exist in nature?
- How many proteins are synthesized per burst of mRNA translation?
- How long are noncoding RNAs?
- What is the length of sequence required for homologous recombination?
- What are the rates of somatic recombination and transposition?

- What is the error rate in antibody recognition?
- How many neurons are in the brain?
- What proportion of the ribosome is rRNA?
- How many cell types are there in the human body?

We leave our readers with the hope that they will find these or other questions inspiring and will set off on their own path to biological numeracy.

Index

Roman page numbers refer to the *Preface* or *The Path to Biological Numeracy*. The suffixes 'F', 'T' and 'n' indicate mentions in Figures, Tables and footnotes when these are on different pages from any text treatment. When coverage in the text has already been indexed, non-text material visible on the same page is not distinguished.